AF805979

Advances in Experimental Medicine and Biology

Volume 959

Editorial Board
IRUN R. COHEN, *The Weizmann Institute of Science, Rehovot, Israel*
ABEL LAJTHA, N.S. *Kline Institute for Psychiatric Research, Orangeburg, NY, USA*
JOHN D. LAMBRIS, *University of Pennsylvania, Philadelphia, PA, USA*
RODOLFO PAOLETTI, *University of Milan, Milan, Italy*

More information about this series at http://www.springer.com/series/5584

Robert M. Tanguay
Editor

Hereditary Tyrosinemia

Pathogenesis, Screening and Management

Springer

Editor
Robert M. Tanguay
Department of Molecular Biology, Medical Biochemistry
and Pathology, Medical School
Université Laval
Québec, QC, Canada

ISSN 0065-2598 ISSN 2214-8019 (electronic)
Advances in Experimental Medicine and Biology
ISBN 978-3-319-55779-3 ISBN 978-3-319-55780-9 (eBook)
DOI 10.1007/978-3-319-55780-9

Library of Congress Control Number: 2017946312

© Springer International Publishing AG 2017
This work is subject to copyright. All rights are reserved by the Publisher, whether the whole or part of the material is concerned, specifically the rights of translation, reprinting, reuse of illustrations, recitation, broadcasting, reproduction on microfilms or in any other physical way, and transmission or information storage and retrieval, electronic adaptation, computer software, or by similar or dissimilar methodology now known or hereafter developed.
The use of general descriptive names, registered names, trademarks, service marks, etc. in this publication does not imply, even in the absence of a specific statement, that such names are exempt from the relevant protective laws and regulations and therefore free for general use.
The publisher, the authors and the editors are safe to assume that the advice and information in this book are believed to be true and accurate at the date of publication. Neither the publisher nor the authors or the editors give a warranty, express or implied, with respect to the material contained herein or for any errors or omissions that may have been made. The publisher remains neutral with regard to jurisdictional claims in published maps and institutional affiliations.

Printed on acid-free paper

This Springer imprint is published by Springer Nature
The registered company is Springer International Publishing AG
The registered company address is: Gewerbestrasse 11, 6330 Cham, Switzerland

Foreword

Hereditary tyrosinemia type 1 (HT1), the most severe inborn error of the tyrosine degradation pathway, is due to a deficiency in fumarylacetoacetate hydrolase (FAH). The worldwide frequency of HT1 is around one per 100,000 births, but some regions have a significantly higher incidence due to founder effects. Thus the high incidence of HT1 (1:1,800) in the northeastern part of Quebec known as Saguenay-Lac-Saint-Jean (SLSJ) has led to continued interest in the community as the defective gene carrier rate within this region is 1:22. The disease was discovered by a remarkable paediatrician in the region, Dr. Jean Larochelle, in the mid-1960s (see Chap. 1).

The FAH defect results in the accumulation of toxic metabolites, mainly in the liver. If left untreated, HT1 is usually fatal before the age of 2. HT1 patients develop several chronic complications including cirrhosis with a high risk of hepatocellular carcinoma (HCC) and neuropsychological impairment (Chaps. 6, 9 and 10). Since 1992, thanks to Sven Lindstedt in Gothenburg, treatment with an inhibitor of the pathway, nitisinone, combined with a strict dietary treatment has largely replaced liver transplantation (Chaps. 16, 17 and 18). However, early treatment is important to avoid HCC. The book includes the latest developments on the molecular basis of HT1 and its pathology and the screening and diagnosis and management of the disease written by leading scientists, geneticists, hepatologists and clinicians in the field.

In October 1994, I organized the First International Symposium on Hereditary Tyrosinemia in Lac-Delage, a small resort outside Québec (Figure 1). All the scientists working on this rare disease were present. Subsequently, Eli Anne Kvittingen, who was working on another cluster of HT1 patients in Norway, uncovered the most unusual property of this disease: the reversal and correction of the mutation in some hepatocytes from some of the patients examined (1993 and 1994). In the autumn of 2015, GAETQ (Groupe d'Aide aux Enfants Tyrosinémiques du Québec) organized a meeting about the disease to mark the 50th anniversary of the discovery of HT1 by Dr. Larochelle. The meeting was attended by a number of clinicians, research scientists and experts on various features of HT1 patients. But the interesting part was the participation of over 200 parents of HT1 patients coming from various parts of the world (Québec, Canada, USA, Europe and Australia). For

Fig. 1 Participants of the 1994 First International Symposium on Hereditary Tyrosinemia, Lac-Delage, Québec, Canada

Fig. 2 Eli Anne Kvittingen

Fig. 3 Luc Bélanger

me and my research assistant Dr. Geneviève Morrow, this was an unsuspected and very pleasant feature of the meeting as we met the parents and children in whom we had diagnosed the mutation or the carrier status years before. It was a refreshing change from the usual scientific symposia format of experts discussing "their science" in the absence of the "affected patients". It humanized the exchanges and took us away from our mouse research models of HT1 and back to real life.

On the basis of the newer knowledge on various features of HT1, I thought that it might be time to assemble a book dedicated to hereditary tyrosinemia. I presented this proposal to Meran Owen of Springer with whom I had worked with on another book on small heat shock proteins the year before. Meran gave me his support and Springer accepted the proposal. I assembled the experts on HT1 from around the world and convinced them to make their contributions. I would like to thank the authors of the 21 chapters for their nice contributions and for their patience in answering my queries and suggestions before the final version. The assembly of the chapters would not have been possible without the kind and patient help of Cynthia Kroonen at Springer. She was very kind in reminding me that the book was overdue (without mentioning the word), but I guess that she is used to working with scientists who are always late.

Finally I would like pay a special tribute to Dr. Luc Bélanger (Université Laval) (Figure 3) who showed the same interest in HT1 as his father who was the clinical biochemist at l'Hôpital de Chicoutimi where he collaborated with Dr. Larochelle. Luc died a few months ago and I dedicate this book to him.

Université Laval
Québec, QC, Canada

Robert M. Tanguay

GAETQ: A Success Story in Community Help

When I think about the history of hereditary tyrosinemia type 1 (HT1), I can't help but think about my own story. I have been witnessing the evolution of this disease for a good 37 years. I know it may seem like 37 years is not a lot, but considering I wasn't expected to live more than a couple of years, it's a lot!

I was born affected with HT1 at the end of the year 1979. My future could have been a lot different, read non-existent, if my parents hadn't been Gérard Tremblay and Marielle Gravel. They're the ones who first decided to fight for my life and then to fight for the life of others like me. When they founded the **Groupe d'Aide aux Enfants Tyrosinémiques du Québec (GAETQ)** in 1983, I don't think they knew the kind of impact that they would have on my life and on the life of hundreds of families.

The first actions that were conducted by GAETQ merely allowed families to get together and share their stories and their fears. Back in the 1980s, there wasn't much hope for babies who were born affected, but still my parents did their best and, most importantly, never gave up. They raised money and awareness and worked with the medical teams who didn't know much at the time. Dr. Jean Larochelle, who was part of the group that identified the disease in 1965, would most certainly agree that they were very tenacious! Dr. Larochelle became a very strong ally for my parents, and this relationship most certainly explains why GAETQ still works very closely with medical teams and researchers interested in HT1.

Then the 1980s were over and a new era came for HT1 patients. Against all odds, I had survived with a low-protein diet as the sole treatment option. But I wasn't alone; some of my friends had also survived. Of course, all patients had a damaged liver, and doctors started to fear that we would develop liver cancer. That's when GAETQ and I began a new chapter in our lives. Liver transplant was made available. It was a risky surgery, but it was our only hope. I clearly remember getting the phone call saying my friend Simon had received his new liver. I remember all the phone calls. Each time, it was both scary and exciting. Then the scariest and most exciting phone call of all came, and it was my turn. During that period, GAETQ played a crucial role. Families were getting financial help from GAETQ to support them during the long periods that they had to spend in Montreal to get the new liver and recover from that challenging surgery. The emotional support was also very important. My parents supported families through job loss, divorce situations and everything else that comes with a sick child. Sadly, some of us didn't make it. But still, GAETQ, i.e. my parents, was there to help and support.

In 1993, shortly after I had received my new liver, another great revolution was made available for HT1 patients who had not yet received a transplant or were just being born. It was the kind of news that all patients suffering from an orphan disease wish for. The miracle that my parents had been waiting for all those years had finally arrived. It was a little too late for me, but it was going to help so many families. This miracle was called NTBC better known as nitisinone. We couldn't pronounce it right, but it sure accomplished miracles. Again, GAETQ played a very important role. We fought hard so that the medicine would be available to all affected children and covered by the Quebec government. Again, financial support from GAETQ was very helpful for families who had to travel to Montreal several times a year to be part of the NTBC protocol. Some 25 years later, close to a hundred patients are being treated using both the medicine and the low-protein diet. Although cooking remains a daily challenge for affected families, there finally is a proper treatment for HT1. Children who first started being treated with nitisinone grew up and became inspiring adults.

Time passed and I became the happy mother of two amazing children. I did more than just surviving; I gave birth twice! Now don't go thinking that I didn't have to fight for this. Getting pregnant was the fun part! Getting doctors to agree to follow my pregnancies was the tough part. This wasn't my first time scaring doctors away, and I knew I had to fight for my dream. A high-risk pregnancy protocol was finally established for my pregnancies and was used for two other females in my region with whom I share the same story.

Time also passed for GAETQ and we started feeling that families were doing good and didn't need our help so much. Kids are growing up, the medical services are so well organized, and the treatment is well established. But then we realized that we could still improve the quality of life of our members. We also realized that the Quebec situation was unique. Being born with HT1 in Quebec or in the rest of the world will not necessarily bring you the same life. And that's when the idea of organizing an international meeting came to our minds. It was a crazy idea, but for some reason, crazy people decided to live this journey with me. **The Tyrosinemia 2015 International Conference** was born. I acted as president of the organizing committee, and I will never thank enough the people who spent numerous hours supporting me in this experience. Thanks to Gérard Tremblay, Marielle Gravel, Francine Gauthier, Jean Harvey, Sylvie St-Pierre, Anne Vigneault and Dr. Guy Parizeault, we were able to get together 250 patients, researchers, health professionals and doctors interested in tyrosinemia. Speakers from Montreal, Québec, France, the United States, the Netherlands and England travelled to Saguenay (Quebec) to share their knowledge and expertise. Patients travelled from British Columbia, Ontario, the United States, Africa, India and Australia to meet with us and share their reality. This book, very much like Tyrosinemia 2015, shows how the HT1 community is always eager to share and learn. This most certainly is a great advantage for patients all over the world but also for experts who are oblivious of distance and cultural barriers.

Now as much as my future is still to be written, the future of GAETQ is unknown. But there is one thing that I know for sure: just like my parents taught me, I will never give up. I will stand up for people affected with HT1 around the world as long as I can. Fortunately, I'm not alone. GAETQ will advocate for its members as long as necessary, i.e. as long as is needed to find a cure. Hopefully, I will be there to witness this moment.

GAETQ
Québec, Canada
anicktremblay@yahoo.ca
http://gaetq.org/, http://tyrosinemie2015.com/

Janick Tremblay

Contents

Part I

Hereditary Tyrosinemia Type I

Discovery of Hereditary Tyrosinemia in Saguenay- Lac St-Jean

1

Jean Larochelle

Keywords

Liver cirrhosis • HT1 • Tyrosyluria • Alpha foetoprotein • Diagnosis • Charlevoix

Given the interest of many people and families directly or indirectly affected by hereditary tyrosinemia (HT1), I have tried to give my view on the history of the disease from 1965 to 2015 (Fig. 1.1).

Even if I was a significant actor in this story, I would like first to point the important contributions of a number of persons who participated in the identification, treatment and diagnosis of HT1. I had the precious support and help of Dr. Maurice Bélanger, a biochemist, in the identification of the metabolic problem and of Dr. Lucien Privé, pathologist for the anatomo-pathological analysis of the first patient at Chicoutimi's hospital. Dr. Charles Scriver (Montréal) was my first mentor (Scriver et al. 1967a, b) together with Dr. Claude Laberge (CHUL, Québec) who initially showed the founder effect of HT1 in our region (Laberge 1969; Laberge and Dallaire 1967). Dr. Luc Bélanger, an MD-biochemist interested in alpha-foetoprotein (AFP) (Belanger et al. 1973) also made substantial contributions to the various therapeutic assays used to improve the "state" of HT1 patients. Through the years many researchers joined us in our search for diagnostic methods for HT1.

Hereditary tyrosinemia type I is a recessive hereditary disease caused by a deficiency of the hepatic enzyme fumarylacetoacetate hydrolase (FAH), which is involved in the catabolism of tyrosine. The deficiency causes an accumulation of metabolic products with a high toxicity in liver, kidneys and peripheral nerves. Tyrosine is an amino acid present in animal proteins and to a lesser degree in plant proteins.

The incidence of the disease in Saguenay-Lac St-Jean (SLSJ) is high with 1 case for 1846 births (De Braekeleer and Larochelle 1990) (Fig. 1.2).

On a population of 280,000, it is estimated that 14,200 persons are carriers thus one person on 21. Since 1954, approximately 170 babies were born with HT1 in the region. One hundred

J. Larochelle (✉)
2855 Le Noblet, Apt 309, Québec, QC G1V 0C6, Canada
e-mail: larochellejean1@videotron.ca

© Springer International Publishing AG 2017
R.M. Tanguay (ed.), *Hereditary Tyrosinemia*, Advances in Experimental Medicine and Biology 959, DOI 10.1007/978-3-319-55780-9_1

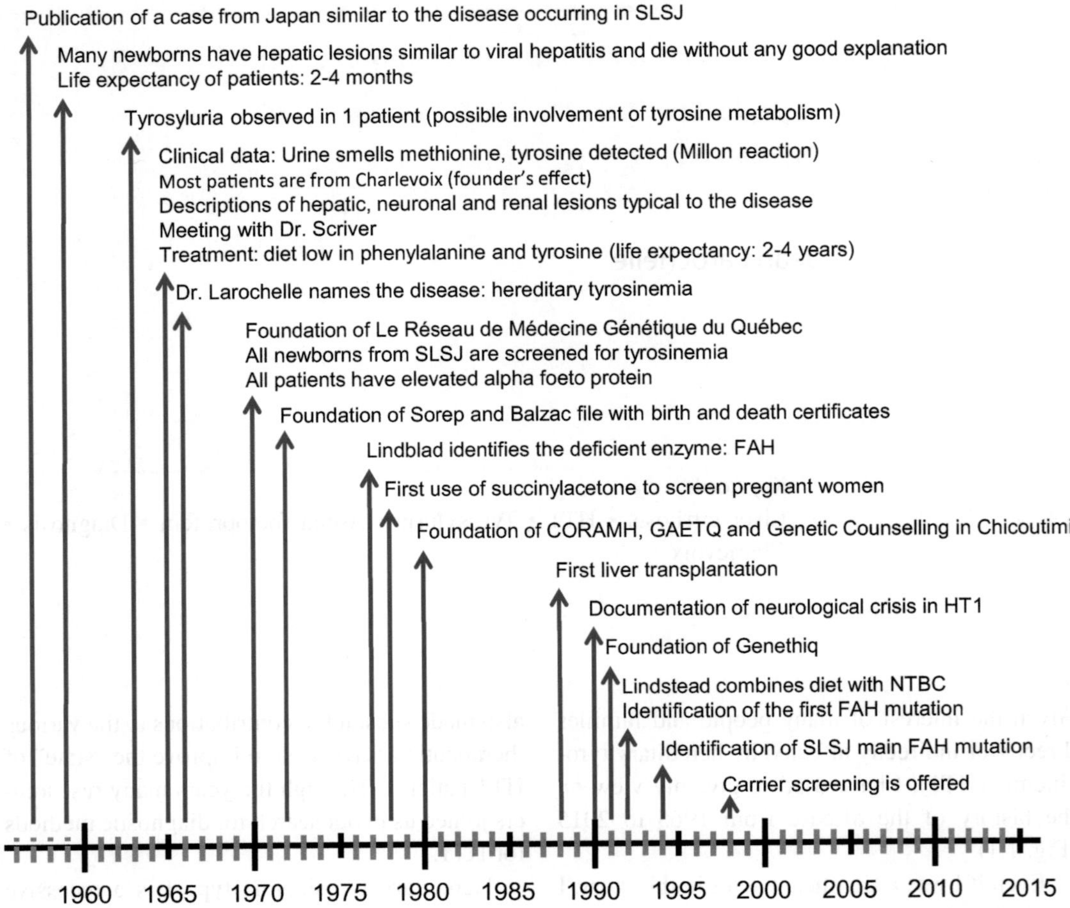

Fig. 1.1 Time line of HT1 marking events

forty of them died from the disease while the others survived as a result of liver transplantation or due to a special diet in combination with NTBC (2-(2-nitro-trifluoromethylbenzoyl)-1,3-cyclohexanedione) a drug that was introduced in 1992 (Lindstedt et al. 1992).

In 1959 and 1960 I was resident at Chicoutimi's Hospital and I could observe many newborns that were hospitalized for fulminant cirrhosis. The patients had hepato- and splenomegaly, ecchymosis, icteral bleeding suggesting a diagnosis of viral hepatitis. Urine also had a peculiar smell, which Dr. Maurice Bélanger attributed to methionine in 1965. All patients died between 2 and 4 months of age without any logical explanations.

Transitory tyrosinemia of the newborn and premature was already known. Oculo cutaneous tyrosinemia (tyrosinemia type II) resulting from a defect in tyrosine transaminase (TAT) had been identified and a low tyrosine diet had been successfully used as treatment. Phenylketonuria was treated with a low phenylalanine diet.

In 1957 two Japanese clinical researchers Dr. K. Sakai and T. Kitagawa described in a medical journal a patient with cirrhosis, tubulopathy, rickets and tyrosyluria (Sakai and Kitagawa 1957a, b, 1959), features that were similar to those observed in our patients. A few other observations followed in the 1960s. In 1963 Dr. Maurice Tremblay (pediatrician) referred a patient in a particular bad condition to Dr. Andrew Sass-Kortsak who was working at the Hospital for Sick Children in Toronto. The patient died soon after arrival in Toronto. In line with the observed tyrosyluria, Dr. Sass-Kortsak wrote in his report that a trouble in tyrosine metabolism should be investigated. At

Fig. 1.2 Localization of SLSJ region in Quebec province (Map modified from https://en.wikipedia.org/wiki/File:Canada_Quebec_relief_location_map.jpg#filelink)

the end of 1964, that is by the end of my training in pediatrics at St-Justine Hospital (Montréal), I asked Dr. Maurice Bélanger (Chicoutimi) to order an amino acid auto-analyzer to measure amino acids when another patient showed up. On January 2, 1965, the first child hospitalized under my supervision had all the characteristics of this terrible disease. All amino acids were high and the Millon reaction, which detects tyrosine, was positive. This was the ideal situation to look more deeply at the disease. I decided to examine the files of all patients deceased at a young age with a diagnosis of cirrhosis or hepatitis. I also went through the Index Medicus hoping to find a description of a disease similar to that observed in our SLSJ patients. No hits. I followed up with a familial and environmental study of the families of patients. In this way I was able to eliminate a problem of marriage of first cousins in patients but I noted that most of them came from Charlevoix (Fig. 1.2). I asked Dr. Lucien Privé, pathologist, to have a look at the autopsy reports of newborns deceased with hepatic damages; he was able to describe the hepato-renal damages and neurological problems typical of the disease. With all this information in hand I presented these cases at the Canadian Association of Pediatry meeting in June of 1965. I described the history, biochemistry and anatomo-pathological features of 29 patients diseased with cirrhosis and hypermethioninemia. Many researchers present at the meeting expressed their interests in the presentation. This is when I met Dr. Charles R Scriver, a world authority in genetic and metabolic diseases. Charles became my mentor. He informed me that he had given a diagnosis of possible tyrosinemia in April 1965 in the case of two patients from the same family that were born in Roberval (SLSJ region, Fig. 1.2). The diagnosis and diet were recommended to him by a Norwegian colleague Dr. Gjessing. In August of the same year, I referred a patient to Dr. Scriver in Montréal. The patient rapidly died and the exams suggested that tyrosinemia could be the cause of death. Since no one had yet been able to prove the cause of the disease, Dr. Scriver suggested that we measure amino acids as soon as a newborn baby was susceptible to have the disease. In December 1965, a baby girl, sister of a patient who had died earlier, was born in Chicoutimi. We were able to show that tyrosine went up in the first week of life, phenylalanine in the second week and methionine in the third week. A tyrosine defect was therefore likely to be the first cause of tyrosinemia.

The same year Drs Scriver and Sass-Kortsak approached Dr. Claude Laberge to make an epidemiological study of HT1as the topic of his doctoral thesis. Claude established that a founder effect in the SLSJ region was at the origin of rare diseases with a high incidence in the region. From 1675 to 1850, 600 individuals from Perche and Normandie (France) settled in Charlevoix, each family having 10–12 children (Fig. 1.2). From 1838 to 1911, the opening period of SLSJ, 70% of the 5000 migrants came from Charlevoix. Since all these immigrants moved with their family, it explains why the number of carriers of the disease is so high.

In March of 1966, I was invited to give a keynote address in Toronto in which I could present all the data I had and suggested that this new disease could be called hereditary tyrosinemia. From 1965 to 1970, all newborns at Chicoutimi's hospital were screened for tyrosinemia by Dr. Maurice Bélanger. This was probably the first screening test for HT1 in the world. Later Dr. Scriver suggested to apply the test to the other hospitals of the region and even elsewhere in the province where patients started to be reported. In 1970 Dr. Laberge (Université Laval) and Dr. Scriver (McGill Universty) founded the Réseau de Médecine Génétique du Québec (RMGQ).

From 1965, HT1 patients were treated using a low tyrosine-phenylalanine diet with encouraging results. Unfortunately most of them died of cirrhosis or neurological crises between 2 and 4 years of age with a few exceptions. In the 1970s, Dr. Luc Bélanger devised a number of therapeutic trials and asked us to contact Prof. Robert M. Tanguay (CHUL, Québec) to accelerate the identification of the tyrosinemia gene. Dr. Bélanger also studied the immunological system of patients in order to evaluate the possibilities of liver transplantation in the future. In 1970 Dr. Bélanger discovered the presence of alpha foeto protein (AFP) in the blood of HT1 patients (Belanger et al. 1973). This discovery was applied for neonatal diagnosis in the RMGQ network program. In 1972, Gérard Bouchard (UQAC) set up the Balzac file that now includes over three million births, weddings and death certificates from the beginning of the Nouvelle-France up to now.

In 1977, Berndt Lindblad (Gothenburg, Sweden) identified the defective enzyme as fumarylacetoacetate hydrolase (FAH) (Lindblad et al. 1977). The next year, Dr. Richard Gagné (CHUL, Québec) described the use of succinylacetone in amniotic fluids to diagnose HT1 at 16 weeks of pregnancy (Gagne 1978). In 1980, CORAMH (Corporation de recherche et d'action sur les maladies héréditaires) was created. This corporation's role is to foster research and inform the population on hereditary diseases. The same year, Claude Prévost created genetic counceling in Chicoutimi and Gerard Tremblay founded the Fondation du Groupe d'Aide des Enfants tyrosinémiques (GAETQ). Gerard is still the president of GAETQ. Given the lack of results of the diet treatment alone, a first liver transplantation was done with success. Seventeen of our 20 transplanted patients are still doing well. However they must take anti-rejection drugs and be regularly followed. Many efforts and pressures were needed to implant the transplantation program at Ste.-Justine Hospital (Montréal).

In 1990, Dr. Grant Mitchell reports the features of neurological crises in a seminal paper in the New England Journal of Medicine (Mitchell et al. 1990). The next year Dr. Richard Gagné creates Genethiq, an inter-university group on ethics in genetics.

1992 is a key year in HT1 as Sven Lindsted (Gothenborg) reports that NTBC, a strange compound used for biological warfare, corrects with a proper diet most of the symptoms of HT1 and prevents neurological crisis (Lindstedt et al. 1992). The same year Daniel Phaneuf and Robert M Tanguay identify the first mutation of FAH in a French Canadian patient (Phaneuf et al. 1992). Then in 1994, Markus Grompe (Portland, Oregon) in collaboration with Prof. Robert Tanguay (Québec) discover the so-called SLSJ mutation that maps to chromosome 15 (Grompe et al. 1994). 95.4% of our patients have the same mutation, which is exceptional and greatly facilitates a non-costly and easy screening test. Claude Prévost in 1997 presents a project to study the impact of screening of carriers in SLSJ families. To our surprise 90% of the SLSJ population are in favor of such screening. The next year the Conseil d'Évaluation des Technologies de la Santé du Québec recommends that carrier tests should be offered to the affected families and even to other citizens from the region requesting the test.

Thus in 50 years, while a full cure of HT1 has yet to be reached, dozens of children have benefited from liver transplantation and NTBC treatment combined with a low tyrosine-phenylalanine diet.

I would like to thank all who participated in the research on HT1, in screening and on treatment of HT1, the various centers in Montréal (Hospital St-Luc and Ste.-Justine), Québec and Saguenay for the follow-up of HT1 patients and the geneticist Dr. Grant A Mitchell who has supervised the NTBC treatment and distribution of this orphan drug since its implementation (Larochelle et al. 2012). I also thank Dr. Guy Parizeault (Hôpital de Chicoutimi) who has taken charge of my patients since 2001 in SLSJ and our GAETQ president Gerard Tremblay whose tyrosinemic daughter, who has received liver transplantation, gave birth to two beautiful children.

And finally I thank Dr. Robert M. Tanguay and Dr. Luc Bélanger (Université Laval, Faculté de medicine, Québec) for the translation and editing of my recollection of the discovery of HT1 in SLSJ.

References

Belanger L, Belanger M, Prive L, Larochelle J, Tremblay M, Aubin G (1973) Hereditary tyrosinemia and alpha-1-fetoprotein. I. Clinical value of alpha-fetoprotein in hereditary tyrosinemia. Pathol Biol 21(5):449–455

De Braekeleer M, Larochelle J (1990) Genetic epidemiology of hereditary tyrosinemia in Quebec and in Saguenay-Lac-St-Jean. Am J Hum Genet 47(2):302–307

Gagne R (1978) Genetic counseling: experience of 4 years. L'union Med Canada 107(4):391–393

Grompe M, St-Louis M, Demers SI, al-Dhalimy M, Leclerc B, Tanguay RM (1994) A single mutation of the fumarylacetoacetate hydrolase gene in French Canadians with hereditary tyrosinemia type I. N Engl J Med 331(6):353–357. doi:10.1056/NEJM199408113310603

Laberge C (1969) Hereditary tyrosinemia in a French Canadian isolate. Am J Hum Genet 21(1):36–45
Laberge C, Dallaire L (1967) Genetic aspects of tyrosinemia in the Chicoutimi region. Can Med Assoc J 97(18):1099–1101
Larochelle J, Alvarez F, Bussieres JF, Chevalier I, Dallaire L, Dubois J, Faucher F, Fenyves D, Goodyer P, Grenier A, Holme E, Laframboise R, Lambert M, Lindstedt S, Maranda B, Melancon S, Merouani A, Mitchell J, Parizeault G, Pelletier L, Phan V, Rinaldo P, Scott CR, Scriver C, Mitchell GA (2012) Effect of nitisinone (NTBC) treatment on the clinical course of hepatorenal tyrosinemia in Quebec. Mol Genet Metab 107(1–2):49–54
Lindblad B, Lindstedt S, Steen G (1977) On the enzymic defects in hereditary tyrosinemia. Proc Natl Acad Sci U S A 74(10):4641–4645
Lindstedt S, Holme E, Lock EA, Hjalmarson O, Strandvik B (1992) Treatment of hereditary tyrosinaemia type I by inhibition of 4-hydroxyphenylpyruvate dioxygenase. Lancet 340(8823):813–817
Mitchell G, Larochelle J, Lambert M, Michaud J, Grenier A, Ogier H, Gauthier M, Lacroix J, Vanasse M, Larbrisseau A et al (1990) Neurologic crises in hereditary tyrosinemia. N Engl J Med 322(7):432–437
Phaneuf D, Lambert M, Laframboise R, Mitchell G, Lettre F, Tanguay RM (1992) Type 1 hereditary tyrosinemia. Evidence for molecular heterogeneity and identification of a causal mutation in a French Canadian patient. J Clin Invest 90(4):1185–1192
Sakai K, Kitagawa T (1957a) An atypical case of tyrosinosis (1-para-hydroxyphenyllactic aciduria) Part 2. A research on the metabolic block. Jikei Med J 4:11–15
Sakai K, Kitagawa T (1957b) An atypical case of tyrosinosis (1-Parahydroxyphenyl-lactic aciduria) Part 1. Clinical and laboratory findings. Jikei Med J 2:1–5
Sakai K, Kitagawa T (1959) An atypical case of tyrosinosis (1-Parahydroxyphenyl-lactic aciduria) Part 3. The outcome of the patient; pathological and biochemical observations of the organ tissues. Jikei Med J 6:15–18
Scriver CR, Larochelle J, Silverberg M (1967a) Hereditary tyrosinemia and tyrosyluria in a French Canadian geographic isolate. Am J Dis Child 113(1):41–46
Scriver CR, Silverberg M, Clow CL (1967b) Hereditary tyrosinemia and tyrosyluria: clinical report of four patients. Can Med Assoc J 97(18):1047–1050

2 Biochemical and Clinical Aspects of Hereditary Tyrosinemia Type 1

Geneviève Morrow and Robert M. Tanguay

Abstract

Inborn errors of metabolism (IEMs) are a group of diseases involving a genetic defect that alters a metabolic pathway and that presents usually during infancy. The tyrosine degradation pathway contains five enzymes, four of which being associated with IEMs. The most severe metabolic disorder associated with this catabolic pathway is hereditary tyrosinemia type 1 (HT1; OMIM 276700). HT1 is an autosomal recessive disease caused by a deficiency of fumarylacetoacetate hydrolase (FAH), the last enzyme of the tyrosine catabolic pathway. Although a rare disease worldwide, HT1 shows higher incidence in certain populations due to founder effects. The acute form of the disease is characterized by an early onset and severe liver failure while the chronic form appears later and also involves renal dysfunctions. Until 1992 the only treatment for this disease was liver transplantation. Since then, NTBC/Nitisone (a drug blocking the pathway upstream of FAH) is successfully used in combination with a diet low in tyrosine and phenylalanine, but patients are still at risk of developing hepatocellular carcinoma. This chapter summarizes the biochemical and clinical features of HT1.

Keywords

Hereditary tyrosinemia type 1 (HT1) • Fumarylacetoacetate hydrolase (FAH) • Liver • Kidney • Hepatocellular carcinoma

G. Morrow • R.M. Tanguay (✉)
Laboratoire de génétique cellulaire et développementale, Département de biologie moléculaire, biochimie médicale et pathologie, Faculté de médecine, Institut de biologie intégrative et des systèmes (IBIS) and PROTEO, Université Laval, 1030 avenue de la médecine, Québec, QC G1V 0A6, Canada
e-mail: genevieve.morrow@fmed.ulaval.ca; robert.tanguay@fmed.ulaval.ca

Abbreviations

ALA	δ-aminolevulinic acid
ALAD	δ-aminolevulinic acid dehydratase
BER	Base excision repair
FAA	Fumaryl acetoacetate

© Springer International Publishing AG 2017
R.M. Tanguay (ed.), *Hereditary Tyrosinemia*, Advances in Experimental Medicine and Biology 959,
DOI 10.1007/978-3-319-55780-9_2

FAH	Fumarylacetoacetate hydrolase
GSH	Glutathione
HCC	Hepatocellular carcinoma
HGA	Homogentisic acid
HGO	Homogentisic acid oxidase
HPD	*p*-hydroxyphenylpyruvate dioxygenase
HT1	Hereditary tyrosinemia
IEM	Inborn errors of metabolism
MAA	Maleyl acetoacetate
MAAI	Maleyl acetoacetate isomerase also known as (ζ) 1 GSTZ1
OLT	Orthotopic liver transplantation
PAH	Phenylalanine hydroxylase
SAA	Succinylacetone
TAT	Tyrosine aminotransferase
TCA	Trichloroacetic acid cycle

2.1 Introduction

Inborn errors of metabolism (IEMs) are a group of diseases (more than 400) in which a single gene defect causes a block in a metabolic pathway resulting either in the accumulation of unwanted metabolites or in product deficiency. Most IEMs disorders present early in life although milder forms may remain undetected until adulthood. Individually they are rare, but collectively they are common. In Canada, the incidence of IEMs is estimated at 40/100,000 live births (Applegarth et al. 2000).

IEMs are classified according to their clinical features, the type of enzyme involved and their pattern of inheritance. They include genetic defects that alter amino acids, lipid and carbohydrate metabolisms in addition to mucopolysaccharidoses, purine and pyrimidine disorders and porphyrias.

Amino acids are the building blocs of proteins and are a source of nitrogen for biologically important compounds such as hormones and neurotransmitters. Many IEMs alter amino acid metabolism such as alkaptonuria, maple syrup urine disease and homocystinuria. This chapter focusses on the most severe metabolic disorder of the tyrosine catabolic pathway, hereditary tyrosinemia type 1 (HT1; OMIM 276700) (Mitchell et al. 2001; Sniderman King et al. 2006) and is aimed at describing the general biochemical and clinical aspects of this disease.

2.2 Phenylalanine and Tyrosine

Phenylalanine and tyrosine only differ by the presence of an – OH group attached to the benzene ring (Fig. 2.1). Hence, tyrosine is also known as 4-hydroxyphenylalanine. Phenylalanine normally has two fates in cells: incorporation into polypeptide chains and hydroxylation to tyrosine via phenylalanine hydroxylase (PAH). Therefore phenylalanine degradation follows the tyrosine catabolic pathway.

Tyrosine is involved in protein biosynthesis and is also a precursor for neurotransmitters and hormones. Indeed, tyrosine is involved in epinephrine, melanin and thyroxine synthesis among others. Together with phenylalanine, they are considered glucogenic and ketogenic amino acids due to the products yield by their degradation.

2.2.1 Tyrosine Catabolism

The tyrosine degradation pathway involves five enzymes that ultimately lead to the conversion of tyrosine into fumarate and acetoacetate (Fig. 2.2).

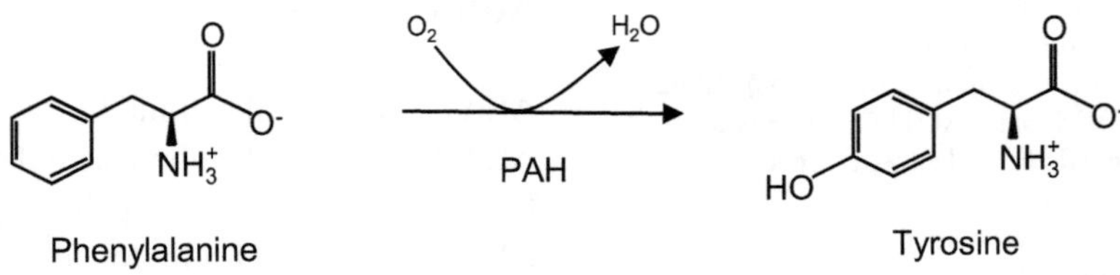

Fig. 2.1 Conversion of phenylalanine to tyrosine by phenylalanine hydroxylase. The transformation of phenylalanine to tyrosine is done by phenylalanine hydroxylase (PAH)

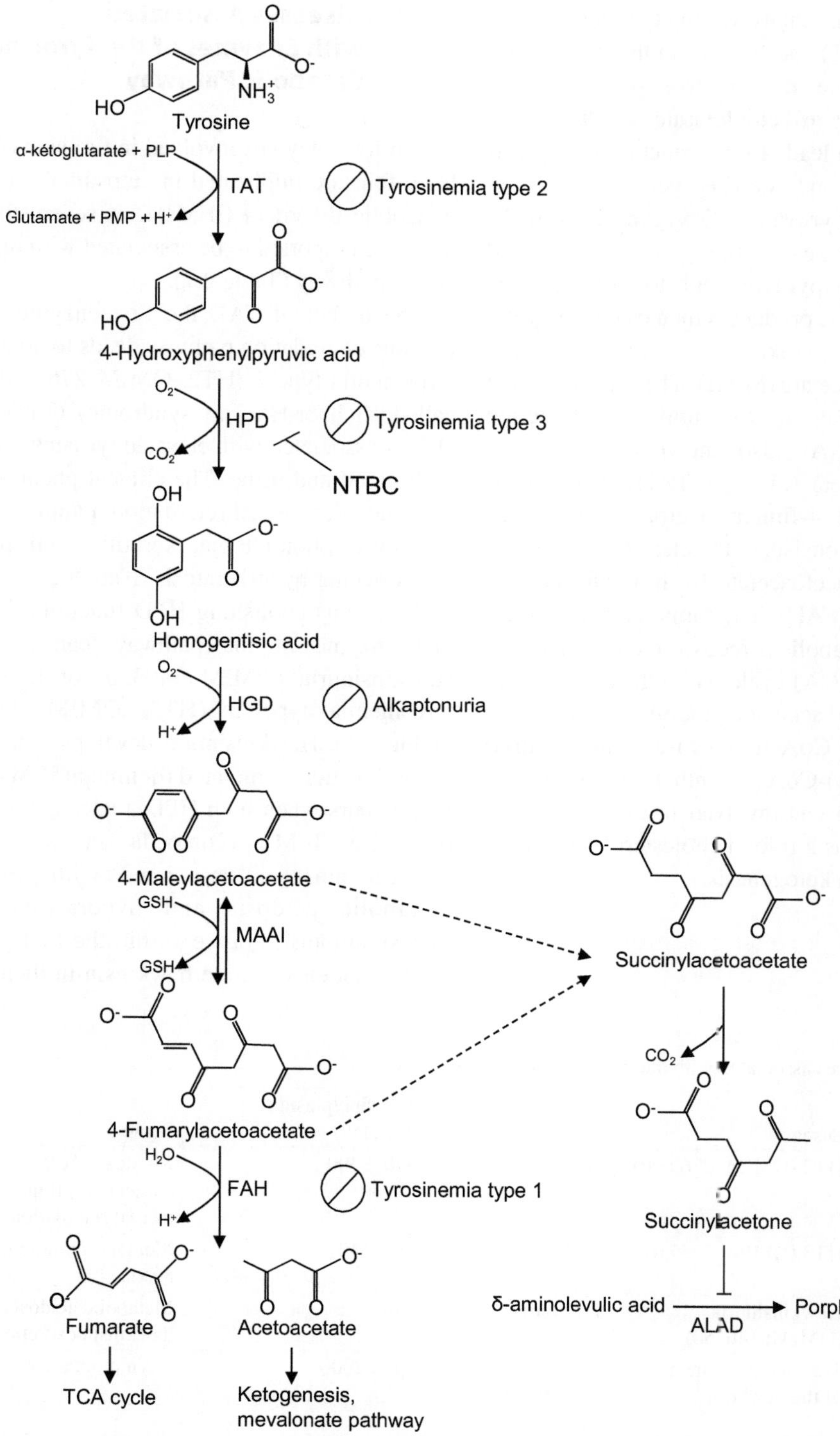

Fig. 2.2 Tyrosine degradation pathway. Enzymatic steps of the tyrosine catabolic pathway. Each enzyme is abbreviated and the corresponding IEM is written near the "no" symbol. The conversion of 4-maleylacetoacetate (MAA) and 4-fumarylacetoacetate (FAA) into succinylacetoacetate (SAA) and then succinylacetone (SA) is depicted as well as the inhibitory effect of SA on ALAD. The site of action of NTBC is also indicated. *TAT* tyrosine aminotransferase, *HPD* *p*-hydroxyphenylpyruvate dioxygenase, *HGO* homogentisate oxidase, *MAAI* maleylacetoacetate isomerase, *FAH* fumarylacetoacetate hydrolase, *TCA cycle* tricarboxylic acid cycle, *ALAD* δ-aminolevulic acid dehydratase

The first step is catalyzed by tyrosine aminotransferase (TAT) and consists in the transformation of tyrosine in 4-hydroxyphenylpyruvic acid. TAT uses α-ketoglutarate as the amino acceptor, which leads to the generation of glutamate. The second step is performed by *p*-hydroxyphenylpyruvate dioxygenase (HPD), which catalyzes the oxidation of 4-hydroxyphenylpyruvic acid to homogentisic acid (HGA). This product is then oxidized by the homogentisate oxidase (HGO) into 4-maleylacetoacetate (MAA). The fourth enzyme of the pathway is the maleylacetoacetate isomerase (MAAI, also known as glutathione S-transferase zeta (ζ) 1 (GSTZ1)), which converts MAA in 4-fumarylacetoacetate (FAA). The final step consists in the cleavage of FAA in fumarate and acetoacetate by fumarylacetoacetate hydrolase (FAH). The fumarate end product of tyrosine catabolism feeds directly the tricarboxylic acid (TCA) cycle while the acetoacetate is activated to acetoacetyl-CoA via succinyl-CoA:3-oxoacid-CoA transferase (mitochondria) or acetoacetyl-CoA synthetase (cytosol). Acetoacetyl-CoA is involved in the mevalonate pathway and has a role in cholesterol biosynthesis as well as in ketogenesis.

2.2.2 Diseases Associated with Enzymes of the Tyrosine Catabolic Pathway

Of the five enzymes involved in tyrosine catabolism, four are implicated in autosomal recessive metabolic disorders (IEMs) and seven diseases have been reported to be associated with this catabolic pathway (Table 2.1).

The defect of TAT, the first enzyme of the tyrosine degradation pathway, leads to hereditary tyrosinemia type 2 (HT2, OMIM 276600) (also called Richner-Hanhart syndrome) (Table 2.1). HT2 is associated with elevated tyrosine levels in both blood and urine. The clinical phenotype of HT2 includes mental retardation, painful corneal eruptions, photophobia, keratitis, and painful palmoplantar hyperkeratosis (Natt et al. 1992).

Mutations abolishing HPD function, the second enzyme of the pathway, can result in Hawkinsinuria (OMIM 140350) or hereditary tyrosinemia type 3 (HT3, OMIM 276710) (Table 2.1). Hawkinsinuria develops when only one HPD allele is mutated (dominant IEM) while HT3 occurs when both HPD alleles are affected (recessive IEM) (Tomoeda et al. 2000). Hawkinsinuria is associated with transient metabolic acidosis and hypertyrosinemia. The symptoms improve within the first year of life but patients excrete hawkinsin in their urine

Table 2.1 Diseases associated with the tyrosine catabolic pathway

Enzyme	Disease	Incidence	Tyrosine/plasma (pmole/l)	Signs
TAT	HT2 (OMIM: 276,600)	Rare	370–3300	Keratosis, keratitis, corneal eruptions, mental retardation
HPD	HT3 (OMIM: 276710)	Very rare	355–640	Mental retardation, ataxia
	Hawkinsinuria (OMIM: 140350)	Very rare	196	Metabolic acidosis, hawkinsin excretion
	Transient tyrosinemia of the newborn	30–50% of premature baby, transient	up to 2000	Asymptomatic
HGO	Alkaptonuria (OMIM: 203500)	Frequent	–	Arthritis, ochronosis, dark urine
MAAI	HT1b	Very rare	1000	Undefined
FAH	HT1 (OMIM: 276700)	Rare to frequent	400–800	Liver and kidney dysfunctions

throughout life (Niederwieser et al. 1977; Wilcken et al. 1981). HT3 is a very rare disease with clinical phenotype including mild mental retardation and/or convulsions, and noteworthy, the absence of liver damage (Tomoeda et al. 2000). Of note, a delay in HPD maturation results in transient tyrosinemia of the newborn. This condition is benign and disappears spontaneously with no sequelae (Mitchell et al. 2001; Russo et al. 2001).

A defect of HGO, the third enzyme of the pathway, leads to the first IEM ever recognized, alkaptonuria (AKU, OMIM 203500) (Garrod 1902). This disease is not life-threatening and the usual consequences are ochronosis (bluish-black discoloration of the tissues) and arthritis (reviewed in (Vilboux et al. 2009)). The accumulation of HGA also causes the urine to darken on exposure to air.

Up to now, a deficiency in MAAI has been suggested twice in literature but it has not been clearly confirmed (Karnik et al. 2004; Fernandez-Canon et al. 2002) (OMIM gene ID: 603758). In both cases, patient developed liver failure and kidney dysfunction similar to hereditary tyrosinemia type 1 (HT1, see below) but with absence of SA accumulation and the corresponding disease was therefore called HT1 type b (HT1b). It remains unclear if MAAI deficiency actually exists in human patients, especially since there is an enzyme-independent bypass that allows the isomerization of MAA in FAA (Fernandez-Canon et al. 2002).

Finally, the defect of the last enzyme of the pathway (FAH) causes HT1 (OMIM 276700), which is a severe progressive liver disease coupled with renal tubular dysfunction (Lindblad et al. 1977; Mitchell et al. 2001; Russo et al. 2001). HT1 is the most severe disease associated to the tyrosine catabolism pathway.

2.3 HT1 Incidence

The incidence of HT1 is around 1/100,000 births worldwide but can be much higher in certain regions due to founder effects (reviewed in (Angileri et al. 2015)). The highest incidence of HT1 is found in the region of Saguenay-Lac-St-Jean (SLSJ) (Quebec Province, Canada), where 1:1846 children has HT1 and 1:22 individual is a carrier of a disease allele (De Braekeleer and Larochelle 1990; Grompe et al. 1994; Poudrier et al. 1996) (See Larochelle, Chap. 1). The splice mutation c.1062+5G>A (IVS12+5G→A) is the most frequently found mutation within this population (~90% of the HT1 reported allele) and accounts for approximately one third of all HT1 reported alleles worldwide (Grompe et al. 1994; Poudrier et al. 1996; St-Louis and Tanguay 1997; Angileri et al. 2015).

A second mutation cluster of HT1 is found in Scandinavia, most precisely in the Finnish population of Pohjanmaa where 1/5000 individual is affected by HT1 while the overall incidence of HT1 in Finland is 1/60,000 (Kvittingen et al. 1981; Mustonen et al. 1997; St-Louis et al. 1994). The most frequent HT1 reported allele in Finland is c.786G>A (p.W262X) which represents ~88% of the reported alleles in nine out of ten alleles in this country (St-Louis et al. 1994). A third cluster of HT1 occurs in an immigrant population from Pakistan living predominantly in Birmingham (United Kingdom) (Hutchesson et al. 1998). HT1 alleles have been reported all over the world except in Central America and on the Oceania continent (Angileri et al. 2015) (See Morrow et al. Chap. 3).

2.4 FAH, the Deficient Enzyme in HT1

The *FAH* gene is located on the long arm of chromosome 15 at position 25.1 (15q25.1; base pairs 80,152,890 to 80,186,581) (Phaneuf et al. 1991). The corresponding protein (FAH, E.C. 3.7.1.2) is a metalloenzyme for which the structure has been determined by crystallography (Timm et al. 1999; Tanguay 2002; Bateman et al. 2007; Bateman et al. 2001). FAH is a cytosolic homodimeric enzyme of two 46 kDa subunits, which is mainly expressed in liver and kidneys. It is also expressed in other tissues, such as fibroblasts, amniocytes, chorionic villi, erythrocytes and oligodendrocytes, albeit at a lower level. As the last enzyme of the tyrosine catabolic pathway,

FAH catalyzes the conversion of FAA into fumarate and acetoacetate (Phaneuf et al. 1992; Tanguay et al. 1990) (Fig. 2.2).

Ninety-eight mutations have been reported to cause HT1 and their occurrence worldwide has been recently compiled (Angileri et al. 2015). The most common *FAH* mutation causing HT1 are c.1062+5G>A (IVS12+5G>A) followed by c.554-1G>T (IVS6-1G > T) and c.786G > A (p.W262X) (See Morrow et al. Chap. 3).

2.5 Biochemical Features of HT1

The deficiency of FAH was originally associated with increased levels of hepatic transaminases, as well as increased plasma levels of tyrosine, methionine and phenylalanine and with urinary elevated concentrations of tyrosine metabolites (p-hydroxyphenylpyruvate, p-hydroxyphenyllactate and p-hydroxyphenylacetate) (Table 2.2). Hypertyrosinemia can be caused by numerous other conditions affecting the liver and is also a feature of transient tyrosinemia of the newborn, a condition that resolves spontaneously without significant damages (Mitchell et al. 2001; Russo et al. 2001) (Table 2.1).

HT1 patients usually present with high levels of plasma α-fetoprotein (AFP) and most importantly with high levels of succinylacetone in plasma and urine (SA) the only valid prognosis marker of HT1. The latter metabolite is widely used for HT1 screening and its presence can be directly linked to the lack of FAH activity. Indeed, the absence of FAH results in the accumulation of FAA and MAA, which are then reduced in succinylacetoacetate (SAA) (Fig. 2.2). The subsequent decarboxylation of SAA is responsible of the observed succinylacetone (SA).

FAA and SA are the most damaging metabolites resulting from FAH deficiency and considerable efforts have been made to find by which molecular mechanisms these compounds act to produce the severe phenotype seen in HT1 (See Tanguay et al. Chap. 4).

2.5.1 Overview of FAA Toxicity

FAA is an electrophilic compound that has been suggested to damage DNA. While a direct effect on DNA remains to be demonstrated, it was shown to be mutagenic in a cell assay, albeit at a much lower level than classical mutagens (Tanguay et al. 1996). FAA induces genome instability through activation of the ERK pathway (Jorquera and Tanguay 2001). Moreover, it induces cell cycle arrest and apoptosis through glutathione (GSH) depletion (Jorquera and Tanguay 1997, 2001). Hence, GSH, a major actor of redox homeostasis, was shown to reduce FAA mutagenicity in cultured cells (Jorquera and

Table 2.2 Clinical characteristics of HT1

Symptoms	Biochemistry	Pathology
Vomiting	Tyrosinemia	Liver cirrhosis
Diarrhea	Methioninemia	Fanconi syndrome
Muscle weakness	Hyperbilirubinemia	Peripheral neuropathy
Hepato-splenomegaly	Hypoglycemia	Hepatocarcinoma (HCC)
Jaundice	Hypoproteinemia	Renal tubule dilatation
Cabbage smell	Hypothrombinemia	Rickets
Auto-mutilation	Aminoaciduria	Hypertrophy pancreas
Fever	Succinylacetonuria	
Anemia	Tyrosyluria	
Irritability	Glucosuria	
Ascites	Phosphaturia	
Bleeding	Alphafoetoprotein	
Paralysis	Anemia	
Hypotonia	Hematuria	

Tanguay 1997, 2001) and to rescue neonatal death in the *fah* knockout model of HT1 (Langlois et al. 2008). At the cellular level, the stress caused by FAA accumulation was also shown to induce the unfolding protein response (UPR) in the endoplasmic reticulum (Bergeron et al. 2006).

In an interesting study, FAA was shown to inhibit 6 of the 7 human DNA glycosylases involved in DNA base removal during base excision repair (BER) (Bliksrud et al. 2013). This is in agreement with a previous report showing that the expression of DNA glycosylase OGG1 and the nucleotide excision repair protein ERCC1, was reduced in lymphocytes of two HT1 patients (van Dyk et al. 2010). Preventing DNA repair by the BER pathway, would favor accumulation of oxidative damage to DNA resulting in increased potent mutagenic lesions. Noteworthy, GSH depletion due to FAA accumulation is likely to affect DNA repair indirectly as redox homeostasis is also important for this process (Langie et al. 2007; Storr et al. 2012).

2.5.2 Overview of SA Toxicity

Contrary to FAA, SA has no mutagenic effect on DNA (Tanguay et al. 1996) nor was it shown to have any inhibiting effect on human DNA glycosylases (Bliksrud et al. 2013).

The toxicity of SA mostly relies on its ability to be a competitive inhibitor of δ-aminolevulinic acid dehydratase (ALAD), the enzyme responsible of the conversion of δ-aminolevulinic acid (ALA) in porphobilinogen, which is precursor of heme synthesis (Fig. 2.2). This inhibition results in the accumulation of ALA and its excretion in urine. ALA has been associated with mitochondrial toxicity, liver toxicity, liver cancer, and neuropshychiatric problems. Indeed, it lowers heme levels for cytochrome/hemoproteins synthesis and increases mitochondrial iron levels, resulting in heme deficiency, down regulation of mitochondrial cytochrome oxidase and overall mitochondrial toxicity (reviewed in (Lee and O'Brien 2010)). Accordingly, SA is widely used as a heme synthesis inhibitor to create a mitochondrial iron-loading model that is similar to the mitochondrial iron loading found in Friedreich's ataxia (Lee and O'Brien 2010; Richardson et al. 2001).

At the tissue level, SA causes dysfunction of kidney membrane transport by altering membrane fluidity and possibly disrupting normal structure. Indeed, it was shown to cause renal tubular dysfunction in normal rat kidneys, mimicking human Fanconi syndrome (Roth et al. 1991; Wyss et al. 1992; Tanguay et al. 2009).

2.6 Clinical Features of HT1

HT1 is characterized by progressive liver disease and renal tubular dysfunction leading to hypophosphatemic rickets. Moreover, it is the IEM with the highest incidence of hepatocellular carcinoma (HCC) (Schady et al. 2015) (Table 2.3). HT1 is categorized into three main clinical types (acute, subacute and chronic) based on the age of onset and the clinical manifestations (Tanguay et al. 1990; Mitchell et al. 2001; van Spronsen et al. 1994).

2.6.1 Acute, Subacute and Chronic Forms of HT1

The acute form of HT1 has an onset before 2 months of age and is mainly characterized by severe liver failure associated with cirrhosis, hepato- and spleno-megaly, abnormal blood coagulation and hypoglycemia leading to death in the first months of life (Table 2.3). Renal tubular dysfunctions such as Fanconi syndrome and rickets have also been considered hallmarks of HT1 (Mitchell et al. 2001; Russo et al. 2001). The subacute form is similar to the acute form but symptoms appear between 2 and 6 months (van Spronsen et al. 1994) (Table 2.3).

The chronic form is initially less aggressive and presents after 6 months of age. While its onset is insidious and progressive, renal manifestations, such as proximal tubulopathy, are prominent and may even be the presenting problem

Table 2.3 Distinctive features of each HT1 forms

	Acute	Subacute	Chronic
Age at manifestation	0–2 months	2–6 months	After 6 months
Progression	Fast	Fast	Slow
Life expectancy when untreated	0–1 year	0–1 year	0–10 years
Characteristic symptoms	Severe hepatic dysfunction	Rickets	Cirrhosis
	Hepatomegaly	Failure to thrive	Renal tubular Dysfunction
		Easy bruising	Growth retardation
	Abnormal blood coagulation	Hepatomegaly	
Main cause of death	Liver failure	HCC	HCC
	Recurrent bleeding	Liver failure	Porphyria
Extent of mutation reversion	Low (1.6%)	Low to moderate (22%)	Moderate (36%)
FAH activity	Absent	Absent to residual	Residual

Data in the table are based on the following papers: (Demers et al. 2003; Tanguay et al. 1990; van Spronsen et al. 1994).

(Table 2.3). Patients show impaired renal tubular reabsorption functions leading to Fanconi syndrome, renal tubular acidosis, generalized aminoaciduria, hypophosphatemic vitamin D-resistant rickets and growth retardation (Paradis et al. 1990; Russo and O'Regan 1990; Fernandez-Lainez et al. 2014; Forget et al. 1999).

2.6.2 High Incidence of HCC in HT1

HT1 is characterized by gradual liver alterations leading to cirrhosis and HCC development. In fact, the risk of developing HCC in HT1 is considered the highest among all metabolic disorders (Russo et al. 2001; Schady et al. 2015). In an early study, HCC was reported in 37% of HT1 patients over 2 years of age (Weinberg et al. 1976) but subsequent studies in Scandinavia (van Spronsen et al. 1989) and in Quebec (Russo et al. 2001) showed a lower frequency of HCC (~15%) likely due to the advent of transplantation and improved treatment. In addition to be at high risk of developing HCC, HT1 patients develop them earlier then patients having other diseases (often before 5 years of age) (Schady et al. 2015) (See van Ginkel et al. Chap. 9).

2.7 Diagnosis of HT1

As mentioned above, the deficiency of FAH gives rise to elevated plasma concentrations of amino acids such as tyrosine and methionine as well as excretion of unusual tyrosine metabolites like SA (Mitchell et al. 2001; Russo et al. 2001) (Table 2.2). Although elevated levels of tyrosine and plasma AFP are indicative of HT1, the most reliable biochemical diagnostic marker consists in the presence of SA in urine, blood and amniotic fluid (Grenier et al. 1976; Grenier et al. 1982). In Quebec, where a high incidence of HT1 is observed, a neonatal screening program for the disease has been established in 1970 and consists of measuring SA levels in dried blood spots. Tandem mass spectrometry (MS/MS) is now used as a sensitive and rapid method, for screening HT1 by measuring the level of SA (Allard et al. 2004). Prenatal biochemical diagnosis can also be done by measuring the level of SA in amniotic fluid sampled between the 14th and 16th weeks of pregnancy (Grenier et al. 1996; Jakobs et al. 1990). However, some false-positives have been reported using this method.

An enzymatic assay based on FAH activity measurements on cultured fibroblasts, blood, or liver specimen has also been used for diagnosis (Kvittingen et al. 1981, 1983). However, this method is less reliable on liver specimen as some HT1 patients have mosaic expression of FAH in their liver due to reversion of the mutation (Demers et al. 2003; Kvittingen et al. 1993, 1994; Poudrier et al. 1998).

Genetic diagnosis tests are also performed when the family history or the origin of the patient suggests that one parent may be a carrier of the disorder. This kind of test has been facilitated by the technological progress achieved in the past decade and the identification of predominant mutations in certain ethnic groups (Angileri et al. 2015). The best example of this is the genetic screening test that was designed to detect the most common mutation found in HT1 (c.1062+5G>A, IVS12+5G→A) (Grompe and al-Dhalimy 1995). The method is based on the amplification of the genomic DNA region containing the mutation by PCR followed by enzymatic cleavage of the amplified sequence in order to distinguish the mutated allele from the wild-type sequence. Similar molecular tests have since been developed for most mutations and can be performed on blood, chorionic villi or cultured amniocytes. However, with the improvement of new sequencing technologies, it is becoming very simple to perform the *FAH* gene sequencing.

2.8 Treatment of HT1

2.8.1 Restrictive Diet

Before 1990s, there was no treatment available for HT1 except liver transplantation. Patients were following a restrictive diet with low phenylalanine and tyrosine intake. While this was beneficial at the beginning, it was not fully preventing ulterior liver damage and renal dysfunction.

2.8.2 Orthotopic Liver Transplantation

Orthotopic liver transplantation (OLT) is performed in the most severe HT1 cases due to the risks associated to the surgery. OLT is essentially curative but does not fully correct metabolic perturbations in HT1 since kidneys continue to excrete SAA, SA and ALA in urine (Fernandez-Lainez et al. 2014; Tuchman et al. 1987; Pierik et al. 2005; Bartlett et al. 2013). Only half of liver transplanted patients show partial improvement of renal function but still, altered kidney size and architecture persists (Paradis et al. 1990; Fernandez-Lainez et al. 2014; Forget et al. 1999) (See Alvarez and Mitchell Chap. 5 and McKiernan Chap. 7 on liver transplantation).

2.8.3 NTBC

NTBC (2-(2-nitro-4-trifluoro-methylbenzyol)-1,3 cyclohexanedione, Nitisinone) was first used in 1992 (Lindstedt et al. 1992). It acts by inhibiting the second enzyme of the tyrosine catabolic pathway, HPD (Fig. 2.2). The advantage of blocking the pathway at this step is that there is no accumulation of FAA and MAA and therefore no accumulation of SA either. Moreover, as mentioned in Sect. 2.2 no liver damages are associated with inhibition of HPD in HT3.

The use of NTBC combined to the low tyrosine/phenylalanine diet has proven to be very efficient in preventing HT1 progression, by curing both liver and kidney dysfunctions (Larochelle et al. 2012; Bartlett et al. 2014). It is very efficient particularly when introduced early in life (Larochelle et al. 2012), but it is still unsure if it will be sufficient to prevent problems on a long-term basis. For instance, failure to respond to NTBC has been reported in one child (Mohan et al. 1999) and high risks of HCC have been reported when NTBC is introduced after 2 years of age (McKiernan 2006; van Spronsen et al. 2005) (for more information on NTBC see Maiorana and Dionisi-Vici Chap. 8 and Lock Chap. 16).

2.9 Concluding Remarks

HT1 is a severe liver disease that should be detected at the earliest to be treated effectively. The management of this disease has been revolutionised by the introduction of NTBC (de Laet et al. 2013), but the venue of alternative/complementary treatments will be of upmost importance due to difficulties to fully comply to the restrictive diet for some patients and the high cost of NTBC. It is also important to favor newborn screening programs for the early detection of HT1 patients whenever possible. Such screening is inexpensive and should prevent late intervention when liver damage has already been done. Finally additional basic research is still needed to unveil the pathogenic mechanisms involved in HT1.

Acknowledgements The work on HT1 in the RMT's lab was supported by grants from the Canadian Institutes of Health Research (CIHR) and Fondation du Grand Défi Pierre Lavoie.

References

Allard P, Grenier A, Korson MS, Zytkovicz TH (2004) Newborn screening for hepatorenal tyrosinemia by tandem mass spectrometry: analysis of succinylacetone extracted from dried blood spots. Clin Biochem 37(11):1010–1015. doi:S0009-9120(04)00202-4

Angileri F, Bergeron A, Morrow G, Lettre F, Gray G, Hutchin T, Ball S (2015) Geographical and ethnic distribution of mutations of the fumarylacetoacetate hydrolase gene in hereditary tyrosinemia type 1. JIMD Rep 19:45–58. doi:10.1007/8904_2014_363

Applegarth DA, Toone JR, Lowry RB (2000) Incidence of inborn errors of metabolism in British Columbia, 1969–1996. Pediatrics 105(1):e10

Bartlett DC, Preece MA, Holme E, Lloyd C, Newsome PN, McKiernan PJ (2013) Plasma succinylacetone is persistently raised after liver transplantation in tyrosinaemia type 1. J Inherit Metab Dis 36(1):15–20. doi:10.1007/s10545-012-9482-1

Bartlett DC, Lloyd C, McKiernan PJ, Newsome PN (2014) Early nitisinone treatment reduces the need for liver transplantation in children with tyrosinaemia type 1 and improves post-transplant renal function. J Inherit Metab Dis 37(5):745–752. doi:10.1007/s10545-014-9683-x

Bateman RL, Bhanumoorthy P, Witte JF, McClard RW, Grompe M, Timm DE (2001) Mechanistic inferences from the crystal structure of fumarylacetoacetate hydrolase with a bound phosphorus-based inhibitor. J Biol Chem 276(18):15284–15291. doi:10.1074/jbc.M007621200

Bateman RL, Ashworth J, Witte JF, Baker LJ, Bhanumoorthy P, Timm DE, Hurley TD, Grompe M, McClard RW (2007) Slow-onset inhibition of fumarylacetoacetate hydrolase by phosphinate mimics of the tetrahedral intermediate: kinetics, crystal structure and pharmacokinetics. Biochem J 402(2):251–260. doi:10.1042/BJ20060961

Bergeron A, Jorquera R, Orejuela D, Tanguay RM (2006) Involvement of endoplasmic reticulum stress in hereditary tyrosinemia type I. J Biol Chem 281(9):5329–5334. doi:M506804200

Bliksrud YT, Ellingsen A, Bjoras M (2013) Fumarylacetoacetate inhibits the initial step of the base excision repair pathway: implication for the pathogenesis of tyrosinemia type I. J Inherit Metab Dis 36(5):773–778. doi:10.1007/s10545-012-9556-0

De Braekeleer M, Larochelle J (1990) Genetic epidemiology of hereditary tyrosinemia in Quebec and in Saguenay-Lac-St-Jean. Am J Hum Genet 47(2):302–307

de Laet C, Dionisi-Vici C, Leonard JV, McKiernan P, Mitchell G, Monti L, de Baulny HO, Pintos-Morell G, Spiekerkotter U (2013) Recommendations for the management of tyrosinaemia type 1. Orphanet J Rare Dis 8:8. doi:1750-1172-8-8

Demers SI, Russo P, Lettre F, Tanguay RM (2003) Frequent mutation reversion inversely correlates with clinical severity in a genetic liver disease, hereditary tyrosinemia. Hum Pathol 34(12):1313–1320. doi:S0046817703004064

Fernandez-Canon JM, Baetscher MW, Finegold M, Burlingame T, Gibson KM, Grompe M (2002) Maleylacetoacetate isomerase (MAAI/GSTZ)-deficient mice reveal a glutathione-dependent nonenzymatic bypass in tyrosine catabolism. Mol Cell Biol 22(13):4943–4951

Fernandez-Lainez C, Ibarra-Gonzalez I, Belmont-Martinez L, Monroy-Santoyo S, Guillen-Lopez S, Vela-Amieva M (2014) Tyrosinemia type I: clinical and biochemical analysis of patients in Mexico. Ann Hepatol 13(2):265–272. doi:1090941

Forget S, Patriquin HB, Dubois J, Lafortune M, Merouani A, Paradis K, Russo P (1999) The kidney in children with tyrosinemia: sonographic, CT and biochemical findings. Pediatr Radiol 29(2):104–108

Garrod AE (1902) About alkaptonuria. Med-Chir Trans 85:69–78

Grenier A, Belanger L, Laberge C (1976) alpha1-Fetoprotein measurement in blood spotted on paper: discriminating test for hereditary tyrosinemia in neonatal mass screening. Clin Chem 22(7):1001–1004

Grenier A, Lescault A, Laberge C, Gagne R, Mamer O (1982) Detection of succinylacetone and the use of its measurement in mass screening for hereditary tyrosinemia. Clin Chim Acta 123(1–2):93–99

Grenier A, Cederbaum S, Laberge C, Gagne R, Jakobs C, Tanguay RM (1996) A case of tyrosinaemia type

I with normal level of succinylacetone in the amniotic fluid. Prenat Diagn 16(3):239–242. doi:10.1002/(SICI)1097-0223(199603)16:3<239::AID--PD829>3.0.CO;2-W

Grompe M, Al-Dhalimy M (1995) Rapid nonradioactive assay for the detection of the common French Canadian tyrosinemia type I mutation. Hum Mutat 5(1):105. doi:10.1002/humu.1380050117

Grompe M, St-Louis M, Demers SI, al-Dhalimy M, Leclerc B, Tanguay RM (1994) A single mutation of the fumarylacetoacetate hydrolase gene in French Canadians with hereditary tyrosinemia type I. N Engl J Med 331(6):353–357. doi:10.1056/NEJM199408113310603

Hutchesson AC, Bundey S, Preece MA, Hall SK, Green A (1998) A comparison of disease and gene frequencies of inborn errors of metabolism among different ethnic groups in the West Midlands, UK. J Med Genet 35(5):366–370

Jakobs C, Stellaard F, Kvittingen EA, Henderson M, Lilford R (1990) First-trimester prenatal diagnosis of tyrosinemia type I by amniotic fluid succinylacetone determination. Prenat Diagn 10(2):133–134

Jorquera R, Tanguay RM (1997) The mutagenicity of the tyrosine metabolite, fumarylacetoacetate, is enhanced by glutathione depletion. Biochem Biophys Res Commun 232(1):42–48. (47). doi:http://dx.doi.org/10.1006/bbrc.1997.6220

Jorquera R, Tanguay RM (2001) Fumarylacetoacetate, the metabolite accumulating in hereditary tyrosinemia, activates the ERK pathway and induces mitotic abnormalities and genomic instability. Hum Mol Genet 10(17):1741–1752

Karnik D, Thomas N, Eapen CE, Jana AK, Oommen A (2004) Tyrosinemia type I: a clinico-laboratory case report. Indian J Pediatr 71(10):929–932

Kvittingen EA, Jellum E, Stokke O (1981) Assay of fumarylacetoacetate fumarylhydrolase in human liver-deficient activity in a case of hereditary tyrosinemia. Clin Chim Acta 115(3):311–319

Kvittingen EA, Halvorsen S, Jellum E (1983) Deficient fumarylacetoacetate fumarylhydrolase activity in lymphocytes and fibroblasts from patients with hereditary tyrosinemia. Pediatr Res 17(7):541–544. doi:10.1203/00006450-198307000-00005

Kvittingen EA, Rootwelt H, Brandtzaeg P, Bergan A, Berger R (1993) Hereditary tyrosinemia type I. Self-induced correction of the fumarylacetoacetase defect. J Clin Invest 91(4):1816–1821. doi:10.1172/JCI116393

Kvittingen EA, Rootwelt H, Berger R, Brandtzaeg P (1994) Self-induced correction of the genetic defect in tyrosinemia type I. J Clin Invest 94(4):1657–1661. doi:10.1172/JCI117509

Langie SA, Knaapen AM, Houben JM, van Kempen FC, de Hoon JP, Gottschalk RW, Godschalk RW, van Schooten FJ (2007) The role of glutathione in the regulation of nucleotide excision repair during oxidative stress. Toxicol Lett 168(3):302–309. doi:S0378-4274(06)01335-X

Langlois C, Jorquera R, Orejuela D, Bergeron A, Finegold MJ, Rhead WJ, Tanguay RM (2008) Rescue from neonatal death in the murine model of hereditary tyrosinemia by glutathione monoethylester and vitamin C treatment. Mol Genet Metab 93(3):306–313. doi:S1096-7192(07)00441-6

Larochelle J, Alvarez F, Bussieres JF, Chevalier I, Dallaire L, Dubois J, Faucher F, Fenyves D, Goodyer P, Grenier A, Holme E, Laframboise R, Lambert M, Lindstedt S, Maranda B, Melancon S, Merouani A, Mitchell J, Parizeault G, Pelletier L, Phan V, Rinaldo P, Scott CR, Scriver C, Mitchell GA (2012) Effect of nitisinone (NTBC) treatment on the clinical course of hepatorenal tyrosinemia in Quebec. Mol Genet Metab 107(1–2):49–54. doi:S1096-7192(12)00211-9

Lee O, O'Brien PJ (2010) Modifications of mitochondrial function by toxicants. In: CA MQ (ed) Comprehensive toxicology. Elsevier, Oxford, pp 411–445

Lindblad B, Lindstedt S, Steen G (1977) On the enzymic defects in hereditary tyrosinemia. Proc Natl Acad Sci U S A 74(10):4641–4645

Lindstedt S, Holme E, Lock EA, Hjalmarson O, Strandvik B (1992) Treatment of hereditary tyrosinaemia type I by inhibition of 4-hydroxyphenylpyruvate dioxygenase. Lancet 340(8823):813–817. doi:0140-6736(92)92685-9

McKiernan PJ (2006) Nitisinone in the treatment of hereditary tyrosinaemia type 1. Drugs 66(6):743–750. doi:6662

Mitchell GA, Grompe M, Lambert H, Tanguay RM (2001) Hypertyrosinemia. In: Scriver C, Beaudet A, Sly WSJ, Valle D (eds) The metabolic and molecular bases of inherited diseases, vol II, 8th edn. McGrawHill, New York, pp 1777–1805

Mohan N, McKiernan P, Preece MA, Green A, Buckels J, Mayer AD, Kelly DA (1999) Indications and outcome of liver transplantation in tyrosinaemia type 1. Eur J Pediatr 158(Suppl 2):S49–S54. doi:9158S049.431

Mustonen A, Ploos van Amstel HK, Berger R, Salo MK, Viinikka L, Simola KO (1997) Mutation analysis for prenatal diagnosis of hereditary tyrosinaemia type 1. Prenat Diagn 17(10):964–966. doi:10.1002/(SICI)1097-0223(199710)17:10<964::AID--PD164>3.0.CO;2-6

Natt E, Kida K, Odievre M, Di Rocco M, Scherer G (1992) Point mutations in the tyrosine aminotransferase gene in tyrosinemia type II. Proc Natl Acad Sci U S A 89(19):9297–9301

Niederwieser A, Matasovic A, Tippett P, Danks DM (1977) A new sulfur amino acid, named hawkinsin, identified in a baby with transient tyrosinemia and her mother. Clin Chim Acta 76(3):345–356

Paradis K, Weber A, Seidman EG, Larochelle J, Garel L, Lenaerts C, Roy CC (1990) Liver transplantation for hereditary tyrosinemia: the Quebec experience. Am J Hum Genet 47(2):338–342

Phaneuf D, Labelle Y, Berube D, Arden K, Cavenee W, Gagne R, Tanguay RM (1991) Cloning and expression of the cDNA encoding human fumarylacetoacetate hydrolase, the enzyme deficient in hereditary tyrosinemia: assignment of the gene to chromosome 15. Am J Hum Genet 48(3):525–535

Phaneuf D, Lambert M, Laframboise R, Mitchell G, Lettre F, Tanguay RM (1992) Type 1 hereditary tyrosinemia. Evidence for molecular heterogeneity and identification of a causal mutation in a French Canadian patient. J Clin Invest 90(4):1185–1192. doi:10.1172/JCI115979

Pierik LJ, van Spronsen FJ, Bijleveld CM, van Dael CM (2005) Renal function in tyrosinaemia type I after liver transplantation: a long-term follow-up. J Inherit Metab Dis 28(6):871–876. doi:10.1007/s10545-005-0059-0

Poudrier J, St-Louis M, Lettre F, Gibson K, Prevost C, Larochelle J, Tanguay RM (1996) Frequency of the IVS12 + 5G-->A splice mutation of the fumarylacetoacetate hydrolase gene in carriers of hereditary tyrosinaemia in the French Canadian population of Saguenay-Lac-St-Jean. Prenat Diagn 16(1):59–64. doi:10.1002/(SICI)1097-0223(199601)16:1<59::AID-PD810>3.0.CO;2-D

Poudrier J, Lettre F, Scriver CR, Larochelle J, Tanguay RM (1998) Different clinical forms of hereditary tyrosinemia (type I) in patients with identical genotypes. Mol Genet Metab 64(2):119–125. doi:S1096-7192(98)92695-6

Richardson DR, Mouralian C, Ponka P, Becker E (2001) Development of potential iron chelators for the treatment of Friedreich's ataxia: ligands that mobilize mitochondrial iron. Biochim Biophys Acta 1536(2–3):133–140

Roth KS, Carter BE, Higgins ES (1991) Succinylacetone effects on renal tubular phosphate metabolism: a model for experimental renal Fanconi syndrome. Proc Soc Exp Biol Med 196(4):428–431

Russo P, O'Regan S (1990) Visceral pathology of hereditary tyrosinemia type I. Am J Hum Genet 47(2):317–324

Russo PA, Mitchell GA, Tanguay RM (2001) Tyrosinemia: a review. Pediatr Dev Pathol 4(3):212–221

Schady DA, Roy A, Finegold MJ (2015) Liver tumors in children with metabolic disorders. Transl Pediatr 4(4):290–303. doi:10.3978/j.issn.2224-4336.2015.10.08

Sniderman King L, Trahms C, Scott CR (2006) Tyrosinemia type 1. In: Pagon RA, Adam MP, Bird TD, Dolan CR, Fong CT, Stephens K (eds). doi:http://www.ncbi.nlm.nih.gov/books/NBK1515/

St-Louis M, Tanguay RM (1997) Mutations in the fumarylacetoacetate hydrolase gene causing hereditary tyrosinemia type I: overview. Hum Mutat 9(4):291–299. doi:10.1002/(SICI)1098-1004(1997)9:4<291::AID-HUMU1>3.0.CO;2-9

St-Louis M, Leclerc B, Laine J, Salo MK, Holmberg C, Tanguay RM (1994) Identification of a stop mutation in five Finnish patients suffering from hereditary tyrosinemia type I. Hum Mol Genet 3(1):69–72

Storr SJ, Woolston CM, Martin SG (2012) Base excision repair, the redox environment and therapeutic implications. Curr Mol Pharmacol 5(1):88–101. doi:EPub-Abstract-CMP-60

Tanguay R (2002) Fumarylacetoacetate hydrolase. Wiley encyclopedia of molecular medicine. Wiley, New York

Tanguay RM, Valet JP, Lescault A, Duband JL, Laberge C, Lettre F, Plante M (1990) Different molecular basis for fumarylacetoacetate hydrolase deficiency in the two clinical forms of hereditary tyrosinemia (type I). Am J Hum Genet 47(2):308–316

Tanguay RM, Jorquera R, Poudrier J, St-Louis M (1996) Tyrosine and its catabolites: from disease to cancer. Acta Biochim Pol 43(1):209–216

Tanguay RM, Bergeron A, Jorquera R (2009) Hepatorenal tyrosinemia. In: Lifton RP, Somlo S, Giebsih D, Seldin DW (eds) Genetic diseases of the kidney, vol Chap 40. Elsevier, Amsterdam, pp 681–691

Timm DE, Mueller HA, Bhanumoorthy P, Harp JM, Bunick GJ (1999) Crystal structure and mechanism of a carbon-carbon bond hydrolase. Structure 7(9):1023–1033. doi:st7915

Tomoeda K, Awata H, Matsuura T, Matsuda I, Ploechl E, Milovac T, Boneh A, Scott CR, Danks DM, Endo F (2000) Mutations in the 4-hydroxyphenylpyruvic acid dioxygenase gene are responsible for tyrosinemia type III and hawkinsinuria. Mol Genet Metab 71(3):506–510. doi:10.1006/mgme.2000.3085

Tuchman M, Freese DK, Sharp HL, Ramnaraine ML, Ascher N, Bloomer JR (1987) Contribution of extrahepatic tissues to biochemical abnormalities in hereditary tyrosinemia type I: study of three patients after liver transplantation. J Pediatr 110(3):399–403

van Dyk E, Steenkamp A, Koekemoer G, Pretorius PJ (2010) Hereditary tyrosinemia type 1 metabolites impair DNA excision repair pathways. Biochem Biophys Res Commun 401(1):32–36. doi:S0006-291X(10)01677-3

van Spronsen FJ, Berger R, Smit GP, de Klerk JB, Duran M, Bijleveld CM, van Faassen H, Slooff MJ, Heymans HS (1989) Tyrosinaemia type I: orthotopic liver transplantation as the only definitive answer to a metabolic as well as an oncological problem. J Inherit Metab Dis 12(Suppl 2):339–342

van Spronsen FJ, Thomasse Y, Smit GP, Leonard JV, Clayton PT, Fidler V, Berger R, Heymans HS (1994) Hereditary tyrosinemia type I: a new clinical classification with difference in prognosis on dietary treatment. Hepatology 20(5):1187–1191. doi:S027091399400340X

van Spronsen FJ, Bijleveld CM, van Maldegem BT, Wijburg FA (2005) Hepatocellular carcinoma in hereditary tyrosinemia type I despite 2-(2 nitro-4-3 trifluoro- methylbenzoyl)-1, 3-cyclohexanedione treatment. J Pediatr Gastroenterol Nutr 40(1):90–93. doi:00005176-200501000-00017

Vilboux T, Kayser M, Introne W, Suwannarat P, Bernardini I, Fischer R, O'Brien K, Kleta R, Huizing M, Gahl WA (2009) Mutation spectrum of homogentisic acid oxidase (HGD) in alkaptonuria. Hum Mutat 30(12):1611–1619. doi:10.1002/humu.21120

Weinberg AG, Mize CE, Worthen HG (1976) The occurrence of hepatoma in the chronic form of hereditary tyrosinemia. J Pediatr 88(3):434–438

Wilcken B, Hammond JW, Howard N, Bohane T, Hocart C, Halpern B (1981) Hawkinsinuria: a dominantly inherited defect of tyrosine metabolism with severe effects in infancy. N Engl J Med 305(15):865–868. doi:10.1056/NEJM198110083051505

Wyss PA, Boynton SB, Chu J, Spencer RF, Roth KS (1992) Physiological basis for an animal model of the renal Fanconi syndrome: use of succinylacetone in the rat. Clin Sci 83(1):81–87

Part II

Molecular Basis of HT1

Molecular Aspects of the *FAH* Mutations Involved in HT1 Disease

3

Geneviève Morrow, Francesca Angileri, and Robert M. Tanguay

Abstract

Hereditary tyrosinemia type 1 (HT1) is caused by the lack of fumarylacetoacetate hydrolase (FAH), the last enzyme of the tyrosine catabolic pathway. Up to now, around 100 mutations in the *FAH* gene have been associated with HT1, and despite many efforts, no clear correlation between genotype and clinical phenotype has been reported. At first, it seems that any mutation in the gene results in HT1. However, placing these mutations in their molecular context allows a better understanding of their possible effects. This chapter presents a closer look at the *FAH* gene and its corresponding protein in addition to provide a complete record of all the reported mutations causing HT1.

Keywords

Hereditary tyrosinemia type 1 (HT1) • Fumarylacetoacetate hydrolase (FAH) • Mutations

3.1 Introduction

Amino acid catabolism provides nitrogen for the synthesis of biologically important compounds like hormones and neurotransmitters as well as energy for the cell. This process occurs mainly in the liver and kidneys and the enzymatic pathway involved depends on the nature of the amino acid.

Phenylalanine and tyrosine are important for protein biosynthesis and intermediates in the biosynthesis of catecholamines. They are both catabolized through the tyrosine degradation pathway, which converts tyrosine into fumarate and acetoacetate,

G. Morrow • F. Angileri • R.M. Tanguay (✉)
Laboratoire de génétique cellulaire et développementale, Département de biologie moléculaire, biochimie médicale et pathologie, Faculté de médecine, Institut de biologie intégrative et des systèmes (IBIS) and PROTEO, Université Laval, 1030 avenue de la médecine, Québec, QC G1V 0A6, Canada
e-mail: genevieve.morrow@fmed.ulaval.ca; francesca.angileri.1@ulaval.ca; robert.tanguay@fmed.ulaval.ca

© Springer International Publishing AG 2017
R.M. Tanguay (ed.), *Hereditary Tyrosinemia*, Advances in Experimental Medicine and Biology 959,
DOI 10.1007/978-3-319-55780-9_3

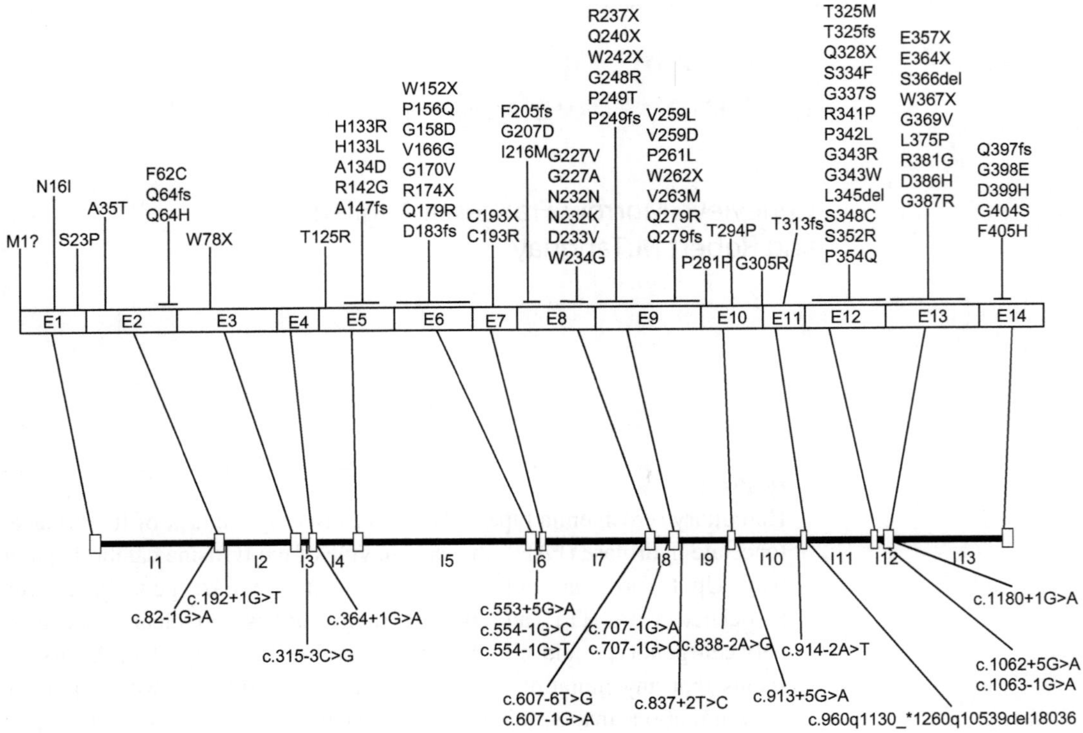

Fig. 3.1 Schematic representation of the *FAH* gene region and the corresponding protein. The location of the 98 mutations identified on the *FAH* gene is presented. *E* exon, *I* intron

two substrates of the mitochondrial tricarboxilic cycle (TCA cycle) (see Chap. 2).

Fumarylacetoacetate hydrolase (FAH, E.C. 3.7.1.2) is the last enzyme of the tyrosine degradation pathway containing 420 amino acids in a homodimeric form (Mahuran et al. 1977; Phaneuf et al. 1991). Up to now, close to 100 mutations in the *FAH* gene have been associated with hereditary tyrosinemia type 1 (HT1; HGMD® Professional 2016.1, accessed on April 2016) (Fig. 3.1) (Angileri et al. 2015). Despite multiple efforts, no clear link between mutation (genotype) and HT1 phenotype has been found. This chapter will focus on molecular aspects of the *FAH* gene and its corresponding protein in addition to give a complete listing of all the mutations identified to date.

3.2 Architecture of the fah Gene

The *FAH* gene is localized on chromosome 15 (15q23–25) and consists of 14 exons spanning over 35 kb of DNA (Awata et al. 1994; Labelle et al. 1993; Phaneuf et al. 1991) (Fig. 3.1). All exon-intron junctions possess the 5′ splice donor (gt) and 3′ splice acceptor (ag) consensus sequence and the major splicing product of the *FAH* gene results in an mRNA of 1260 coding nucleotides (Labelle et al. 1993). Table 3.1 summarizes the length of exons and introns and the number of HT1 mutations found in each of these elements while Fig. 3.1 is a schematic representation of the *FAH* gene region and protein. Exons 9 and 12 have the largest clusters of HT1 disease-causing *FAH* mutations. Interestingly both of these exons contain metal and substrate binding sites.

Table 3.1 Characteristics of *FAH* gene and its corresponding protein

Region	Number of nucleotides	Number of amino acids	Protein features[a]	Number of mutations
Exon 1	81	27	N-terminal domain (FF09298) N-acetylserine in position 2	3
Intron 1	4924	–	–	1
Exon 2	111	37	N-terminal domain (FF09298)	4
Intron 2	1585	–	–	1
Exon 3	122	41	N-terminal domain (FF09298) Phosphoserine in position 92	2
Intron 3	532	–	–	1
Exon 4	50	16	N-terminal domain (FF09298)	0
Intron 4	1786	–	–	1
Exon 5	91	31	FAH C-terminal domain (PF01557) Active site, Hxx… region Substrate binding Dimerization	7
Intron 5	5715	–	–	0
Exon 6	98	32	FAH C-terminal domain (PF01557) Dimerization	8
Intron 6	114	–	–	3
Exon 7	53	18	FAH C-terminal domain (PF01557) Metal binding Dimerization	2
Intron 7	3832	–	–	2
Exon 8	100	33	FAH C-terminal domain (PF01557) Metal binding	9
Intron 8	765	–	–	2
Exon 9	131	44	FAH C-terminal domain (PF01557) Metal binding Substrate binding	13
Intron 9	1871	–	–	2
Exon 10	76	25	FAH C-terminal domain (PF01557)	3
Intron 10	2445	–	–	2
Exon 11	47	16	FAH C-terminal domain (PF01557) Phosphoserine in position 309	1
Intron 11	2540	–	–	1
Exon 12	102	34	FAH C-terminal domain (PF01557) Substrate binding	13
Intron 12	816	–	–	2

(continued)

Table 3.1 (continued)

Region	Number of nucleotides	Number of amino acids	Protein features[a]	Number of mutations
Exon 13	118	39	FAH C-terminal domain (PF01557)	9
			Active site, …xxE region	
Intron 13	4970	–	–	1
Exon 14	213	26	FAH C-terminal domain (PF01557)	5
			Phosphotyrosine in position 395	

[a]Based on Uniprot, PFAM and (Ran et al. 2013; Timm et al. 1999; Bateman et al. 2001, 2007)

3.2.1 Fah Alternative Transcripts

Two minor alternative splicing products of the *FAH* gene have also been found in normal fibroblasts, namely *del100* and *del231* (Dreumont et al. 2005).

The *del100* transcript lacks exon 8 and, as a consequence, the reading frame is shifted and a premature termination codon (PTC) appears in 3′ end of exon 10. While this transcript is subjected to nonsense-mediated mRNA decay (NMD), a small part of it is transcribed into a protein that shares the first 202 amino acids with FAH and as a stretch of 67 amino acids completely different in the C-terminal. The pattern of DEL100 expression differs from the one of FAH and its function remains unknown (Dreumont et al. 2005).

The *del231* transcript is less abundant than *del100* and lacks exons 8 and 9. The corresponding protein should be similar to FAH except for the 77 amino acids encoded by both exons. While this transcript is not subjected to NMD, the corresponding protein has never been observed (Dreumont et al. 2005). The identification of this transcript has led to the hypothesis that intron 8 would be removed before introns 7 and 9 during normal *FAH* splicing (Dreumont et al. 2005).

The biological relevance of these two minor transcripts has not been demonstrated further. Interestingly, the abundance of *del100* and *del231* transcripts changes in presence of mutations affecting splicing donors/acceptors sites (Q279R, c.707-1G>A, c.707-1G>C) or enhancer elements (N232N, V259L) and in presence of the nonsense mutation W262X (Table 3.2) (Dreumont et al. 2001, 2004; Perez-Carro et al. 2014; Morrow et al. submitted).

3.2.2 Splicing Mutations

Up to now, 25 *FAH* mutations associated with HT1 phenotype have been reported to affect splicing (Table 3.2). Among these, four are located at the exon side of the exon/intron junction (p.Q64H, p.Q279R. p.P281P and p.G305R) and three others are located within exons 8, 9 and 12 (p.N232N, p.V259L and p.G337S). While these later mutations do not alter core sequence elements of splicing, they are probably modifying exonic splicing enhancers (ESE) or silencers (ESS) sites. The importance of these sites in splicing efficiency is gaining increasing support (reviewed in Ward and Cooper 2010) and it was recently shown that 20–45% of pathogenic single nucleotide polymorphisms (SNPs) affect splicing (Wu and Hurst 2016). It is therefore likely that other HT1 causing mutations may affect splicing.

3.3 FAH Protein

FAH forms a homodimer that catalyzes the hydrolytic cleavage of a carbon-carbon bound in fumarylacetoacetate (FAA) to yield fumarate and acetoacetate. It is the first member of an expanding family of metalloenzymes characterized by a unique α/β fold and involving a Glu-His-water catalytic triad (Timm et al. 1999). Orthologs of

Table 3.2 Splicing mutations involved in HT1 disease

	cDNA/protein location	Information	HT1 alleles	References
1	c.82-1G>A/ – I1/E2 junction	mRNA absent, probably degraded by NMD	1	Perez-Carro et al. (2014)
2	c.192G>T/p.Q64H E2/I2 junction	Decreased level of mRNA, retention of 94 pb from intron 2 and PTC after 9 missense amino acids, absence of protein	41	Rootwelt et al. (1994a, 1996), Angileri et al. (2015), and Ijaz et al. (2016)
3	c.192 + 1G>T (IVS2 + 1G>T)/ – E2/I2 junction	No experimental data on mRNA or protein	1	Bergman et al. (1998)
4	c.315-3C>G (IVS3-3C>G)/ – I3/E4 junction	No experimental data on mRNA or protein	8	Dursun et al. (2011)
5	c.364 + 1G>A[a] (IVS4 + 1G>A)/ – E4/I4 junction	No experimental data on mRNA or protein Predicted to completely abolish splicing donor site	8	Imtiaz et al. (2011)
6	c.553 + 5G>A (IVS6 + 5G>A)/ – E6/I6 junction	Abnormal mRNA, absence of protein	1	Timmers and Grompe (1996)
7	c.554-1G>C (IVS6-1G>C)/ – I6/E7 junction	No experimental data on mRNA Protein absent	1	Bergman et al. (1998)
8	c.554-1G>T (IVS6-1G>T) I6/E7 junction	3 distinct mRNA all lacking 5 nt of exon 7; additionally the first 13 nt are lost in one transcript and all exon 8 in the other transcript	155	Angileri et al. (2015), Arranz et al. (2002), Bergman et al. (1998), Couce et al. (2011), Dursun et al. (2011), Kim et al. (2000), la Marca et al. (2011), Laszlo et al. (2013), Ploos van Amstel et al. (1996), Poudrier et al. (1999), Rootwelt et al. (1996), Timmers and Grompe (1996), and Vondrackova et al. (2010)
9	c.607-1G>A (IVS7-1G>A)/ – I7/E8 junction	mRNA bearing a deletion of a single G at the beginning of exon 8 due to a shift of the acceptor splice site	6	Ploos van Amstel et al. (1996), GQET
10	c.607-6T>G (IVS7-6T>G)/– I7/E8 junction	No experimental data on mRNA or protein	?	Sniderman King et al. (2006)
11	c.696C>T/p.N232N 11 bp before E8/I8	mRNA lacking exon 8 (del100 transcript) Creation and alteration of ESE predicted[b]	1	Ploos van Amstel et al. (1996)
12	c.707-1G>A (IVS8-1G>A)/ – I8/E9 junction	mRNA lacking exons 8 and 9 (del231 transcript) Alteration of the splice acceptor site of intron 8	13	Arranz et al. (2002), Couce et al. (2011), and Imtiaz et al. (2011)
13	c.707-1G>C (IVS8-1G>C)/ – I8/E9 junction	mRNA lacking exons 8 and 9 (del231 transcript) Alteration of the splice acceptor site of intron 8	16	Bergman et al. (1998) and Elpeleg et al. (2002)
14	c.775G>C/p.V259L 63 bp before E9/I9	Slightly affect splicing of exons 8 and 9 (del231 transcript) Recombinant protein: same solubility and activity as wt, decreases FAH activity when combined with G398E	1	Angileri et al. (2015) and Morrow et al. (submitted)

(continued)

Table 3.2 (continued)

	cDNA/protein location	Information	HT1 alleles	References
15	c.836A>G/p.Q279R E9/I9 junction	mRNA lacking exons 8 and 9 (del231 transcript)	2	Dreumont et al. (2001), Kim et al. (2000), and Perez-Carro et al. (2014)
		Recombinant protein: same solubility and activity as wt		
16	c.837 + 2T>C (IVS9 + 2T>C)/ – E9/I9 junction	No experimental data on mRNA or protein	4	Dursun et al. (2011)
17	c.838-2A>G (IVS9 -2A>G) I9/E10 junction	No experimental data on mRNA or protein	2	Angileri et al. (2015) and Heath et al. (2002)
18	c.843 C>A/p.P281P[c] I9/E10 junction	No experimental data on mRNA or protein	2	Imtiaz et al. (2011)
		Creation and alteration of ESE predicted[b]		
19	c.913G>C/p.G305R E10/I10 junction	mRNA lacking exon 10	1	Perez-Carro et al. (2014)
20	c.913 + 5G>A/ – E10/I10 junction	No experimental data on mRNA or protein	1	Choi et al. (2014)
21	c.914-2A>T (IVS10-2A>T)/ – I10/E11 junction	No experimental data on mRNA or protein	1	Arranz et al. (2002)
		Possible alteration of exon 11 splicing		
22	c.1009G>A/p.G337S Middle of E12	3 distinct mRNA	30	Bergman et al. (1998), Bliksrud et al. (2005, 2012), Haghighi-Kakhki et al. (2014), Prieto-Alamo and Laval (1998), Rootwelt et al. (1994b, 1996), and St-Louis et al. (1995)
		One lacking first 50 nucleotides of exon 12		
		One having all exon 12 plus 105 nucleotides from intron 12		
		One normal transcript		
		No protein detected in patients		
23	c.1062 + 5G>A (IVS12 + 5G>A)/ – E12/I12 junction	3 distinct mRNA	305	Angileri et al. (2015), Arranz et al. (2002), Bergman et al. (1998), Couce et al. (2011), Dursun et al. (2011), Grompe and al-Dhalimy (1993, 1994), Hahn et al. (1995), Heath et al. (2002), Ijaz et al. (2016), Imtiaz et al. (2011), Perez-Carro et al. (2014), Ploos van Amstel et al. (1996), Poudrier et al. (1996), Rootwelt et al. (1994d, 1996), Timmers and Grompe (1996), and Vondrackova et al. (2010), GQET
		One lacking exon 12		
		One lacking exons 12 and 13		
		One with a 105 base pair retention of intron 12		
		No protein detected in patients		
24	c.1063-1G>A (IVS12-1G>A)/ – I12/E13 junction	No experimental data on mRNA or protein	2	Mak et al. (2013)
		Suggested to abolish the acceptor site by the authors following in silico analysis		
25	c.1180 + 1G>A/ – E13/I13 junction	No experimental data on mRNA or protein	2	GQET

GQET Groupe québécois d'étude de la tyrosinémie/Quebec HT1 study group

[a]Reported as c.442-1G>A in (Angileri et al. 2015)

[b]Prediction made with Human Splicing Finder (Desmet et al. 2009)

[c]Reported as T281P in (Imtiaz et al. 2011)

FAH are found in different species and share a high degree of homology (Fig. 3.2). The protein can be separated in two distinct domains; the FAH N-terminal domain and the FAH C-terminal domain.

3.3.1 FAH N-Terminal Domain

The N-terminal domain of FAH consists of ~100 residues that are encoded by exons 1–4 (Timm et al. 1999) (Pfam: PF09298, InterPro: IPR015377) (Fig. 3.3a). Little is known about this structural domain except that it forms a structure consisting of an SH3-like barrel. This domain is not involved in dimerization or in active site formation, but it could have a regulatory function given its contacts with the C-terminal domain (Bateman et al. 2001, 2007; Timm et al. 1999).

The FAH N-terminal domain contains two identified post-transcriptionally modified amino acids (N-acetyl-S2 and phospho-S92), but the reasons/effects of these modifications have not been investigated (UniProt: P16930) (Huttlin et al. 2010; Vaca Jacome et al. 2015).

Among the nine exonic *FAH* mutations found in this domain (Fig. 3.1), five can be linked to aberrant mRNA processing and two results in the p.W78X nonsense mutation yielding a protein lacking the entire FAH catalytic C-terminal domain (Figs. 3.1 and 3.3b) (Table 3.3). Among the five mutations that can be linked to aberrant mRNA processing, the c.1A>G (p.M1?/exon 1) mutation results in the start codon loss, while p.Q64H (exon 2) has been shown to affect splicing by promoting the retention of 94 nucleo-

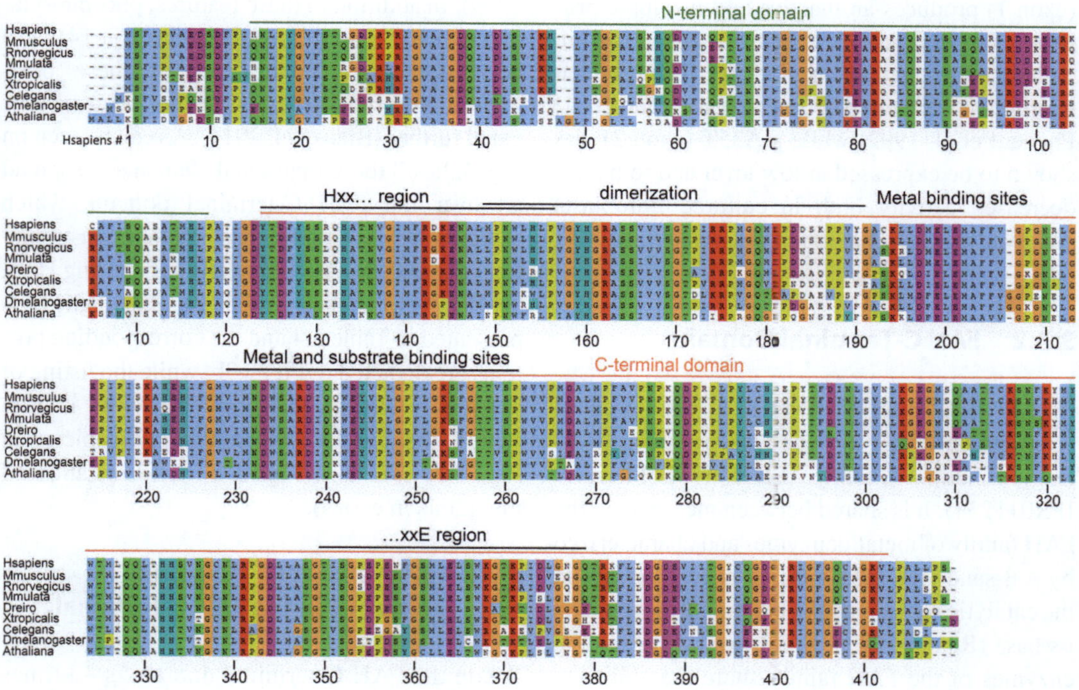

Fig. 3.2 Alignment of FAH orthologs. The protein sequence were retrieved from the NCBI website; *Homo sapiens* (NP_000128.1), *Mus musculus* (NP_034306.2), *Rattus Norvegicus* (NP_058877.1), *Drosophila melanogaster* (NP_524830.2), *Arabidopsis thaliana* (NP_172669.2), *Danio rerio* (NP_955895.1 with N-terminal of Fisher et al. 2008), *Caenorhabditis elegans* (NP_509083.1), *Xenopus tropicalis* (NP_001107523.1), *Macaca mulatta* (NP_001244458.1). The multiple sequence alignment was performed with MUSCLE tool (http://www.ebi.ac.uk/Tools/msa/muscle/) from the EMBL-EBI website. Numbers at the bottom of the figure correspond to *H. sapiens* amino acids sequence. N-terminal and C-terminal domains of FAH are presented, as well as region of interest for FAH activity

tides from intron 2 and resulting in the apparition of a PTC after nine missense amino acids. In addition to these two documented mutations, p. S23P (exon 1), p.F62C (exon 2) and p.Q64fs (exon 2) are likely to affect splicing due to the predicted alteration of an ESE site (S23P), activation of an exonic cryptic donor site (F62C) and direct alteration of the wild-type cryptic donor site (p.Q64fs) (Desmet et al. 2009). However, no experimental data on mRNA and protein are available for the later three mutations preventing a conclusion on their real effect on *FAH* mRNA, protein stability or activity. Of note, S23P was proposed to possibly affect FAH dimerization (Heath et al. 2002) and recombinant F62C was found to be an insoluble/inactive protein (Bergeron et al. 2001). The two remaining disease-causing mutations of the N-terminal domain give rise to normal *FAH* mRNA and are therefore likely to have a structural effect. Indeed, p.N16I (exon 1) produces an inactive and insoluble protein as shown from expression analysis of FAH in patient liver extract and from recombinant expression in cultured cells (Bergeron et al. 2001; Phaneuf et al. 1992) while p.A35T (exon 2) was shown to be expressed at low level and to have a decreased activity both in cultured fibroblasts and liver extracts (Cassiman et al. 2009).

3.3.2 FAH C-Terminal Domain

The FAH C-terminal domain is composed of ~300 residues that are encoded by exons 5–14 (Fig. 3.3a) (Pfam: PF01557, InterPro: IPR011234). It is shared between members of the FAH family of metalloenzymes and characterized by a ß-sandwich fold forming a deep pocket in the catalytic domain and containing a metal ion at its base (Ran et al. 2013; Timm et al. 1999). All enzymes of the FAH family share the ability to cleave C-C bond of their substrate through a Glu-His-water triad involving either a HxxE or Hxx…xxE motif. For extensive alignments between members of the FAH family please refer to (Ran et al. 2013).

The FAH C-terminal domain has functional roles in metal-ion binding, catalysis and dimerization. The dimer formation is needed to form the pocket of the active site, while multiple residues have been shown to be involved in FAH dimerization, the longest stretch of amino acids located at the dimer interface is spanning from the end of exon 5 through exon 6 and the first half of exon 7 (Timm et al. 1999). From the crystal structure of mouse FAH, it was shown that the metal ion is coordinated by residues present in exons 5, 7, 8 and 9 (Bateman et al. 2001, 2007; Timm et al. 1999). Moreover, based on the study of Ran and co-workers, the FAH active site corresponds to the Hxx…xxE motif (Ran et al. 2013) (Figs. 3.2 and 3.3a). The first part of this motif (Hxx…), which would correspond to the lid domain of the active site, is mainly located in exon 5 and the last part of the motif (…xxE) in exon 13. Also based on mouse FAH crystal structure, substrate binding sites are located in exon 5, 9 and 12 (Bateman et al. 2001, 2007; Timm et al. 1999). In addition to these features, phosphorylation of S309 (exon 11) and Y395 (exon 14) have been observed but the role of these post-translational modifications has not been investigated further (Bian et al. 2014). As can be seen on Fig. 3.3a, all these functional elements are spread all over the FAH C-terminal domain, which explains why even nonsense and deletions mutations located in *FAH* last exons are causing HT1. The listing of nonsense mutations causing HT1 is presented in Table 3.4 and the corresponding proteins are depicted in Fig. 3.3b, while the listing of deletion mutations causing frameshift is presented in Table 3.5 and proteins are depicted in Fig. 3.3c (see also Fig. 3.1 for localization of mutations in exons).

3.3.2.1 Missense HT1 Mutations Located in the FAH C-Terminal Domain

While the FAH C-terminal domain is ~3 times larger then the FAH N-terminal domain, it contains nearly eight times more disease causing mutations (69 versus 9) due to its importance for FAH function. As mentioned above, the two exons containing the most disease-causing mutations are exon 9 and 12 (Table 3.1 and Fig. 3.1). Based on the fact that it was recently

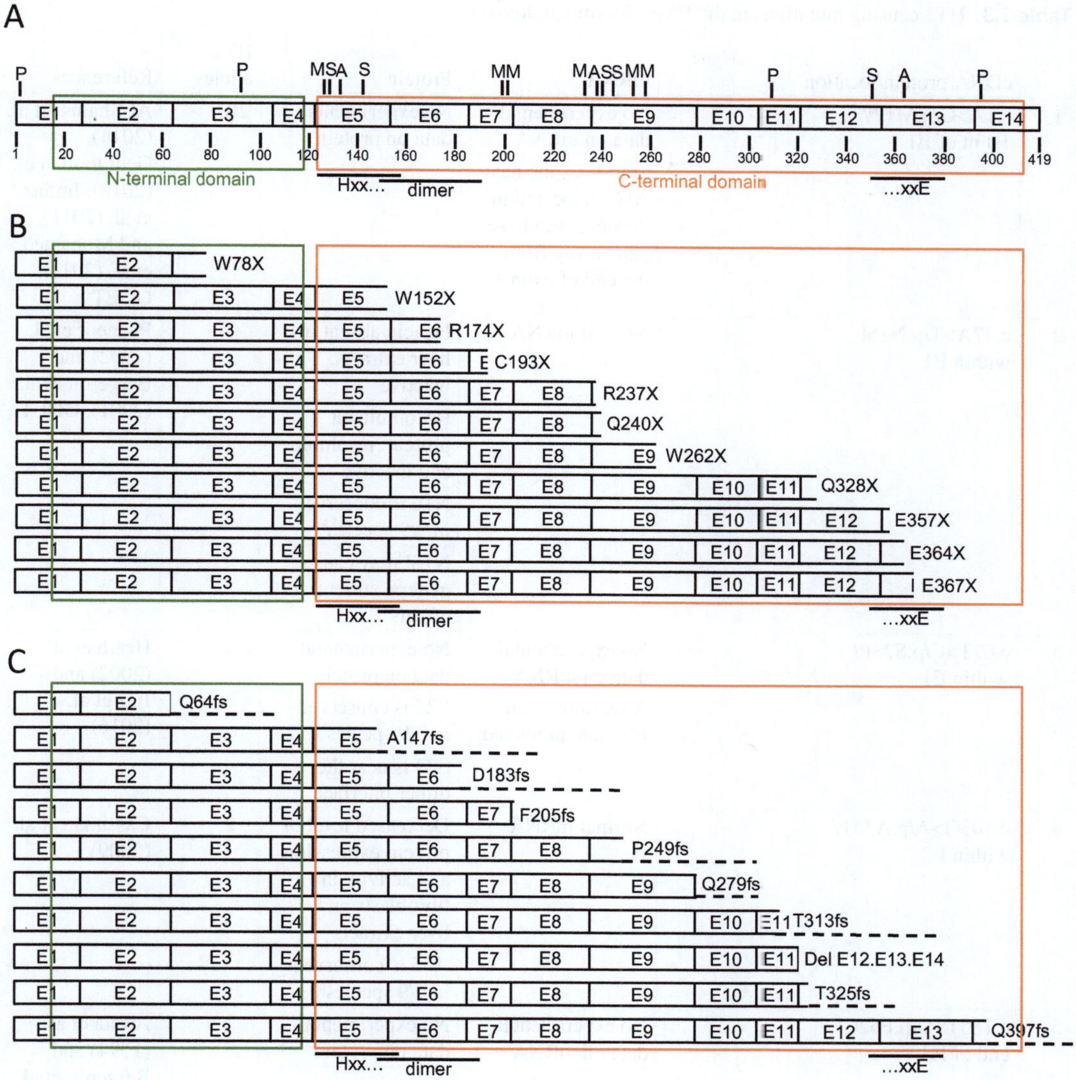

Fig. 3.3 FAH region of interest and effect of nonsense mutations as well as deletion mutations causing frameshift. (**a**) Scaled schematic representation of *FAH* exons. FAH N-terminal domain (Pfam: PF09298), FAH C-terminal doamin (PF01557), Hxx… and …xxE: conserved motif of the active site (Ran et al. 2013). *P* post-translational modifications (Bian et al. 2014; Huttlin et al. 2010; Vaca Jacome et al. 2015), *M* metal binding sites, *S* substrate binding sites, *A* active site (Bateman et al. 2001, 2007; Timm et al. 1999). Most of these features can also be found on the *Homo sapiens* FAH UniProt entry (P16930). (**b**) Schematic representation of the nonsense mutations causing HT1. The corresponding mutations can be found in Tables 3.3 and 3.4. (**c**) Schematic representation of the *FAH* deletion mutations causing frameshift. The corresponding mutations can be found in Tables 3.3 and 3.5

shown that 20–45% of pathogenic SNPs affect splicing (Wu and Hurst 2016), *FAH* missense mutations of the C-terminal domain were separated according to experimental data and to their potential effect on splicing as determined by the Human Splicing Finder website (Desmet et al. 2009). Nineteen mutations were found to potentially affect splicing (Table 3.6), while 29 mutations were not (Table 3.7). Not surprisingly all mutations affecting important residues for FAH activity are found in Table 3.7.

Table 3.3 HT1 causing mutations in the FAH N-terminal domain

	cDNA/protein/location	mRNA	Protein	HT1 alleles	References
1	c.1A>G/p.M1?[a]/ 1st nt of E1	No experimental data on mRNA Start loss, the next ATG is located in intron 1, 325 base pairs away from the end of exon 1	No experimental data on protein	20	Al-Shamsi et al. (2014), Georgouli et al. (2010), Imtiaz et al. (2011), and Mohamed et al. (2013), GQET
2	c.47A>T/p.N16I/ within E1	Normal mRNA	Protein absent in liver extracts, inactive Recombinant protein: insoluble and inactive N16 is conserved in 9/9 species[b] N16I may cause general structural effects	1	Phaneuf et al. (1992) and Bergeron et al. (2001), GQET
3	c.67T>C/p.S23P/ within E1	No experimental data on mRNA Alteration of an ESE site predicted[c]	No experimental data on protein S23 is conserved in 8/9 species[b] S23 is near the dimer interface	4	Heath et al. (2002) and Ijaz et al. (2016)
4	c.103G>A/p.A35T/ within E2	Normal mRNA	Decreased level of protein expression and activity in fibroblasts and liver extracts A35 is conserved in 9/9 species[b]	2	Cassiman et al. (2009)
5	c.185T>G/p.F62C/ end of E2	No experimental data on mRNA Activation of an exonic cryptic donor site and alteration of an ESE site predicted[c]	No experimental data on protein Recombinant protein: insoluble and inactive F62 is conserved in 8/9 species[b]	2	Awata et al. (1994) and Bergeron et al. (2001)
6	c.191delA/p.Q64fs/ end of E2	No experimental data on mRNA Alteration of the wild-type donor site, activation of an exonic cryptic donor site and alteration of an ESE site predicted[c]	No experimental data on protein	2	Dursun et al. (2011)

(continued)

Table 3.3 (continued)

	cDNA/protein/location	mRNA	Protein	HT1 alleles	References
7	c.192G>T[d]/p.Q64H/ E2/I2 junction	Decreased level of mRNA retention of 94 pb from intron 2 and PTC after 9 missense amino acids	Absence of protein	41	Rootwelt et al. (1994a, 1996), Angileri et al. (2015) and Ijaz et al. (2016)
8	c.233G>A/p.W78X within E3	Normal mRNA	No experimental data on protein Truncation in the middle of E3	4	Arranz et al. (2002)
9	c.234G>A/p.W78X/ within E3	No experimental data on mRNA	No experimental data on protein Truncation in the middle of E3	1	Couce et al. (2011)

GQET Groupe québécois d'étude de la tyrosinémie/Quebec HT1 study group
[a]*referred to p.M1V in* (Al-Shamsi et al. 2014; Angileri et al. 2015; Mohamed et al. 2013)
[b]based on Fig. 3.2
[c]*prediction made with Human Splicing Finder* (Desmet et al. 2009)
[d]also described in Table 3.2 (splicing mutations)

3.4 Most Frequent FAH Mutations and Their Geographical Localization

In total, more than 98 *FAH* mutations have been reported at this time to cause HT1. This number will likely increase, since SNPs causing the disease are found in SNPs database such as the ones from the Exome Aggregation Consortium (ExAC) (URL: http://exac.broadinstitute.org) and NHLBI Exome Sequencing project (http://evs.gs.washington.edu/EVS/). For example, a W234X SNP is reported on the ExAC website (last access: April 2016). This mutation is located at the end of exon 8 and based on experimental data from Table 3.4 is likely to cause HT1.

The worldwide incidence of HT1 is relatively low, with 1/100,000 affected individual (Hutchesson et al. 1996). The population that possesses the highest incidence of HT1 is the French Canadian population of the Saguenay-Lac-Saint-Jean (SLSJ) region in the province of Quebec (Canada). Not surprisingly, the most frequent *FAH* mutation in SLSJ region (~90% of all the disease causing alleles; c.1062 + 5G>A (IVS12 + 5G>A)) is also the most frequent worldwide (32.3% of the reported alleles) (Table 3.8) (Angileri et al. 2015). Since c.1062 + 5G>A accounts for the third of the HT1 reported allele and due to the fact that it is reported in a wide range of ethnic groups, it is likely to be a very old mutation (Angileri et al. 2015). The second most frequent HT1 mutation encountered worldwide is c.554-1G>T (IVS6-1G>T) with a frequency of 16.4% (Table 3.8). While this mutation is not associated to a specific cluster, it is more prevalent in the Mediterranean region and in southern Europe (Angileri et al. 2015).

Two other clusters of HT1 are found in the world. The first one is in the Finnish population of Pohjanmaa where the c.786G>A (p.W262X) represents ~88% of disease causing alleles (Angileri et al. 2015; Kvittingen 1991; Mustonen et al. 1997). The other cluster is in an immigrant population from Pakistan living in the United Kingdom (predominantly in Birmingham), and for which the c.192G>T (p.Q64H) mutation accounts for 42% of the alleles reported (Angileri et al. 2015; Hutchesson et al. 1998). These two mutations are also among the most frequent mutations worldwide, with frequencies of 5.6% and 4.3% (Table 3.8; third and fourth rank respectively).

Table 3.4 Nonsense mutations found in FAH C-terminal domain

	cDNA/protein	Location	Information regarding mRNA and protein	HT1 alleles	References
1	c.455G>A/p. W152X	End of E5	No experimental data on mRNA and protein	3	Dou et al. (2013) and Yang et al. (2012)
2	c.456G>A/p. W152X	Beginning of E6	No experimental data on mRNA and protein	1	GQET
3	c.520C>T/p. R174X	Middle of E6	No experimental data on mRNA and protein	4	Dursun et al. (2011), Heath et al. (2002), and Timmers and Grompe (1996)
4	c.579C>A/p. C193X	Middle of E7	No experimental data on mRNA and protein	2	Vondrackova et al. (2010)
5	c.709C>T/p. R237X	Beginning of E9	Reduced level of mRNA, the PTC is far from the last exon-exon junction and could therefore lead to NMD	39	Angileri et al. (2015), Cao et al. (2012), Dursun et al. (2011), Heath et al. (2002), Imtiaz et al. (2011), Jitraruch et al. (2011), la Marca et al. (2011), and Ploos van Amstel et al. (1996), GQET
6	c.718 C>T/p. Q240X	Beginning of E9	No experimental data on mRNA and protein	2	Imtiaz et al. (2011)
7	c.726G>A/p. W242X	Middle of E9	No experimental data on mRNA and protein	1	Angileri et al. (2015)
8	c.786G>A/p. W262X	Middle of E9	Drastic reduction of mRNA level No protein	53	Angileri et al. (2015), Mustonen et al. (1997), Rootwelt et al. (1994a, 1996), and St-Louis et al. (1994)
9	c.982C>T/p. Q328X	Beginning of E12	mRNA normal No experimental data on protein	2	Arranz et al. (2002)
10	c.1069G>T/p. E357X	Beginning of E13	From reduced to normal level of mRNA depending on the publication No protein	12	Angileri et al. (2015), Grompe and al-Dhalimy (1993), Heath et al. (2002), Ploos van Amstel et al. (1996), Rootwelt et al. (1994d, 1996), and St-Louis et al. (1995)
11	c.1090G>T/p. E364X	Middle of E13	From reduced to normal level of mRNA depending on the publication No protein	16	Bergman et al. (1998), Grompe and al-Dhalimy (1993), Grompe et al. (1994), Ploos van Amstel et al. (1996), Poudrier et al. (1999), Rootwelt et al. (1994d 1996), and Timmers and Grompe (1996)
12	c.1100G>A/ pW367X	Middle of E13	No experimental data on mRNA and protein	2	Yang et al. (2012)

GQET Groupe québécois d'étude de la tyrosinémie/Quebec HT1 study group

Table 3.5 Deletion mutations causing frameshift in FAH C-terminal domain

	cDNA/protein/location	Information regarding mRNA and protein	HT1 alleles	References
1	c.441_448del8/p.A147fs/end of E5	No experimental data on mRNA and protein	1	Dursun et al. (2011)
2	c.548_553+20del26/p.D183fs/end of E6	No experimental data on mRNA and protein (E6/I6del26)	1	Arranz et al. (2002)
3	c.615delT/p.F205fs/beginning of E8	Frameshift changing F205 in L and the next triplet to a PTC (p.F205lLfsX2)	1	Bliksrud et al. (2012)
4	c.744delG/p.P249fs/middle of E9	Frameshift changing P249 in H and leading to apparition of a PTC at the end of E10 (p.P249HfsX55)	7	Bliksrud et al. (2012)
5	c.835delC/p.Q279fs/end of E9	Frameshift changing Q279 in R and leading to apparition of a PTC at the end of E10 (pQ279RfsX25)	2	Bliksrud et al. (2012)
6	c.938delC/p.T313fs/middle of E11	Frameshift keeping a T in position 313 and introducing a PTC 60 amino acids downstream (p.T313TfsX60)	1	Arranz et al. (2002)
7	c.960q1130_*1260q10539del18036/–/within I11	Large deletion beginning in intron 11 and ending in the intergenic region of FAH-ARNT2 Absence of E12, E13 and E14	4	Park et al. (2009)
8	c.974_976delCGAinsGC/p.T325fs/beginning of E12	No experimental data on mRNA and protein Frameshit will change T325 in S but the has not been investigated further	6	Yang et al. (2012)
9	c.1190delA/pQ397fs/beginning of E14	Frameshift abolishing the wild-type stop codon and resulting in an abnormally prolonged protein with 41 extra amino acids	4	Imtiaz et al. (2011)

The geographical distribution of almost all of the *FAH* mutations has been the subject of a recent review (Angileri et al. 2015).

3.5 Correlation Between FAH Mutations and HT1 Phenotype

HT1 is classified in three different forms depending on the clinical phenotype of patients and the age of onset. The acute form presents before 2 months of age with acute liver failure, while the subacute form presents between 2 and 6 months of age with liver disease and the chronic form presents after 6 months of age with slowly progressive liver cirrhosis and hypophosphatemic rickets (Bergman et al. 1998; Mitchell et al. 2001; van Spronsen et al. 1994) (See Morrow and Tanguay, Chap. 2). However, despite multiple efforts, no clear genotype-phenotype relationships have been unveiled (Arranz et al. 2002; Bergman et al. 1998; Dursun et al. 2011; Rootwelt et al. 1996).

3.5.1 HT1 Pseudodeficiency

To date, one missense mutation (c.1021C>T, p. R341W) has been described as a pseudodeficiency

Table 3.6 FAH C-terminal domain missense mutations that could affect splicing

	cDNA/protein/location	mRNA	Protein	HT1 alleles	References
1	c.509G>T/p.G170V/within E6	No experimental data on mRNA	No experimental data on protein	4	Imtiaz et al. (2011)
		Creation of an ESE predicted[a]	V166 is conserved in 5/9 species[b]		
			V166 is near the active site and at the dimer interface		
2	c.536A>G/p.Q179R/within E6	No experimental data on mRNA	No experimental data on protein	1	Choi et al. (2014)
		Creation and alteration of an ESE predicted[a]	Q179 is conserved in 9/9 species[b]		
			Q179 is at the dimer interface		
3	c.648C>G/p.I216M/within E8	No experimental data on mRNA	No experimental data on protein	4	Sheth et al. (2012)
		Creation of a potential donor splice site suggested by the authors	I216 is conserved in 7/9 species[b]		
4	c.680G>C/p.G227A/within E8	No experimental data on mRNA	No experimental data on protein	2	Imtiaz et al. (2011)
		Alteration of an ESE predicted[a]	G227 is conserved in 9/9 species[b]		
5	c.696C>A/p.N232K/within E8	Alteration of an ESE predicted[a]	No experimental data on mRNA	2	Dursun et al. (2011)
			No experimental data on protein		
			N232 is conserved in 9/9 species[b]		
			N232 is located in the active site		
6	c.775G>C[c]/p.V259L within exon 9	Slightly affect splicing of exons 8 and 9 (del231 transcript)	Recombinant protein: same solubility and activity as wt, decreases activity when combined with G398E	1	Angileri et al. (2015) and Morrow et al. (submitted)
			V259 is conserved in 2/9 species[b]		
			V259 is at the dimer interface		
7	c.836A>G[c]/p.Q279R/E9/I9 junction	mRNA lacking exons 8 and 9 (del231 transcript)	No protein	2	Dreumont et al. (2001), Kim et al. (2000), and Perez-Carro et al. (2014)
			recombinant protein: same solubility and activity as wt		
			Q279 is conserved in 9/9 species[b]		
8	c.913G>C[c]/p.G305R E10/I10 junction	mRNA lacking exon 10	No experimental data on protein	1	Perez-Carro et al. (2014)
			G305 is conserved in 6/9 species[b]		
9	c.974C>T/p.T325M within E12	No experimental data on mRNA	No experimental data on protein	5	Angileri et al. (2015), Couce et al. (2011), and Heath et al. (2002), GQET
		Creation and alteration of an ESE predicted[a]	T325 is conserved in 7/9 species[b]		

10	c.1001 C>T/p.S334F/within E12	No experimental data on mRNA	No experimental data on protein	2	Imtiaz et al. (2011)
		Alteration of an ESE predicted[a]	S334 is conserved in 4/9 species[b]		
11	c.1009G>A[c,d]/p.G337S/middle of E12	3 distinct mRNA	No protein detected in patients	30	Bergman et al. (1998), Bliksrud et al. (2005, 2012), Haghighi-Kakhki et al. (2014), Prieto-Alamo and Laval (1998), Rootwelt et al. (1994b, d, 1996), and St-Louis et al. (1995)
		One lacking first 50 nucleotides of exon 12	G337 is conserved in 9/9 species[b]		
		One having all exon 12 plus 105 nucleotides from intron 12			
		One normal transcript			
12	c.1035_1037del[4] p.L345del/within E12	No experimental data on mRNA	No experimental data on protein	2	Mak et al. (2013)
		Alteration of an ESE site predicted[a]	L345 is conserved in 9/9 species[b]		
			L345 is near the xxE motif region		
13	c. 1061C>A/p.P354Q/end of E12	No experimental data on mRNA	No experimental data on protein	1	Bliksrud et al. (2005) and Bergman et al. (1998)
		Creation and alteration of an ESE predicted[a]	P354 is conserved in 6/9 species[b]		
			P354 is in the xxE motif region		
14	c.1097_1099delCGT[d]/p.S366del/within E13	No experimental data on mRNA	No experimental data on protein	2	
		Creation and alteration of an ESE sites predicted[a]	S366 is conserved in 7/9 species[b]		
			S366 is in the xxE motif region		
15	c.1141A>G/p.R381G/within E13	No experimental data on mRNA	No protein	6	St-Louis et al. (1995)
		Activation of an exonic cryptic donor site and alteration of an ESE site predicted[a]	R381 is conserved in 8/9 species[b]		
			R381 is near the xxE motif region		
16	c.1156G>C/p.D386H/within E13	No experimental data on mRNA	No experimental data on protein	2	Al-Shamsi et al. (2014)
		Activation of an exonic cryptic donor site and alteration of an ESE site predicted[a]	D386 is conserved in 9/9 species[b]		
17	c.1159G>A/p.G387R/within E13	No experimental data on mRNA	No experimental data on protein	4	Sheth et al. (2012)
		Activation of an exonic cryptic donor site, creation and alteration of an ESE site predicted[a]	G387 is conserved in 8/9 species[b]		

(continued)

Table 3.6 (continued)

	cDNA/protein/location	mRNA	Protein	HT1 alleles	References
18	c.1195G>C/p.D399H[e]/within E14	No experimental data on mRNA	No experimental data on protein	2	Imtiaz et al. (2011)
		Alteration of an ESE site predicted[a]	D399 is conserved in 5/9 species[b]		
			D399 is between the N- and C-terminal domain (quaternary structure)		
19	c.1213_1214delTTinsCA/p.F405H/ within E14	No experimental data on mRNA	No experimental data on protein	1	Bergman et al. (1998)
		creation of an ESS site predicted[a]	F405 is conserved in 9/9 species[b]		
			F405 is between the N- and C-terminal domain (quaternary structure)		

GQET Groupe québécois d'étude de la tyrosinémie/Quebec HT1 study group
[a]Prediction made with Human Splicing Finder (Desmet et al. 2009)
[b]Based on Fig. 3.2.
[c]Proven splicing mutations, also presented in Table 3.2
[d]Also reported as c.961_1010del50 in (Angileri et al. 2015; Prieto-Alamo and Laval 1998; Rootwelt et al. 1994d)
[e]Also reported as p.N400H in (Angileri et al. 2015; Imtiaz et al. 2011)
[f]In frame deletion of one codon

Table 3.7 Missense mutations located in the FAH C-terminal domain and affecting protein function

	cDNA/protein/location	Information regarding mRNA and protein	HT1 alleles	References
1	c.374C>G/p.T125R/ exon 5	No experimental data on mRNA and protein	2	Imtiaz et al. (2011)
		T125 is conserved in 9/9 species[a]		
		T125 is in the active site (Hxx… motif region)		
2	c.398A>G/p.H133R/ exon 5	No experimental data on mRNA and protein	2	Heath et al. (2002)
		H133 is conserved in 9/9 species[a]		
		H133 is in the catalytic triad		
3	c.398A>T/p.H133L/ exon 5	No experimental data on mRNA and protein	2	Couce et al. (2011)
		H133 is conserved in 9/9 species[a]		
		H133 is in the catalytic triad		
4	c.401C>A/p.A134D/ exon 5	normal mRNA, low level of protein, no activity	3	Bergeron et al. (2001), Labelle et al. (1993), and Rootwelt et al. (1994c, 1996)
		Recombinant protein: same solubility as wild-type, inactive		
		A134 is conserved in 9/9 species[a]		
		A134D is in the active site, in the Hxx… motif		
5	c.424A>G/p.R142G/ Exon 5	No experimental data on mRNA and protein	1	GQET
		R142 is conserved in 9/9 species[a]		
		R142 is in the active site, is involved in substrate binding		
6	c.467C>A/p.P156Q/ exon 6	No experimental data on mRNA and protein	1	Heath et al. (2002)
		P156 is conserved in 9/9 species[a]		
		P156 is in the largest contiguous sequence of contacts between monomers		
7	c.473G>A/p.G158D/ exon 6	No experimental data on mRNA and protein	1	Bergman et al. (1998)
		G158 is conserved in 8/9 species[a]		
		G158 is in the largest contiguous sequence of contacts between monomers and beside a substrate interaction site		
8	c.497T>G/p.V166G/ exon 6	Normal mRNA, no experimental data on protein	11	Bergman et al. (1998), Dursun et al. (2011), Grompe and al-Dhalimy (1993), and Rootwelt et al. (1996)
		V166 is conserved in 5/9 species[a]		
		V166 is in the largest contiguous sequence of contacts between monomers		
9	c.577 T > C/p.C193R/ Exon 7	normal mRNA, no experimental data on protein	1	Bergeron et al. (2001) and Ploos van Amstel et al. (1996)
		Recombinant protein: insoluble and inactive		
		C193 is conserved in 4/9 species[a]		
		C193 is near the active site		

(continued)

Table 3.7 (continued)

	cDNA/protein/location	Information regarding mRNA and protein	HT1 alleles	References
10	c.620G>A/p.G207D/ exon 8	No experimental data on mRNA and protein	1	Timmers and Grompe (1996)
		G207 is conserved in 9/9 species[a]		
		G207 is near the active site		
11	c.680G>T/p.G227V/ exon 8	No experimental data on mRNA and protein	4	Vondrackova et al. (2010)
		G227 is conserved in 9/9 species[a]		
		G227 is beside F226 which is involved in metal binding		
12	c.698A>T/p.D233V/ exon 8	Normal mRNA, low level of protein, no activity	15	Dursun et al. (2011), Rootwelt et al. (1994a, 1996)
		Recombinant protein: same solubility as wild-type, inactive		
		D233 is conserved in 9/9 species[a]		
		D233 is involved in metal binding		
13	c.700T>G/p.W234G/ exon 8	Normal mRNA, inactive protein	1	Hahn et al. (1995), Rootwelt et al. (1996), and Bergeron et al. (2001)
		Recombinant protein: insoluble and inactive		
		W234 is conserved in 9/9 species[a]		
		W234 is within a metal cation pocket		
14	c.742G>A/p.G248R/ exon 9	No experimental data on mRNA and protein	2	GQET
		G248 is conserved in 9/9 species[a]		
		G248 is beside P249 which is involved in substrate binding		
15	c.745C>A/p.P249T/ exon 9	No experimental data on mRNA and protein	1	Timmers and Grompe (1996)
		P249 is conserved in 9/9 species[a]		
		P249 is involved in substrate binding and is located at the dimer interface		
16	c.776T>A/p.V259D/ exon 9	No experimental data on mRNA and protein	2	Dursun et al. (2011)
		V259 is conserved in 2/9 species[a]		
		V259 is in the active site, at the dimer interface		
17	c.782C>T/p.P261L/ exon 9	No experimental data on mRNA and protein	16	Bergman et al. (1998), Imtiaz et al. (2011), Elpeleg et al. (2002), and Angileri et al. (2015), GQET
		P261 is conserved in 9/9 species[a]		
		P261 occur between the N- and C-terminal domains (quaternary structure)		
18	c.787G>A/p.V263M/ exon 9	No experimental data on mRNA and protein	4	Imtiaz et al. (2011)
		V263 is conserved in 8/9 species[a]		
		V263 is near metal binding site		
19	c.880A>C/p.T294P/ exon 10	No experimental data on mRNA and protein	2	Bergman et al. (1998) and Timmers and Grompe (1996)
		T294 is conserved in 7/9 species[a]		

(continued)

Table 3.7 (continued)

	cDNA/protein/location	Information regarding mRNA and protein	HT1 alleles	References
20	c.1022G>C/p.R341P/ exon 12	No experimental data on mRNA and protein	2	Imtiaz et al. (2011)
		R341 is conserved in 9/9 species[a]		
		R341 is near the active site		
21	c.1025C>T/p.P342L/ exon 12	Normal mRNA, absence of protein	5	Bergman et al. (1998, 1994c) and Rootwelt et al. (1996)
		P342 is conserved in 8/9 species[a]		
		P342 is near the active site, between the N- and C-terminal domains (quaternary structure)		
22	c.1027G>T/p.G343W/ exon 12	Normal mRNA, no experimental data on protein	2	Arranz et al. (2002)
		G343 is conserved in 9/9 species[a]		
		G343 is near the active site		
23	c.1027G>C/p.G343R/ exon 12	No experimental data on mRNA and protein	9	Dou et al. (2013) and Imtiaz et al. (2011)
		G343 is conserved in 9/9 species[a]		
24	c.1043C>G/p.S348C[b]/ exon 12	No experimental data on mRNA and protein	?	Prieto-Alamo and Laval (1998)
		S348 is conserved in 7/9 species[a]		
		S348 is near the active site		
25	c.1056C>A/p.S352R/ exon 12	No experimental data on mRNA and protein	2	Heath et al. (2002)
		S352 is conserved in 9/9 species[a]		
		S352 is in the …xxE motif region		
26	c.1106G>T/p.G369V/ exon 12	Normal level of mRNA, no experimental data on protein	1	Ploos van Amstel et al. (1996)
		G369 is conserved in 8/9 species[a]		
		G369 is in the …xxE motif region		
27	c.1124T>C/p.L375P/ exon 12	Normal level of mRNA	2	Cao et al. (2012)
		L375 is conserved is 6/9 species[a]		
		L375 is in the …xxE motif region		
28	c.1193G>A/p.G398E/ exon 14	Normal mRNA, no protein	1	Morrow et al. (submitted)
		Recombinant protein: soluble, decreased activity		
		G398 is conserved in 7/9 species[a]		
29	c.1210G>A/p.G404S/ exon 14	No experimental data on mRNA and protein	2	Vondrackova et al. (2010)
		G404 is conserved in 9/9 species[a]		

GQET Groupe québécois d'étude de la tyrosinémie/Quebec HT1 study group
[a]Based on Fig. 3.2
[b]Reported as p.S348G in (Angileri et al. 2015; Prieto-Alamo and Laval 1998)

Table 3.8 HT1 most frequent mutations

cDNA/protein	Location	Type of mutation	HT1 alleles	Frequency[1] (%)
c.1062 + 5G>A (IVS12 + 5G>A)/–	E12/I12 junction	Splicing (Table 3.2)	305	32.3
c.554-1G>T (IVS6-1G>T)/–	I6/E7 junction	Splicing (Table 3.2)	155	16.4
c.786G>A/p.W262X	Exon 9	Nonsense (Table 3.4)	53	5.6
c.192G>T/p.Q64H	E2/I2 junction	Splicing (Table 3.2)	41	4.3
c.709C>T/p.R237X	Exon 9	Nonsense (Table 3.4)	39	4.1
c.1009G>A/p.G337S	Exon 12	Splicing (Table 3.2)	30	3.2
c.1A>G/p.M1?/	1st nt of E1	Start loss (Table 3.3)	20	2.1
c.707-1G>C (IVS8-1G>C)/–	I8/E9 junction	Splicing (Table 3.2)	16	1.7
c.782C>T/p.P261L/	Exon 9	Missense (Table 3.7)	16	1.7
c.1090G>T/p.E364X	Exon 13	Nonsense (Table 3.4)	16	1.7
c.698A>T/p.D233V	Exon 8	Missense (Table 3.7)	15	1.6
c.707-1G>A (IVS8-1G>A)/–	I8/E9 junction	Splicing (Table 3.2)	13	1.4
c.1069G>T/p.E357X	Exon 13	Nonsense (Table 3.4)	12	1.3
c.497T>G/p.V166G	Exon 6	Missense (Table 3.7)	11	1.2

[1]Frequency is calculated on the 944 alleles reported in this chapter

variant since individuals homozygous for this mutation are healthy due to residual FAH activity while compound heterozygotes with another *FAH* mutation develop HT1 (Rootwelt et al. 1994b). This mutation does not change the mRNA level nor its size but it results in decreased amount of FAH protein with less activity than the wild-type protein (Bergeron et al. 2001; Rootwelt et al. 1994b), suggesting that a minimal requirement of FAH activity is needed to prevent HT1 disease.

3.5.2 Reversion of FAH Mutation

A mosaic pattern of FAH expression in liver of HT1 patients has been reported for four splicing mutations; c.192G>T (p.Q64H), c.836A>G (p.Q279R), c.1009G>A (p.G337S) and c.1062 + 5G>A (IVS12 + 5G>A) (Demers et al. 2003; Dreumont et al. 2001; Kvittingen et al. 1993, 1994; Poudrier et al. 1998). The presence of FAH positive nodules was shown to be due to the reversion of the primary point mutation (Demers et al. 2003; Kvittingen et al. 1993, 1994) and favored by the selective advantage that the reversion would provide (Demers et al. 2003). Interestingly, it was also shown that the severity of the disease is directly correlated with the extent of HT1 mutation reversion in the liver of HT1 patients (Demers et al. 2003).

3.6 Concluding Remarks

This report summarizes the available information for each of the *FAH* mutations reported in the literature and places them back in their molecular context. While it does not provide explanations for the effect of all mutations on *FAH* mRNA and protein, it does suggest new ways to look at them in addition to highlight the importance of splicing mutations in HT1.

Acknowledgements The work on HT1 in RMT's laboratory is supported by grants from the Canadian Institutes of Health Research (CIHR) and Fondation du Grand Défi Pierre Lavoie.

References

Al-Shamsi A, Hertecant JL, Al-Hamad S, Souid AK, Al-Jasmi F (2014) Mutation spectrum and birth prevalence of inborn errors of metabolism among emiratis: a study from Tawam Hospital Metabolic Center, United Arab Emirates. Sultan Qaboos Univ Med J 14(1):e42–e49

Angileri F, Bergeron A, Morrow G, Lettre F, Gray G, Hutchin T, Ball S (2015) Geographical and ethnic distribution of mutations of the fumarylacetoacetate hydrolase gene in hereditary tyrosinemia type 1. JIMD Rep 19:45–58. doi:10.1007/8904_2014_363

Arranz JA, Pinol F, Kozak L, Perez-Cerda C, Cormand B, Ugarte M, Riudor E (2002) Splicing mutations, mainly IVS6-1(G>T), account for 70% of fumarylacetoacetate hydrolase (FAH) gene alterations, including 7 novel mutations, in a survey of 29 tyrosinemia type I patients. Hum Mutat 20(3):180–188. doi:10.1002/humu.10084

Awata H, Endo F, Tanoue A, Kitano A, Nakano Y, Matsuda I (1994) Structural organization and analysis of the human fumarylacetoacetate hydrolase gene in tyrosinemia type I. Biochim Biophys Acta 1226(2):168–172. doi:0925-4439(94)90025-6

Bateman RL, Bhanumoorthy P, Witte JF, McClard RW, Grompe M, Timm DE (2001) Mechanistic inferences from the crystal structure of fumarylacetoacetate hydrolase with a bound phosphorus-based inhibitor. J Biol Chem 276(18):15284–15291. doi:10.1074/jbc.M007621200

Bateman RL, Ashworth J, Witte JF, Baker LJ, Bhanumoorthy P, Timm DE, Hurley TD, Grompe M, McClard RW (2007) Slow-onset inhibition of fumarylacetoacetate hydrolase by phosphinate mimics of the tetrahedral intermediate: kinetics, crystal structure and pharmacokinetics. Biochem J 402(2):251–260. doi:10.1042/BJ20060961

Bergeron A, D'Astous M, Timm DE, Tanguay RM (2001) Structural and functional analysis of missense mutations in fumarylacetoacetate hydrolase, the gene deficient in hereditary tyrosinemia type 1. J Biol Chem 276(18):15225–15231. doi:10.1074/jbc.M009341200

Bergman AJ, van den Berg IE, Brink W, Poll-The BT, Ploos van Amstel JK, Berger R (1998) Spectrum of mutations in the fumarylacetoacetate hydrolase gene of tyrosinemia type 1 patients in northwestern Europe and Mediterranean countries. Hum Mutat 12(1):19–26. doi:10.1002/(SICI)1098-1004(1998)12:1<19::AID-HUMU3>3.0.CO;2-3

Bian Y, Song C, Cheng K, Dong M, Wang F, Huang J, Sun D, Wang L, Ye M, Zou H (2014) An enzyme assisted RP-RPLC approach for in-depth analysis of human liver phosphoproteome. J Proteome 96:253–262. doi:10.1016/j.jprot.2013.11.014

Bliksrud YT, Brodtkorb E, Andresen PA, van den Berg IE, Kvittingen EA (2005) Tyrosinaemia type I – de novo mutation in liver tissue suppressing an inborn splicing defect. J Mol Med (Berl) 83(5):406–410. doi:10.1007/s00109-005-0648-2

Bliksrud YT, Brodtkorb E, Backe PH, Woldseth B, Rootwelt H (2012) Hereditary tyrosinaemia type I in Norway: incidence and three novel small deletions in the fumarylacetoacetase gene. Scand J Clin Lab Invest 72(5):369–373. doi:10.3109/00365513.2012.676210

Cao YY, Zhang YL, Du J, Qu YJ, Zhong XM, Bai JL, Song F (2012) Compound mutations (R237X and L375P) in the fumarylacetoacetate hydrolase gene causing tyrosinemia type I in a Chinese patient. Chin Med J 125(12):2132–2136

Cassiman D, Zeevaert R, Holme E, Kvittingen EA, Jaeken J (2009) A novel mutation causing mild, atypical fumarylacetoacetase deficiency (Tyrosinemia type I): a case report. Orphanet J Rare Dis 4:28. doi:1750-1172-4-28

Choi HJ, Bang HI, Ki CS, Lee SY, Kim JW, Song J, Shin MR, Lee YW, Lee DH, Park HD (2014) Two novel FAH gene mutations in a patient with hereditary tyrosinemia type I. Ann Clin Lab Sci 44(3):317–323. doi:44/3/317

Couce ML, Dalmau J, del Toro M, Pintos-Morell G, Aldamiz-Echevarria L (2011) Tyrosinemia type 1 in Spain: mutational analysis, treatment and long-term outcome. Pediatr Int 53(6):985–989. doi:10.1111/j.1442-200X.2011.03427.x

Demers SI, Russo P, Lettre F, Tanguay RM (2003) Frequent mutation reversion inversely correlates with clinical severity in a genetic liver disease, hereditary tyrosinemia. Hum Pathol 34(12):1313–1320. doi:S0046817703004064

Desmet FO, Hamroun D, Lalande M, Collod-Beroud G, Claustres M, Beroud C (2009) Human splicing finder: an online bioinformatics tool to predict splicing signals. Nucleic Acids Res 37(9):e67. doi:10.1093/nar/gkp215

Dou LM, Fang LJ, Wang XH, Lu W, Chen R, Li LT, Zhao J, Wang JS (2013) Mutation analysis of FAH gene in patients with tyrosinemia type 1. Zhonghua Er Ke Za Zhi 51(4):302–307

Dreumont N, Poudrier JA, Bergeron A, Levy HL, Baklouti F, Tanguay RM (2001) A missense mutation (Q279R) in the fumarylacetoacetate hydrolase gene, responsible for hereditary tyrosinemia, acts as a splicing mutation. BMC Genet 2:9

Dreumont N, Maresca A, Khandjian EW, Baklouti F, Tanguay RM (2004) Cytoplasmic nonsense-mediated mRNA decay for a nonsense (W262X) transcript of the gene responsible for hereditary tyrosinemia, fumarylacetoacetate hydrolase. Biochem Biophys Res Commun 324(1):186–192. doi:10.1016/j.bbrc.2004.09.041

Dreumont N, Maresca A, Boisclair-Lachance JF, Bergeron A, Tanguay RM (2005) A minor alternative transcript of the fumarylacetoacetate hydrolase gene produces a protein despite being likely subjected to nonsense-mediated mRNA decay. BMC Mol Biol 6:1. doi:1471-2199-6-1

Dursun A, Ozgul RK, Sivri S, Tokatli A, Guzel A, Mesci L, Kilic M, Aliefendioglu D, Ozcay F, Gunduz M, Coskun T (2011) Mutation spectrum of fumarylacetoacetase gene and clinical aspects of tyrosinemia type I disease. JIMD Rep 1:17–21. doi:10.1007/8904_2011_10

Elpeleg ON, Shaag A, Holme E, Zughayar G, Ronen S, Fisher D, Hurvitz H (2002) Mutation analysis of the FAH gene in Israeli patients with tyrosinemia type I. Hum Mutat 19(1):80–81. doi:10.1002/humu.9001

Fisher AL, Page KE, Lithgow GJ, Nash L (2008) The *Caenorhabditis elegans* K10C2.4 gene encodes a member of the fumarylacetoacetate hydrolase family: a *Caenorhabditis elegans* model of type I tyrosinemia. J Biol Chem 283(14):9127–9135. doi:10.1074/jbc.M708341200

Georgouli H, Schulpis KH, Michelakaki H, Kaltsa M, Sdogou T, Kossiva L (2010) Persistent coagulopathy during *Escherichia coli* sepsis in a previously healthy infant revealed undiagnosed tyrosinaemia type 1. BMJ Case Rep 2010. doi:2010/dec21_1/bcr0720103150

Grompe M, al-Dhalimy M (1993) Mutations of the fumarylacetoacetate hydrolase gene in four patients with tyrosinemia, type I. Hum Mutat 2(2):85–93. doi:10.1002/humu.1380020205

Grompe M, St-Louis M, Demers SI, al-Dhalimy M, Leclerc B, Tanguay RM (1994) A single mutation of the fumarylacetoacetate hydrolase gene in French Canadians with hereditary tyrosinemia type I. N Engl J Med 331(6):353–357. doi:10.1056/NEJM199408113310603

Haghighi-Kakhki H, Rezazadeh J, Ahmadi-Shadmehri A (2014) Identification of a combined missense/splice-site mutation in FAH causing tyrosinemia type 1. J Pediatr Endocrinol Metab. doi:10.1515/jpem-2013-0489

Hahn SH, Krasnewich D, Brantly M, Kvittingen EA, Gahl WA (1995) Heterozygosity for an exon 12 splicing mutation and a W234G missense mutation in an American child with chronic tyrosinemia type 1. Hum Mutat 6(1):66–73. doi:10.1002/humu.1380060113

Heath SK, Gray RG, McKiernan P, Au KM, Walker E, Green A (2002) Mutation screening for tyrosinaemia type I. J Inherit Metab Dis 25(6):523–524

Hutchesson AC, Hall SK, Preece MA, Green A (1996) Screening for tyrosinaemia type I. Arch Dis Child Fetal Neonatal Ed 74(3):F191–F194

Hutchesson AC, Bundey S, Preece MA, Hall SK, Green A (1998) A comparison of disease and gene frequencies of inborn errors of metabolism among different ethnic groups in the West Midlands, UK. J Med Genet 35(5):366–370

Huttlin EL, Jedrychowski MP, Elias JE, Goswami T, Rad R, Beausoleil SA, Villen J, Haas W, Sowa ME, Gygi SP (2010) A tissue-specific atlas of mouse protein phosphorylation and expression. Cell 143(7):1174–1189. doi:10.1016/j.cell.2010.12.001

Ijaz S, Zahoor MY, Imran M, Afzal S, Bhinder MA, Ullah I, Cheema HA, Ramzan K, Shehzad W (2016) Direct sequencing of FAH gene in Pakistani tyrosinemia type 1 families reveals a novel mutation. J Pediatr Endocrinol Metab 29(3):327–332. doi:10.1515/jpem-2015-0289

Imtiaz F, Rashed MS, Al-Mubarak B, Allam R, El-Karaksy H, Al-Hassnan Z, Al-Owain M, Al-Zaidan H, Rahbeeni Z, Qari A, Meyer BF, Al-Sayed M (2011) Identification of mutations causing hereditary tyrosinemia type I in patients of Middle Eastern origin. Mol Genet Metab 104(4):688–690. doi:S1096-7192(11)00206-X

Jitraruch S, Treepongkaruna S, Teeraratkul S, Wattanasirichaigoon D, Leelaudomlipi S, Sornmayura P, Viengteerawat S, Sriphojanart S (2011) Long-term outcome of living donor liver transplantation in a Thai boy with hereditary tyrosinemia type I: a case report. J Med Assoc Thail 94(10):1276–1280

Kim SZ, Kupke KG, Ierardi-Curto L, Holme E, Greter J, Tanguay RM, Poudrier J, D'Astous M, Lettre F, Hahn SH, Levy HL (2000) Hepatocellular carcinoma despite long-term survival in chronic tyrosinaemia I. J Inherit Metab Dis 23(8):791–804

Kvittingen EA (1991) Tyrosinaemia type I – an update. J Inherit Metab Dis 14(4):554–562

Kvittingen EA, Rootwelt H, Brandtzaeg P, Bergan A, Berger R (1993) Hereditary tyrosinemia type I. Self-induced correction of the fumarylacetoacetase defect. J Clin Invest 91(4):1816–1821. doi:10.1172/JCI116393

Kvittingen EA, Rootwelt H, Berger R, Brandtzaeg P (1994) Self-induced correction of the genetic defect in tyrosinemia type I. J Clin Invest 94(4):1657–1661. doi:10.1172/JCI117509

Labelle Y, Phaneuf D, Leclerc B, Tanguay RM (1993) Characterization of the human fumarylacetoacetate hydrolase gene and identification of a missense mutation abolishing enzymatic activity. Hum Mol Genet 2(7):941–946

Laszlo A, Rozsa M, Sallay E, Tiszlavicz L, Janovszky A, Varkonyi A, Karg E, Wittmann G, Turi S, Ugarte M (2013) The fate of tyrosinaemic Hungarian patients before the NTBC aera. Ideggyogy Sz 66(11–12):415–419

Mahuran DJ, Angus RH, Braun CV, Sim SS, Schmidt DE Jr (1977) Characterization and substrate specificity of fumarylacetoacetate fumarylhydrolase. Can J Biochem 55(1):1–8

Mak CM, Lam CW, Chim S, Siu TS, Ng KF, Tam S (2013) Biochemical and molecular diagnosis of tyrosinemia type I with two novel FAH mutations in a Hong Kong chinese patient: recommendation for expanded newborn screening in Hong Kong. Clin Biochem 46(1–2):155–159. doi:S0009-9120(12)00540-1

la Marca G, Malvagia S, Pasquini E, Cavicchi C, Morrone A, Ciani F, Funghini S, Villanelli F, Zammarchi E, Guerrini R (2011) Newborn screening for tyrosinemia type I: further evidence that succinylacetone determination on blood spot is essential. JIMD Rep 1:107–109. doi:10.1007/8904_2011_24

Mitchell GA, Grompe M, Lambert H, Tanguay RM (2001) Hypertyrosinemia. In: Scriver C, Beaudet A, Sly WSJ, Valle D (eds) The metabolic and molecular bases of inherited diseases, vol Volume II, 8th edn. McGrawHill, New York, pp 1777–1805

Mohamed S, Kambal MA, Al Jurayyan NA, Al-Nemri A, Babiker A, Hasanato R, Al-Jarallah AS (2013) Tyrosinemia type 1: a rare and forgotten cause of reversible hypertrophic cardiomyopathy in infancy. BMC Res Notes 6(1):362. doi:1756-0500-6-362

Mustonen A, Ploos van Amstel HK, Berger R, Salo MK, Viinikka L, Simola KO (1997) Mutation analysis for prenatal diagnosis of hereditary tyrosinaemia type 1. Prenat Diagn 17(10):964–966. doi:10.1002/(SICI)1097-0223(199710)17:10<964::AID--PD164>3.0.CO;2-6

Park HD, Lee DH, Choi TY, Lee YK, Kim JW, Ki CS, Lee YW (2009) Clinical, biochemical, and genetic analysis of a Korean neonate with hereditary tyrosinemia type 1. Clin Chem Lab Med 47(8):930–933. doi:10.1515/CCLM.2009.223

Perez-Carro R, Sanchez-Alcudia R, Perez B, Navarrete R, Perez-Cerda C, Ugarte M, Desviat L (2014) Functional analysis and in vitro correction of splicing FAH mutations causing tyrosinemia type I. Clin Genet 86(2):167–171. doi:10.1111/cge.12243

Phaneuf D, Labelle Y, Berube D, Arden K, Cavenee W, Gagne R, Tanguay RM (1991) Cloning and expression of the cDNA encoding human fumarylacetoacetate hydrolase, the enzyme deficient in hereditary tyrosinemia: assignment of the gene to chromosome 15. Am J Hum Genet 48(3):525–535

Phaneuf D, Lambert M, Laframboise R, Mitchell G, Lettre F, Tanguay RM (1992) Type 1 hereditary tyrosinemia. Evidence for molecular heterogeneity and identification of a causal mutation in a French Canadian patient. J Clin Invest 90(4):1185–1192. doi:10.1172/JCI115979

Ploos van Amstel JK, Bergman AJ, van Beurden EA, Roijers JF, Peelen T, van den Berg IE, Poll-The BT, Kvittingen EA, Berger R (1996) Hereditary tyrosinemia type 1: novel missense, nonsense and splice consensus mutations in the human fumarylacetoacetate hydrolase gene; variability of the genotype-phenotype relationship. Hum Genet 97(1):51–59

Poudrier J, St-Louis M, Lettre F, Gibson K, Prevost C, Larochelle J, Tanguay RM (1996) Frequency of the IVS12 + 5G->A splice mutation of the fumarylacetoacetate hydrolase gene in carriers of hereditary tyrosinaemia in the French Canadian population of Saguenay-Lac-St-Jean. Prenat Diagn 16(1):59–64. doi:10.1002/(SICI)1097-0223(199601)16:1<59::AID-PD810>3.0.CO;2-D

Poudrier J, Lettre F, Scriver CR, Larochelle J, Tanguay RM (1998) Different clinical forms of hereditary tyrosinemia (type I) in patients with identical genotypes. Mol Genet Metab 64(2):119–125. doi:S1096-7192(98)92695-6

Poudrier J, Lettre F, St-Louis M, Tanguay RM (1999) Genotyping of a case of tyrosinaemia type I with normal level of succinylacetone in amniotic fluid. Prenat Diagn 19(1):61–63. doi:10.1002/(SICI)1097-0223(199901)19:1<61::AID-PD455>3.0.CO;2-#

Prieto-Alamo MJ, Laval F (1998) Deficient DNA-ligase activity in the metabolic disease tyrosinemia type I. Proc Natl Acad Sci U S A 95(21):12614–12618

Ran T, Gao Y, Marsh M, Zhu W, Wang M, Mao X, Xu L, Xu D, Wang W (2013) Crystal structures of Cg1458 reveal a catalytic lid domain and a common catalytic mechanism for the FAH family. Biochem J 449(1):51–60. doi:10.1042/BJ20120913

Rootwelt H, Berger R, Gray G, Kelly DA, Coskun T, Kvittingen EA (1994a) Novel splice, missense, and nonsense mutations in the fumarylacetoacetase gene causing tyrosinemia type 1. Am J Hum Genet 55(4):653–658

Rootwelt H, Brodtkorb E, Kvittingen EA (1994b) Identification of a frequent pseudodeficiency mutation in the fumarylacetoacetase gene, with implications for diagnosis of tyrosinemia type I. Am J Hum Genet 55(6):1122–1127

Rootwelt H, Chou J, Gahl WA, Berger R, Coskun T, Brodtkorb E, Kvittingen EA (1994c) Two missense mutations causing tyrosinemia type 1 with presence and absence of immunoreactive fumarylacetoacetase. Hum Genet 93(6):615–619

Rootwelt H, Kristensen T, Berger R, Hoie K, Kvittingen EA (1994d) Tyrosinemia type 1 – complex splicing defects and a missense mutation in the fumarylacetoacetase gene. Hum Genet 94(3):235–239

Rootwelt H, Hoie K, Berger R, Kvittingen EA (1996) Fumarylacetoacetase mutations in tyrosinaemia type I. Hum Mutat 7(3):239–243. doi:10.1002/(SICI)1098-1004(1996)7:3<239::AID-HUMU8>3.0.CO;2-5

Sheth JJ, Ankleshwaria CM, Pawar R, Sheth FJ (2012) Identification of novel mutations in FAH gene and prenatal diagnosis of tyrosinemia in Indian family. Case Rep Genet 2012:428075. doi:10.1155/2012/428075

Sniderman King L, Trahms C, Scott CR (2006) Tyrosinemia type 1. In: Pagon RA, Adam MP, Bird TD, Dolan CR, Fong CT, Stephens K (eds). doi:http://www.ncbi.nlm.nih.gov/books/NBK1515/

van Spronsen FJ, Thomasse Y, Smit GP, Leonard JV, Clayton PT, Fidler V, Berger R, Heymans HS (1994) Hereditary tyrosinemia type I: a new clinical classification with difference in prognosis on

dietary treatment. Hepatology 20(5):1187–1191. doi:S027091399400340X
St-Louis M, Leclerc B, Laine J, Salo MK, Holmberg C, Tanguay RM (1994) Identification of a stop mutation in five Finnish patients suffering from hereditary tyrosinemia type I. Hum Mol Genet 3(1):69–72
St-Louis M, Poudrier J, Phaneuf D, Leclerc B, Laframboise R, Tanguay RM (1995) Two novel mutations involved in hereditary tyrosinemia type I. Hum Mol Genet 4(2):319–320
Timm DE, Mueller HA, Bhanumoorthy P, Harp JM, Bunick GJ (1999) Crystal structure and mechanism of a carbon-carbon bond hydrolase. Structure 7(9):1023–1033. doi:st7915
Timmers C, Grompe M (1996) Six novel mutations in the fumarylacetoacetate hydrolase gene of patients with hereditary tyrosinemia type I. Hum Mutat 7(4):367–369. doi:10.1002/(SICI)1098-1004(1996)7:4<367::AID-HUMU14>3.0.CO;2-0
Vaca Jacome AS, Rabilloud T, Schaeffer-Reiss C, Rompais M, Ayoub D, Lane L, Bairoch A, Van Dorsselaer A, Carapito C (2015) N-terminome analysis of the human mitochondrial proteome. Proteomics 15(14):2519–2524. doi:10.1002/pmic.201400617
Vondrackova A, Tesarova M, Magner M, Docekalova D, Chrastina P, Prochazkova D, Zeman J, Honzik T (2010) Clinical, biochemical and molecular characteristics in 11 Czech children with tyrosinemia type I. Cas Lek Cesk 149(9):411–416
Ward AJ, Cooper TA (2010) The pathobiology of splicing. J Pathol 220(2):152–163. doi:10.1002/path.2649
Wu X, Hurst LD (2016) Determinants of the usage of splice-associated cis-motifs predict the distribution of human pathogenic SNPs. Mol Biol Evol 33(2):518–529. doi:10.1093/molbev/msv251
Yang N, Han LS, Ye J, Qiu WJ, Zhang HW, Gong ZW, Zhang YF, Wang Y, Gu XF (2012) Analysis of clinical data and genetic mutations in three Chinese patients with tyrosinemia type I. Zhonghua Yi Xue Yi Chuan Xue Za Zhi 29(6):648–652. doi:940629142

4 Molecular Pathogenesis of Liver Injury in Hereditary Tyrosinemia 1

Robert M. Tanguay, Francesca Angileri, and Arndt Vogel

Abstract

Untreated HT1 rapidly degenerates into very severe liver complications often resulting in liver cancer. The molecular basis of the pathogenic process in HT1 is still unclear. The murine model of FAH-deficiency is a suitable animal model, which represents all phenotypic and biochemical manifestations of the human disease on an accelerated time scale. After removal of the drug 2-(2-N-4-trifluoromethylbenzoyl)-1,3-cyclohexanedione (NTBC), numerous signaling pathways involved in cell proliferation, differentiation and cancer are rapidly deregulated in FAH deficient mice. Among these, the Endoplasmic reticulum (ER) pathway, the heat stress response (HSR), the Nrf2, MEK and ERK pathways, are highly represented. The p21 and mTOR pathways critical regulators of proliferation and tumorigenesis have also been found to be dysregulated. The changes in these pathways are described and related to the development of liver cancer.

R.M. Tanguay (✉) F. Angileri
Laboratoire de génétique cellulaire et développementale, Département de biologie moléculaire, biochimie médicale et pathologie, Faculté de médecine, Institut de biologie intégrative et des systèmes (IBIS) and PROTEO, Université Laval, 1030 avenue de la médecine, Québec, QC G1V 0A6, Canada
e-mail: robert.tanguay@fmed.ulaval.ca; francesca.angileri@lyon.unicancer.fr

A. Vogel
Department of Gastroenterology, Medizinische Hochschule Hannover, Carl-Neuberg Str.1, 30625 Hannover, Germany
e-mail: vogel.arndt@mh-hannover.de

© Springer International Publishing AG 2017

R.M. Tanguay (ed.), *Hereditary Tyrosinemia*, Advances in Experimental Medicine and Biology 959,
DOI 10.1007/978-3-319-55780-9_4

Keywords

Hereditary tyrosinemia type 1 • Metabolic disease • Liver cancer • Hepatocellular Carcinoma (HCC) • Signaling pathways • ER stress • Heat shock response • Nrf2 • MEK pathway • ERK pathway • p21 • NTBC

Abbreviations

AFP	Alpha feto protein
AKT	Protein kinase B
CHOP	C/EBP homologous protein
ER stress	Endoplasmic reticulum stress
ERAD	ER-stress associated degradation
ERK	Extracellular signal-regulated kinase
FAA	Fumarylacetoacetate
FAH	Fumarylacetoacetate hydrolase
GSH	Glutathione
HT1	Hereditary tyrosinemia type 1
MAA	Maleylacetoacetate
MCL-1	Myeloid cell leukemia 1
SAC	Succinylacetone
URP	Unfolded protein response

4.1 Introduction

HT1 is an autosomal-recessive disease caused by a genetic inactivation of the enzyme fumarylacetoacetate hydrolase (FAH), which carries out the last step of the tyrosine catabolism pathway and characterized by an extremely high susceptibility for liver cancer. In order to understand the molecular mechanisms of hepatocarcinogenesis in HT1, a murine model of FAH-deficiency has been developed, which represents all phenotypic and biochemical manifestations of the human disease on an accelerated time scale (Grompe et al. 1995). Previously, it has been thought that the pathophysiology of liver disease in HT1 involves FAA-induced cell death, which would lead to increased hepatocyte turnover and proliferation and thereby cancer. However studies by Vogel and Tanguay's groups in *Fah*$^{-/-}$ mouse model have revealed a very different picture, especially in the chronic liver injury phase. Figure 4.1 summarizes the different protocols that have been used to study the different phases of the pathologic process occurring in the FAH KO mouse upon removal of the drug 2-(2-N-4-trifluoromethylbenzoyl)-1,3-cyclohexanedione (NTBC), which blocks the pathway upstream of the formation of fumarylacetoacetate (FAA), and is used to treat patients with HT1. The first one used by Vogel and his colleagues uses either a low suboptimal dose (0.75 mg/L) to induce moderate liver injury or the completely stopped NTBC treatment to induce severe liver injury (Fig. 4.1a, b) (Marhenke et al. 2008; Vogel et al. 2004). Complete NTBC withdrawal results in a very high mortality with almost all mice dying within 4–6 weeks post-removal. In the case of Tanguay and his colleagues, a complete withdrawal of NTBC was coupled with the addition of a liquid diet to extend the viability from 5 to 15 weeks (Fig. 4.1c) when 100% of mice showed liver cancer (Angileri et al. 2015). This permitted a study of various signaling pathways at the early and late phases of the pathogenic process.

4.2 Liver Injury and Tumorigenesis

4.2.1 HT1 Elicits an ER Stress Response

Amongst the different metabolites accumulating in HT1 (FAA, MAA, SAC), FAA seems to have major effects. For example in cultured cell assays, FAA was shown to be moderately mutagenic while MAA or SAC had no effect on cell transformation (Jorquera and Tanguay 1997; Tanguay et al. 1996). However how FAA affects the liver so severely is not clear. Its mutagenicity was showed potentiated by glutathione (GSH) depletion. ROS are not generated during FAA treatment of GSH-depleted cells (Jorquera and

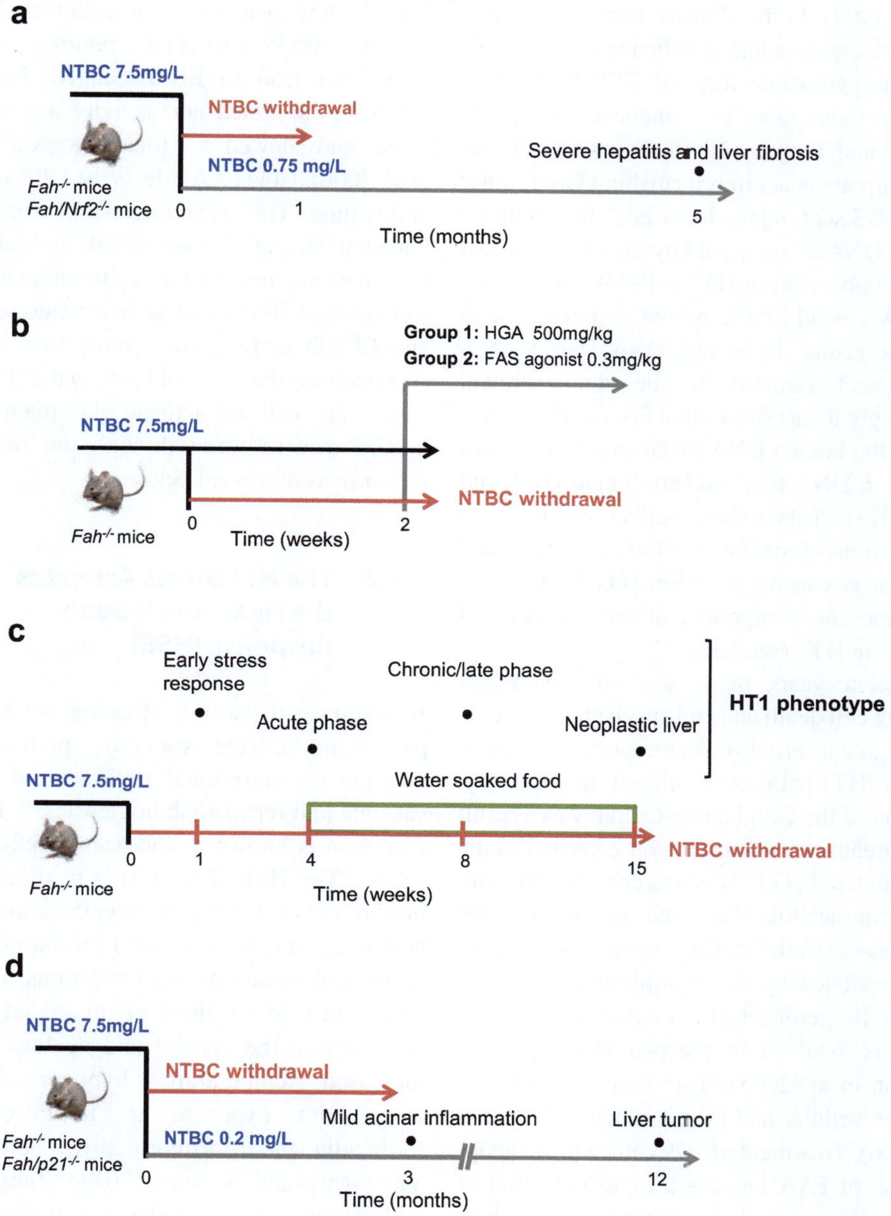

Fig. 4.1 Schematic representation of Vogel and Tanguay's HT1 protocols. *Fah*$^{-/-}$, *Fah*/*p21*$^{-/-}$ and *Fah*/*Nrf2*$^{-/-}$ animals were treated with NTBC-containing drinking water at a concentration of 7.5 mg/L unless otherwise indicated, (**a**) 8 weeks old mice were used in short-term experiments (complete NTBC withdrawal). Surviving mice were killed after 4 weeks. In long-term experiments, NTBC was reduced in the drinking water of 4 weeks old mice and not changed thereafter. All mice receiving 10% NTBC in the drinking water survived until harvest at indicated time points (Marhenke et al. 2008), (**b**) NTBC was withdrawn from *fah*$^{-/-}$ mice, which were kept, on water for 2 weeks. At the end of this period of time, the animals were injected intraperitoneally with HGA (500 mg/kg) or FAS antibody (Jo-2; 0.3 mg/kg) (Vogel et al. 2004), (**c**) 4 months-old *fah*$^{-/-}$ mice were withdrawn from NTBC therapy for periods of 3 days to 15 weeks. Water soaked food treatment was introduced after 4 weeks of NTBC interruption. Mice were weighted three times per week and periodically examined for signs of clinical illness (Angileri et al. 2015). (**d**) Ten-week-old FAH-deficient mice were monitored after NTBC was reduced (2.5%) or withdrawn (0%). *Fah*$^{-/-}$ and *Fah*/*p21*$^{-/-}$ double-knockout mice on 2.5% NTBC were followed for 13 months and *Fah*/*p21*$^{-/-}$ 0% NTBC for 3 months

Tanguay 1997). In the Chinese hamster V79 cell assay, N-acetyl cysteine and the methyl ester cell membrane permeable form of GSH-MEE were found to abrogate FAA-induced apoptosis (Jorquera and Tanguay 1999). This suggests that these compounds act in replenishing GSH rather than as ROS scavengers. Does FAA have a direct effect on DNA as suggested by the chromosomal instability observed in HT1 cells? As an electrophile, FAA would be expected to interact with many molecules including DNA, but such a direct effect remains to be demonstrated. Interestingly it was shown that FAA could inhibit many of the human DNA glycosylases, enzymes involved in DNA base excision repair (Bliksrud et al. 2013). This indirect effect would favor accumulation of oxidative DNA damages and potent cancer-causing mutations. Oxidative stress is an important component of the pathological processes in HT1 (see Sect. 2.2).

In recent years many signaling pathways mediating cell death and cell survival, two events involved in cancer, have been reported regulated during the HT1 pathogenic process. Interestingly, disruption of the Golgi complex likely as a result of FAA treatment was shown on cells and in the mouse model of HT1. This was concomitant with a rise in intracellular Ca^{2+} and activation of the ERK kinase. Cellular Ca^{2+} homeostasis being tightly regulated by the endoplasmic reticulum (ER) led Bergeron et al. to suggest that a ER stress was involved in the pathogenic process (Bergeron et al. 2003). This was further documented in cellular and mouse models (Bergeron et al. 2006). Treatment of V79 cells with an apoptotic dose of FAA induces a rapid induction of ER stress proteins such as GRP78/BiP, phosphorylation of eIF2α and a late induction of CHOP, all events observed in a classical ER stress. This was also observed in the FAH knockout model when mice were taken off NTBC, suggesting an alteration of the Unfolded Protein Response (UPR) in the HT1 process. The mains targets of the UPR are IRE1, PERK and ATF6 via BiP regulation (Fig. 4.2). This was verified by a cell luciferase reporter gene assay that showed that FAA could activate ERSE a cis-element promotor of GRP78/BiP, consistent again with an ER response. In the mouse model, induction of HT1 phenotype by NTBC withdrawal led to upregulation of GRP78/BiP at 3 weeks post-NTBC removal. Since ER stress can lead to ER-associated degradation (ERAD), the proteasome activity was also measured, and showed a 2-fold increase (Bergeron et al. 2006). How FAA affects the UPR is unclear at this time. ATF6 is an interesting target, as it is known to affect Ca^{2+} homeostasis and activate the ERK pathway involved in proliferation and transformation. ATF6 is also as an activator of CHOP and ERAD as discussed above (see Fig. 4.2). Alternatively the effect of FAA on the UPR might signal the cell to activate the main cellular homeostasis response through the heat shock response as discussed below.

4.2.2 The HT1 Stress Activates the Heat Shock Gene Response (HSR)

In response to various stresses, cellular compartments activate signaling pathways that mediate transcriptional programs to promote survival and reestablish homeostasis. This phenomenon is known as the heat shock response (HSR). The HSR (Fig. 4.3) is mediated by the production of a group of proteins known as the heat-shock proteins (HSPs) (Kampinga et al. 2009) and occurs during conditions that cause an increase in unfolded or misfolded proteins primarily in the cytosol and nucleus, such as increased temperature, hypoxia, oxidative stress and exposure to toxic chemicals (Labbadia and Morimoto 2015; Saibil 2013; Vihervaara and Sistonen 2014). Tanguay and collaborators provided the first evidence of an enhanced HSR by FAA toxic action by treating cultured cells with exogenous FAA (Bergeron et al. 2003; Jorquera and Tanguay 2001). In their work, the authors showed that ER stress in HT1 pathophysiology is associated with an early induction of the GRP78/BiP chaperone, a proximal signaling protein, functioning as the key regulator of the three transducers of the unfolded protein response (UPR) during the ER stress (Bergeron et al. 2006).

Mammalian HSPs can be classified in general groups according to their size: high molecular

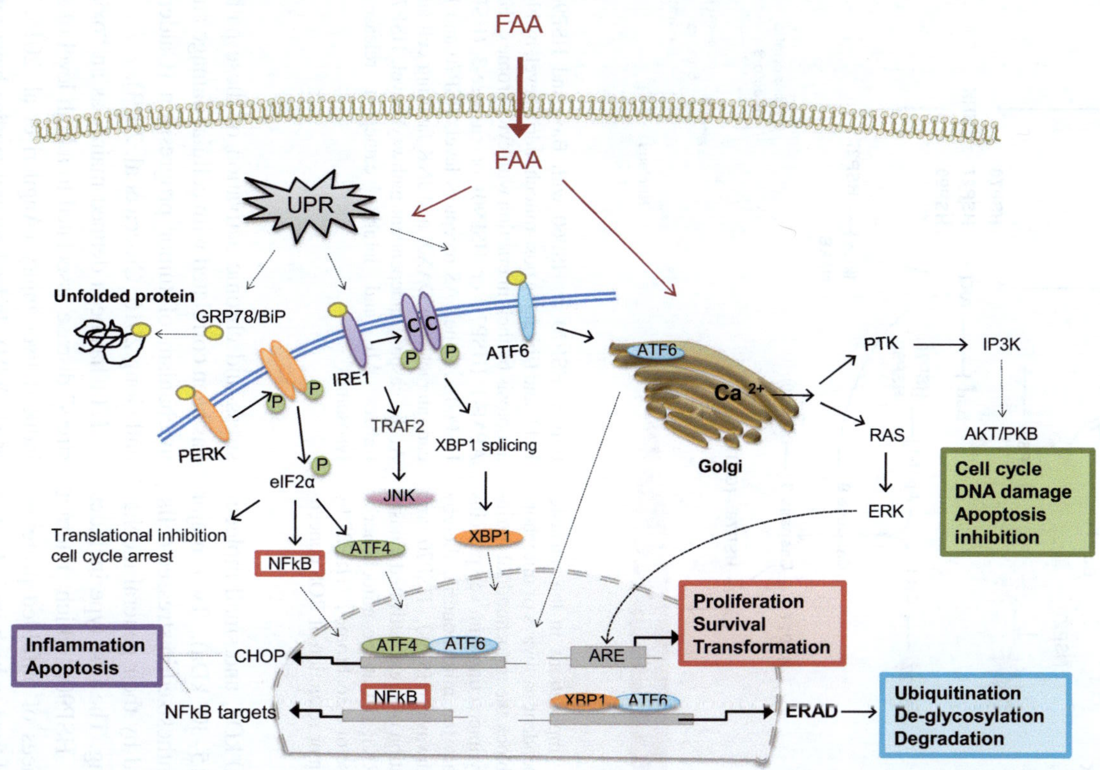

Fig. 4.2 Pathways representing endoplasmic (ER) and unfolded protein responses (UPR) triggered by the toxic metabolite FAA in hepatocytes lacking FAH activity. GRP78/BiP functions as an unfolded protein response (UPR) signaling regulator by binding to and maintaining the ER stress sensors PERK, ATF6 and IRE1 in inactive forms. Under ER stress, the UPR modulators PERK and IRE1 are activated by relocalization of GRP78/BiP and dimerization followed by phosphorylation. ATF6 is cleaved in the Golgi apparatus, inducing the activation of UPR transcription factors ATF6, ATF4 and XBP1s and Ca^{2+} mobilization with activation of ER stress pathways. These transcription factors mainly upregulate the adaptive UPR pathway and normalize ER function via the ATF6, PERK–eIF-2α–ATF4 or IRE1–XBP1 pathways. Under long-term ER stress, the adaptive UPR pathway fails to rescue the cells, and the apoptotic UPR pathway, namely the PERK–eIF-2α–ATF4–CHOP or IRE1–TRAF2–JNK pathway, is induced

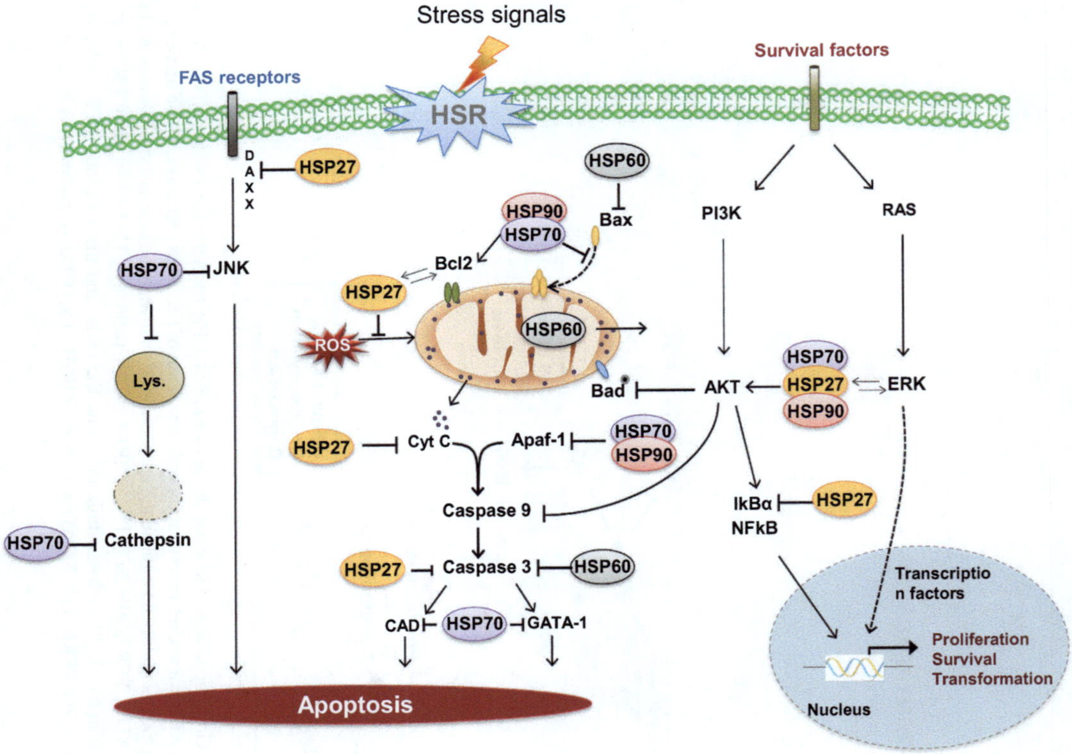

Fig. 4.3 HSPs regulatory implications in the intrinsic, extrinsic and caspase independent pathways of mitochondrial apoptosis. HSPs can block the mitochondrial intrinsic pathway of apoptosis by interacting with key proteins at three levels: (i) upstream the mitochondria, thereby modulating signaling pathways (HSP90, HSP70 and HSP27 modulate the activation of stress-activated kinases such as AKT, JNK or ERK); (ii) at the mitochondrial level, controlling the release of cytochrome c (HSP27 by its interaction with the actin or reduction of ROS concentration, HSP70 or HSP60 with Bax, and HSP90 with Bcl2) and (iii) at the post-mitochondrial level, by blocking apoptosis by their interaction with cytochrome c (HSP27), APAF-1 (HSP70 or HSP90) or caspase-3 (HSP27 or HSP60). At the FAS receptors level, HSP70 and HSP27 can interact with DAXX and JNK favoring cell survival. At the caspase-independent pathways level, HSP70 neutralizes AIF and inhibits cathepsin release from lysosomes

weight HSPs (~60–110 kDa) and small molecular weight HSPs (~15–30 kDa). Two major classes of HSPs are synthesized in stressed cells, which are distinguished by their general mechanisms of protein folding. The first type includes HSP27, HSP70, and HSP90, which interact directly with the surfaces of unfolded proteins. By contrast, the second type (which includes the chaperonin HSP60) assembles into complexes resembling folding chambers, which form privileged environments that exclude bulk cytoplasm and favor recovery of active protein conformations. Manipulation of the magnitude and duration of the activation of stress responses has been proposed as a strategy to prevent or repair the damage associated with stress exposure, aging or in degenerative diseases. However, prolonged stress and chronic activation of these pathways has been correlated with cellular damage and as a mechanism of tumor progression (Calderwood and Gong 2016; Ciocca et al. 2013).

HT1 has been defined mainly as an "oxidative stress" disease associated to a high level of stress-induced liver injury (Angileri et al. 2015; Vogel et al. 2004). Indeed several studies have reported FAA as a mutagenic compound able to cause oxidative damage by forming stable adduct with glutathione (GSH) and sulfhydryl groups of proteins and to induce tumor hypoxia (Dieter et al. 2003; Manabe et al. 1985). Thus it is not surprising that HSPs like HSP70, HSP32 and HSP27 are activated in the stressful environment occurring in HT1 pathogenesis. These observations were further confirmed in the HT1 murine model (*Fah*$^{-/-}$ mouse).

Activation of these proteins was considered to be a response to oxidative stress stimuli during induction of the ER stress response by FAA and toxic metabolites generated as a result of FAH deficiency. Similarly, Vogel and collaborators (2004), observed, a fivefold up-regulation of HSP70 protein levels and a strong increase of HSP27 (200-fold) and HSP32 (58-fold) (Vogel et al. 2004) after taking *Fah*$^{-/-}$ mice off NTBC for 14 days. The activation of the HSR in this model suggests HSPs as contributors to the cell-death resistance phenotype observed in mice that are taken off NTBC, as shown in the same work by testing a wide range of programmed cell death pathways. The cell death resistance observed here in liver damaged by HT1 is mediated not only by the classic apoptotic pathway, modeled by activation of the FAS receptor, but also by other mechanisms, such as the necrosis-like cell death triggered by the drug APAP (acetaminophen), which induces caspase-independent apoptosis.

An important contribution to the overall HSR modulation during HT1 process came from the work of Angileri and colleagues, which were able to prolong the observation of the HT1-induced stress response in a long-term protocol settled on the same murine model (Fig. 4.1c) (Angileri et al. 2014, 2015). For this purpose, the authors built a new protocol of complete NTBC withdrawal, coupled with the addition of a liquid diet to extend the mouse viability up to 15 weeks (Fig. 4.1c). The liver dysfunction could therefore be followed longer and up to the formation of neoplastic liver lesions in 100% of mice under experimentation. In this protocol different endpoints were used as representative of different grades of liver pathology, and HSPs expression was analyzed at transcriptional and translational level. Gene expression by microarray during a long-term HT1 stress characterized by tumor development identified a wide range of genes implicated in cellular growth and proliferation, organismal survival, differentiation, inflammation and cell migration. Interestingly, among them were present several members of the HSPs family (Angileri et al. 2014), corroborating the idea that HSPs induction in HT1 pathology is not just a mechanism of defense against oxidative damage, but might foster an environment that is essential for tumor development. Noteworthy the expression level of HSPs proteins is unevenly modulated in the different phases of liver damage. While the evidence suggests that HSP90 and HSP60 families do not take part in the HT1 liver dysfunction, the other HSPs members are subjected to a variable modulation. The cognate form of the 70 kD heat shock protein, HSC70, remains almost unvaried in *Fah*$^{-/-}$ mice, independently of the NTBC treatment, with a faint reinforcement in the advanced hepatocarcinogenic stage. The inducible HSP70 and the small HSP27 show a significant deregulation along the entire process. Interestingly, HSP70 seems to be involved more in the acute phase of the stress, presenting a normal distribution with a peak of expression in the first week after NTBC interruption (Angileri et al. 2014). On the other hand HSP27 appears as one of the protagonists in the HSR induced by HT1 pathogenic mechanisms, confirming the previous finding of Vogel (Vogel et al. 2004). Indeed, HSP27 is not only highly overexpressed, but also its phosphorylation at Serine 15 is greatly modulated up to liver neoplastic nodules formation. It is well known that phosphorylated HSP27 (Ser15) is involved in the ability of this chaperone to switch between small and large oligomers in order to accomplish its anti-apoptotic functions (Bruey et al. 2000; Oya-Ito et al. 2006). Biochemical and genetic studies have demonstrated the critical role of stress-inducible HSPs in resistance to cell death. The anti-apoptotic effect of HSPs involves protein-protein interactions that seem not to be directly related to their chaperone functions. For example, HSP70 interacts with apoptosis protease activating factor-1 (APAF-1), thereby preventing APAF-1 from interacting with procaspase-9, and HSP27 has been reported to specifically interact with cytochrome c in the cytosol, thereby preventing the activation of caspase 9. HSP70 also has the capacity to interact with the apoptosis-inducing factor to negatively interfere with caspase-independent apoptosis. Furthermore HSP70 has also been implicated in other mechanisms of cell death in addition to programmed cell death (PCD). HSP70 has been shown to increase the lysosome resistance against chemical and physical membrane destabilization and nevertheless it has been suggested that HSP70 could modulate JNK and ERK phosphorylation and promote the stabilization of phosphorylated form of PKC (Gao and Newton 2002; Lee et al. 2005). In addition, these HSPs have been implicated in the protection against APAP-induced necrosis-like liver failure. Therefore, it seems appropriate to affirm that stress-induced malfunction of the death machinery plays a role in the pathophysiology of this disorder and it might be related to the risk of developing HCC, as hypothesized by Angileri et al. (2014).

In the same mechanism of impaired apoptosis and survival reinforcement a key role has been proposed for another small HSP, namely HSP32. In more recent work Angileri and colleagues showed upregulation of this protein concomitantly with factors responsible for its transcriptional regulation (Angileri et al. 2015). HSP32, also known as heme oxigenase -1 (HO-1), is mainly known as a cytoprotective enzyme, sensor of oxidative stress and involved in degradation of heme to form biliverdin, carbon monoxide (CO), and free iron. HO-1 upregulation in HT1 pathology even in the first day of NTBC withdrawal is probably due to its role in the antioxidant response element-dependent gene transcription program (Angileri et al. 2015). It is well known that HO-1 plays an important protective role in the tissues by reducing oxidative injury, attenuating inflammatory response, inhibiting apoptosis, and by regulating angiogenesis and cell proliferation (Banerjee et al. 2011). Recent evidence suggests a role for HO-1 in promoting cancer since its activity has closely been related to growth, apoptosis, angiogenesis, invasiveness, and metastasis of solid tumors (Banerjee et al. 2011). Thus, the HO-1 high protein levels found in the HT1 stress might indicate that it plays a crucial role not just in defense against oxidative stress induced by HT1 toxic metabolites, but also in promoting survival of proliferating cells as hypothesized by Angileri and colleagues. Indeed, since HO-1 induction results in production of CO, that influences glycolytic enzymes activity, this would result in directional glucose utilization ensuring resistance against oxidative stress from which cancer cell survival could take advantage (Angileri et al. 2015) (see Fig. 4.4).

In summary, HSPs induction during the pathogenic process in the HT1 murine model, might participate in a chaperone-induced cytoprotection mechanism able to rescue cells from apoptosis, easing the deleterious consequences of HT1 chronic stress and promoting tumor growth. At this time, it is still unclear if the over-expression of HSPs in cancer plays a role only in supporting malignancy or if it is essential in developing the transformed phenotype.

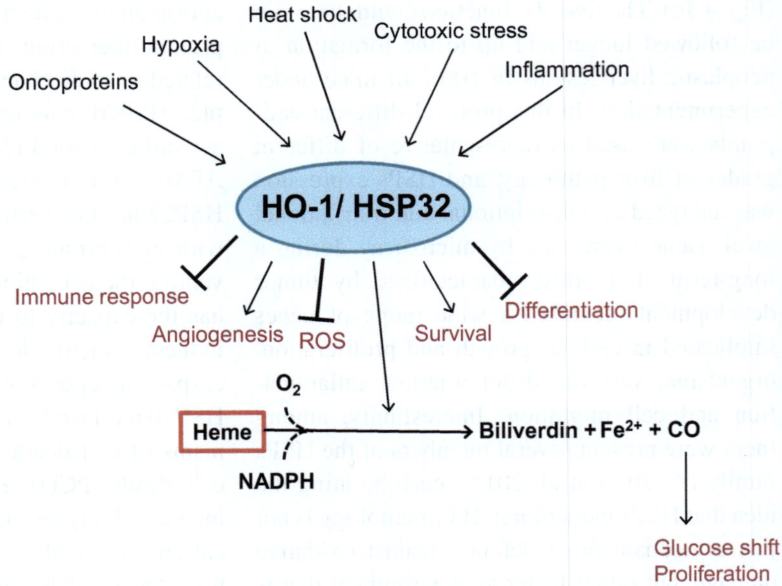

Fig. 4.4 Schematic summary of HO-1 effects on tumor biology. In neoplastic cells, HO-1 expression is induced by various oncoproteins as well as by stress reactions and by (stress-inducing) cytoreductive compounds. HO-1 acts as a survival factor that may rescue tumor cells under various conditions (stress reactions) or when exposed to cytotoxic drugs (HO-1-mediated resistance)

4.2.3 Apoptosis Resistance in HT1 and Activation of AKT/MEK Pathway

The increase in intracellular Ca^{2+} upon treatment of cells with FAA and the occurrence of an ER stress is likely to activate multiple signaling cascades. Among those are the Ras/Raf/MEK/ERK and the PI3K/AKT pathways. Both of these are linked to the mitochondrial intrinsic apoptotic pathways consistent with the physical interaction between mitochondria and ER (Fig. 4.5). These pathways were therefore examined, since resistance to apoptosis had already been reported in HT1 (Jorquera and Tanguay 1999; Kubo et al. 1998; Vogel et al. 2004).

Following the establishment of an HT1 stress by NTBC removal, Orejuela et al. observed a stable activation of the AKT survival pathway along with an important inhibition of mitochondrial apoptosis, associated with a strong induction of BCl-2, BCL-X and a modulation of Bad phosphorylation (Orejuela et al. 2008). BCL-2 a pro survival factor showed a 30-fold rapid increase, which remained high till the 4rth week and then returned to normal level. Phosphorylation of Bad on Serine136 was rapidly induced and was followed by that of Serine 112 3 weeks later (Angileri et al. 2015). Mcl-1 another mitochondrial prosurvival member was induced at 4 weeks post-withdrawal. Moreover, Vogel et al. found that Bid was not dephosphorylated in the FAH mutant mouse taken off NTBC following chal-

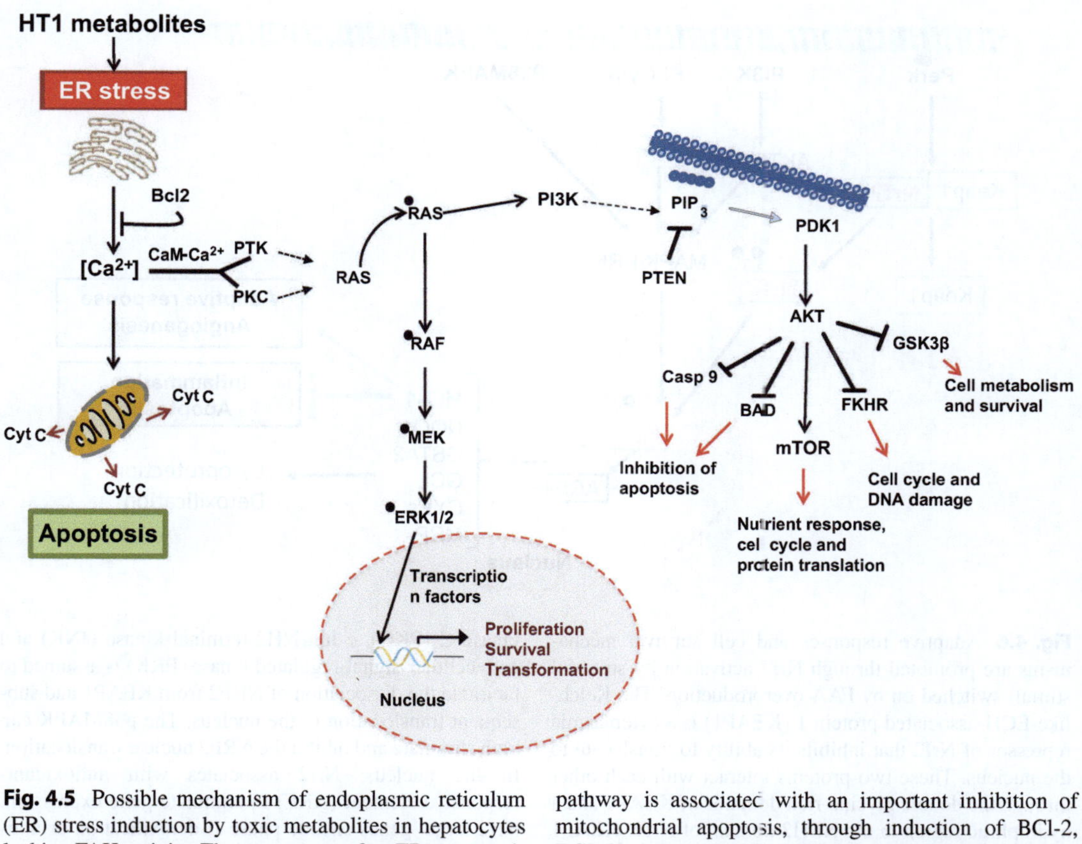

Fig. 4.5 Possible mechanism of endoplasmic reticulum (ER) stress induction by toxic metabolites in hepatocytes lacking FAH activity. The occurrence of an ER stress activates multiple signaling cascades as the PKC, Ras/Raf/MEK/ERK and the PI3K/AKT pathways. Both of these are linked to the mitochondrial intrinsic apoptotic pathways consistent with the physical interaction between mitochondria and ER. Activation of the AKT survival pathway is associated with an important inhibition of mitochondrial apoptosis, through induction of BCl-2, BCL-X and modulation of Bad phosphorylation. AKT signal is also transduced in the inactivation of the glycogen synthase kinase 3 beta (GSK3 β) (phosphorylation of S9) and an activation of the survival transcription factor FKHR through its phosphorylation on S256

lenge with the FAS ligand Jo2 (Vogel et al. 2006) (Fig. 4.1b). The resistance to FAS-induced apoptosis was associated with a sustained phosphorylation of Bid in this model and in another model of chronic liver injury suggesting that the phosphorylation status of Bid may determine the apoptotic threshold of hepatocytes in vivo. Subsequently however, the same group was able to show that in contrast to previous in vitro findings, phosphorylation of Bid does not affect the sensitivity of hepatocytes to FAS receptor-mediated apoptosis in vivo and does therefore not cause the observed apoptosis resistance in *Fah*$^{-/-}$ mice taken off NTBC (Schungel et al. 2009). However, Vogel's data indicate that chronic liver injury induced a cell death resistance in vivo, a process that affects most hepatocytes without requiring genetic mutations. The importance of apoptosis resistance for cancer development is widely appreciated. To date, genetic apoptosis resistance, i.e. mutations in genes such as *Bcl2*, has been implicated in this context. The stress-induced malfunction of the intrinsic death machinery may play a role in the pathophysiology of common hepatic disorders and might be related to the risk of developing HCCs.

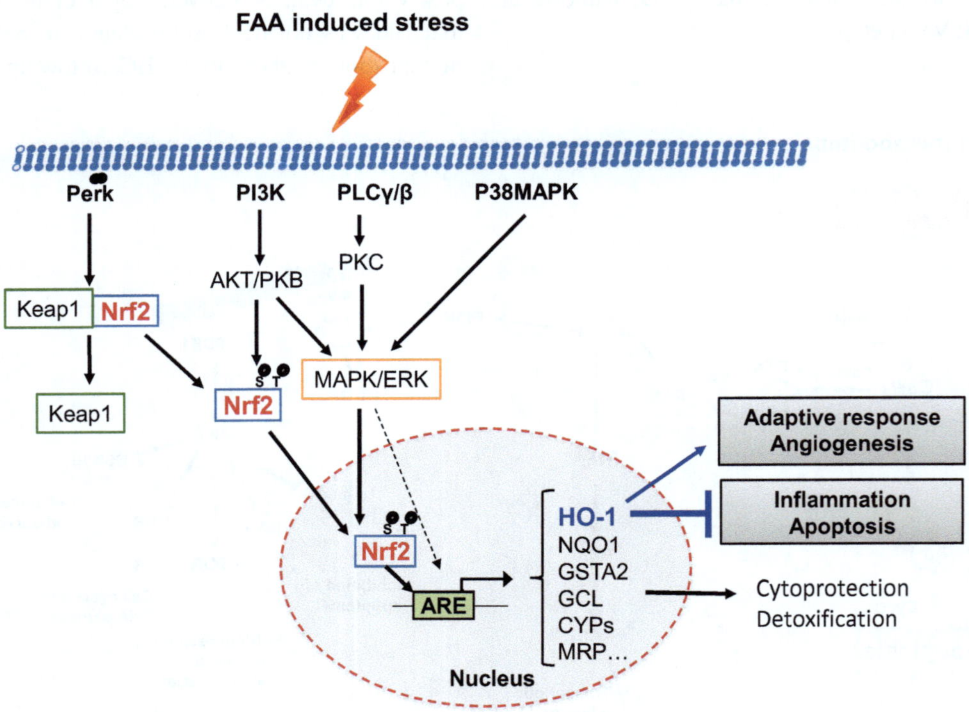

Fig. 4.6 Adaptive responses and cell survival mechanisms are promoted through Nrf2 activation by stressful stimuli switched on by FAA overproduction. The Kelch-like-ECH-associated protein 1 (KEAP1) is a cytoplasmic repressor of Nrf2 that inhibits its ability to translocate to the nucleus. These two proteins interact with each other through the double glycine-rich domains of KEAP1 and a hydrophilic region in the NEH2 domain of Nrf2. KEAP1 contains many cysteine residues. Phase II enzyme inducers and/or prooxidants can cause oxidation or covalent modification of these cysteine residues. As a result, Nrf2 is released from KEAP1. In addition, phosphorylation of Nrf2 at serine (S) and threonine (T) residues by kinases such as phosphatidylinositol 3-kinase (PI3K), protein kinase C (PKC), c-Jun NH2-terminal kinase (JNK) and extracellular-signal-regulated kinase (ERK) is assumed to facilitate the dissociation of NRF2 from KEAP1 and subsequent translocation to the nucleus. The p38MAPK can both stimulate and inhibit the NRF2 nuclear translocation. In the nucleus, Nrf2 associates with antioxidant-responsive element (ARE) to stimulate gene expression. Nrf2 target genes encode phase II detoxification or antioxidant enzymes such as cytochrome P450 (CYP), glutathione S-transferase alpha2 (GSTA2), multidrug resistance proteins (MRP) NAD(P)H:quinone oxidoreductase (NQO1), gamma-glutamate cysteine ligase (gamma-GCLC and gamma-GCLM) and heme oxygenase-1 (HO-1)

In the case of the PI3K/AKT pathway (Fig. 4.5), the AKT signal was also transduced as shown by the inactivation of the glycogen synthase kinase 3 (GSK3 β) (phosphorylation of S9) and an activation of the survival transcription factor FKHR through its phosphorylation on S256 (Orejuela et al. 2008). More recently Angileri et al. also provided evidence for an activation of PKC likely as a result of the damage caused by toxic FAA/MAA/SA tyrosine metabolites (Angileri et al. 2015).

4.2.4 Activation of the Nrf2 Transcription Factor and Its Links with the Hepatocarcinogenic Process in HT1

Nrf2 is a transcriptional factor involved in cancer but its action in the tumorigenic process is still controversial. Nrf2 has been reported to be both tumor suppressive as well as oncogenic (reviewed (Jaramillo and Zhang 2013; Sporn and Liby 2012)). Nrf2 promotes cell survival in stress conditions (Fig. 4.6). Increased Nrf2 has also been found to show high expression in various cancers (lung, ovarian carcinoma) reminding us that cancer cells can also be protected by Nrf2. Moreover Nrf2 has been shown to enhance drug resistance consistent with the increase of the stress protein HSP27 (See 1.2).

The context under which Nrf2 can activate or repress certain types of cancer may be important to understand its positive or negative effects. One should consider the different stages of tumorigenesis, as dysplastic cells differ from more autonomous malignant cells. This important transcription factor is found as a module with KEAP1, a stress sensor with multiple cysteine sulfhydryl groups that can respond to electrophiles and thiol-binding compounds. The binding of KEAP1 to Nrf2 targets the latter to degradation by the proteasome. Binding of xenobiotics, electrophiles, oxidants to KEAP1 cysteines modulate its binding to Nrf2 thereby regulating hundreds of Nrf2 stress-responsive genes. These targets include genes involved in xenobiotics clearance, and genes implicated in anti-oxidant systems. This has been suggested to confer cytoprotection against electrophiles (Sandhu et al. 2015).

FAA and MAA are both theoretically electrophiles and could form adducts with KEAP1 resulting in the activation of Nrf2 by preventing its degradation by the proteasome. Among the genes regulated by Nrf2, induction of glutathione synthesis could be of primary importance in the cancer process in HT1. Glutathione acts not only as a scavenger but also in the process of xenobiotics excretion. It is known that the mutagenicity of FAA increases upon glutathione depletion and that FAA can decrease glutathione in cultured cells (Jorquera and Tanguay 1997). Whether FAA can bind to cysteines in KEAP1 like fumarate is unknown but is surely an attractive concept.

Given its function in inducing cellular antioxidant genes, it was logical to examine this transcription factor's behavior in HT1 since FAA is a strong electrophile. Gene expression profiling was also performed. The microarray analysis revealed a strong activation of the Nrf2 pathway in *Fah*$^{-/-}$ mice. Marhenke et al. initially defined the function of Nrf2 in liver cancer in the HT1 mouse model (Marhenke et al. 2008). They crossed the *Fah*$^{-/-}$ mice with *Nrf2*$^{-/-}$ mice to examine the role of Nrf2 in the pathology of the disease. Upon removal of NTBC the *Fah/Nrf2*$^{-/-}$ mice died rapidly within 5 days as compared to 5 weeks in the *Fah*$^{-/-}$ mouse. Nrf2 did not have any special effects on the other enzymes of the tyrosine catabolism pathway. Thus in acute HT1, loss of Nrf2 increased mortality and accelerated tumor development. In their model of chronic HT1 simulated by a treatment with full or 10% of NTBC the *Fah/Nrf2*$^{-/-}$ mice showed a more severe hepatitis than in the control *Fah*$^{-/-}$ mice. The double KO mice treated with low NTBC had enlarged livers with multiple tumors a situation also observed in the other protocol using total NTBC withdrawal and a liquid diet (Fig. 4.1c) (Angileri et al. 2015). Interestingly the introduction of a chemopreventive drug (CDDO-im: 1[2-cyano-3,12-dioxooleana-1,9(11)-dien-28-oyl]imidazole) inducing Nrf2-regulated genes improved survival from 3 to 8 days in FAH mice placed on a high tyrosine diet inducing high levels of FAA (Marhenke et al. 2008). This was Nrf2-dependent since the double KO mice were not protected by CDDO-im. Finally it was found that Nrf2 is critically important for hepatocyte protection against ethanol-induced toxicity, which is a well-known risk factor for hepatocarcinogenesis in humans (Lamle et al. 2008).

In the second protocol of Angileri et al. (Fig. 4.1c), the levels of Nrf2 were measured at different time points (from 3 days to 15 weeks) chosen to represent different stages of the HT1-induced pathogenic process (Angileri et al. 2015). Nrf2 was induced in the early phase after NTBC withdrawal peaking around 1 week post-NTBC removal (Angileri et al. 2015). A decrease was then observed in Nrf2 levels and this transcription factor was absent in the late stages of the neoplastic process. Interestingly the absence of Nrf2 led to high levels of AFP (alpha-fetoprotein) in the low-NTBC animals of Marhenke et al. (2008). Such an increase in AFP was also observed in the late stages at 15 weeks post-removal (Angileri et al. 2015). All these data support testing of Nrf2 inducing chemoprotective drugs in chronic liver diseases.

HO-1, a target of Nrf2 was also examined and found to be absent in FAH mice treated with NTBC. It was rapidly induced after NTBC removal and its level remained high up to 15 weeks. Thus while in the early stage the level of HO-1 was consistent with its effector Nrf2 (Fig. 4.6), its high level at the late stage of HT1 could be due to its long half-life or a regulation by another pathway. The regulation of HO-1 is complex and goes through many signaling cascades including PKC and MAPK both of which are regulated by the HT1 stress. The increase of HO-1 mRNA was confirmed in microarray data (5.98-fold increase at 8 weeks post withdrawal). A role of HO-1 in promoting cancer has been documented in cancers including renal cancer (Banerjee et al. 2011; Was et al. 2010). Thus its presence at a high level in the late stage of HT1 suggests a role in promoting survival of proliferating cells. In summary, all these studies suggest that Nrf2 plays an important contribution in the neoplastic process, which takes place during the evolution of the HT1 pathology.

Finally it is intriguing that fumarate, one of the products of tyrosine degradation by FAH is a known inhibitor of chemical carcinogenesis in various tissues of rodents including liver. This is related to its action as an inhibitor of KDM2B histone demethylase activity resulting in improved double-strand break repair (Jiang et al. 2015). Whether fumarate levels in HT1 are involved in the cancer process will be an interesting question given its positive implications in DNA repair of DS breaks, a possible site of action of toxic HT1 metabolites.

4.3 Liver Regeneration: Cell Transplantation, p21

Figure 4.7 summarizes the cell cycle pathway and some of the major players in cancer like p53.

Liver regeneration is severely impaired in *Fah*$^{-/-}$ mice taken off NTBC, as they show no signs of hepatocytes proliferation even after partial hepatectomy and invariably die from liver failure within 4 days (Lehmann et al. 2012). Microarray analysis revealed a remarkable p21 up-regulation. The cyclin inhibitor p21 is the primary target of p53 inducing cell cycle arrest and senescence in response to triggers such as DNA damage and telomere shortening (Choudhury et al. 2007) (Fig. 4.7). In agreement with the hypothesis that inhibition of hepatocyte proliferation induced by telomere shortening could delay tumor development, a correlation between increased p21 expression and HCC risk has been observed in human cirrhosis (Plentz et al. 2007). Several studies in the liver have confirmed the importance of p21 for the regulation of liver regeneration and hepatocarcinogenesis (Hui et al. 2008; Willenbring et al. 2008). Interestingly, p21 does not only regulate cell cycle progression in the liver, but also apoptosis, which could facilitate carcinogenesis by reducing the elimination of dysplastic hepatocytes (Zhivotovsky and Kroemer 2004). On the other hand, there is increasing evidence that compensatory proliferation due to increased hepatocyte death actually

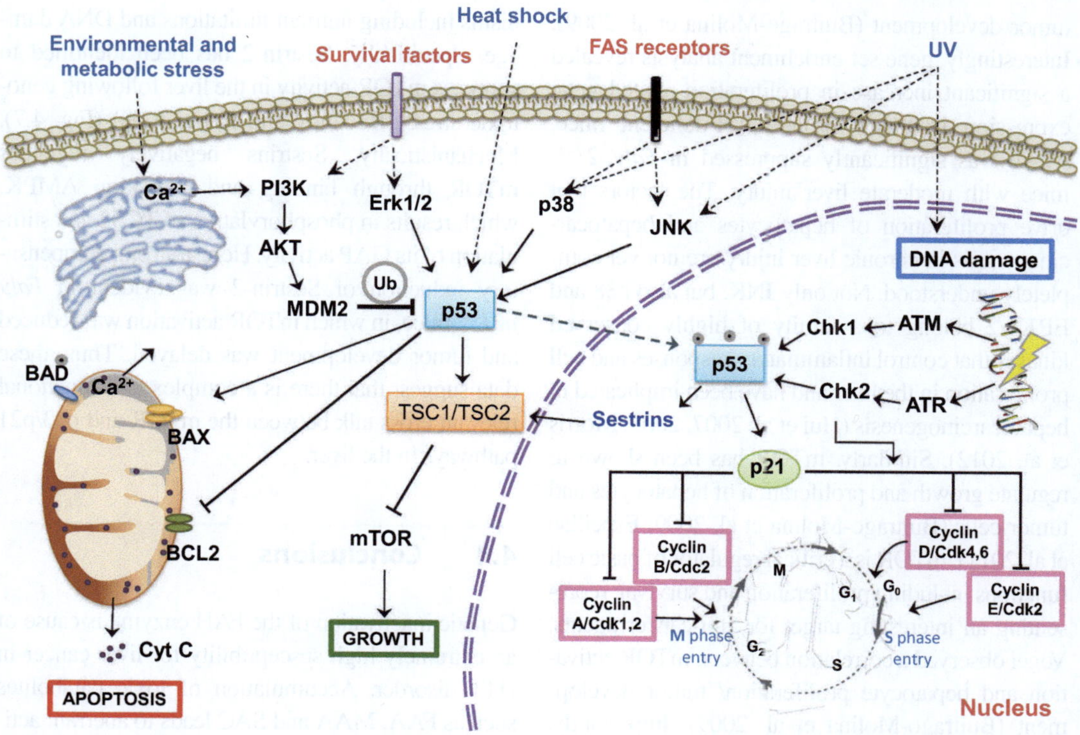

Fig. 4.7 Schematic illustration of the integrative signaling model of cell cycle progression and p53 activation during tumorigenesis. Typical examples of molecules known to act on the cell proliferation, DNA damage response, and cell cycle regulation via the regulatory pathway are shown. At least 50 different enzymes can covalently modify p53 to alter its stability, cellular location or activity. Under normal cellular conditions, MDM2 represses p53 by binding and sequestering p53, and by ubiquitylating (Ub) p53 and targeting it for degradation. DNA damage, oxidative stress and oncogene activation are among the processes that can activate p53 by a range of regulators that disrupt MDM2/p53 interaction. This allows activated p53 to act as a transcription factor, activating or repressing genes involved in apoptosis, cell cycle arrest and senescence. p53 can also move to the mitochondria, where it physically interacts with members of the BCL-2 family to form pores in the mitochondrial membrane

promotes tumor development in the liver (Qiu et al. 2011; Sakurai et al. 2008; Yamaji et al. 2010).

Willenbring et al. were subsequently able to show that in contrast to *Fah*$^{-/-}$ mice, *Fah/ p21*$^{-/-}$ mice showed marked hepatocyte proliferation following complete NTBC withdrawal and significantly improved survival (average life-span 4 months vs. 5.2 weeks) (Willenbring et al. 2008). However, multiple liver tumors were found in *Fah/ p21*$^{-/-}$ mice only 2 months after NTBC withdrawal. Together, these data show that p21 accumulation prevents hepatocyte regeneration and leads to liver failure and death in *Fah*-deficient mice. Loss of p21 in HT1 mice restores the proliferative capabilities of hepatocytes. This growth response compensates cell loss due to increased apoptosis and enables animal survival, but it rapidly leads to HCC suggesting that the p21-mediated cell cycle arrest is crucially important for the prevention of DNA damage-induced hepatocarcinogenesis in the liver.

To examine tumor onset and progression in *Fah*$^{-/-}$ and *Fah/p21*$^{-/-}$ mice under moderate chronic liver injury (see protocol Fig. 4.1d), livers of FAH-deficient mice were examined on low dose 2,5% NTBC treatment. Paradoxically, hepatocytes proliferation appeared significantly reduced and, more importantly, tumor development was profoundly delayed in p21-deficient mice with moderate liver injury, which provides further insight into the complex regulation of cellular processes required for tumor development (Buitrago-Molina et al. 2009). Interestingly, gene set enrichment analysis revealed a significant increase in proliferation-related gene expression in tumor prone FAH-deficient mice, which was significantly suppressed in *Fah/p21*$^{-/-}$ mice with moderate liver injury. The factors that drive proliferation of hepatocytes and hepatocarcinogenesis in chronic liver injury are not yet completely understood. Not only JNK, but also p38 and ERK1/2 belong to a family of highly conserved kinases that control inflammatory responses and cell proliferation in the liver and have been implicated in hepatocarcinogenesis (Hui et al. 2007, 2008; Morris et al. 2012). Similarly, mTOR has been shown to regulate growth and proliferation of hepatocytes and tumor cells (Buitrago-Molina et al. 2009; Espeillac et al. 2011). mTOR is a critical regulator of basic cell functions including proliferation and survival, representing an interesting target for anti-cancer agents. Vogel observed a correlation between mTOR activation and hepatocyte proliferation/ tumor development (Buitrago-Molina et al. 2009). Importantly, mTOR activation was suppressed in *Fah/p21*$^{-/-}$ mice with moderate liver injury, in which tumor development was delayed. In order to analyze the role of mTOR in hepatocyte proliferation, a pharmacological inhibitor of rapamycin (RAD001) was used (Buitrago-Molina et al. 2009). It was found that RAD001 specifically inhibited cell cycle progression of hepatocyte with DNA damage and was significantly less effective in immature or healthy hepatocytes. Interestingly, loss of p53 markedly attenuated the anti-proliferative effects of RAD001 suggesting that the ability of RAD001 to impair cell cycle progression requires the activation of the DNA damage response. In addition to its effect on proliferation, RAD001 sustained the apoptosis sensitivity of hepatocytes during chronic liver injury by inhibiting p53-induced p21 expression. Long-term treatment with RAD001 markedly delayed DNA damage-induced liver tumor development. Together, these data suggested that mTOR activation is critically important for hepatocyte proliferation during liver injury and that it has a substantial effect on hepatocarcinogenesis. Pharmacological inhibition of mTOR may therefore be an effective strategy to delay liver tumor development in patients at risk. mTOR activity can be inhibited by multiple mechanisms including nutrient limitations and DNA damage. Specifically, Sestrin-2 has been identified to suppress mTOR activity in the liver following genotoxic stress (Budanov and Karin 2008) (Fig. 4.7). Mechanistically, Sestrins negatively regulates mTOR through binding and activating AMPK, which results in phosphorylation of TCS2 and stimulation of its GAP activity. Here, a strong compensatory induction of Sestrin-2 was evident in *Fah/p21*$^{-/-}$ mice, in which mTOR activation was reduced and tumor development was delayed. Thus, these data suggest that there is a complex and functional relevant cross talk between the mTOR and p53/p21 pathways in the liver.

4.4 Conclusions

Genetic inactivation of the FAH enzyme is cause of an extremely high susceptibility for liver cancer in HT1 disorder. Accumulation of toxic catabolites such as FAA, MAA and SAC leads to aberrant activation of signaling pathways related to cell death resistance and proliferation. The FAH knockout model, representing all phenotypic and biochemical manifestations of the human disease has helped researchers in a deeper understanding of the mechanisms responsible for cancer onset. This knowledge warrants further investigation to improve prognosis and diagnosis of affected patients. However the major difficulty in HT1 research remains the elusive nature of FAA potential target. We hope this question will be resolved with the help of forthcoming research.

Acknowledgements Work in RMT's lab was supported by the Canadian Institute of Health Research (CIHR) and La Fondation Pierre Lavoie (GO). FA received postdoctoral fellowships from PROTEO.

References

Angileri F, Morrow G, Roy V, Orejuela D, Tanguay RM (2014) Heat shock response associated with hepatocarcinogenesis in a murine model of hereditary tyrosinemia type I. Cancers (Basel) 6(2):998–1019. doi:cancers6020998 [pii]. doi:10.3390/cancers6020998

Angileri F, Roy V, Morrow G, Scoazec JY, Gadot N, Orejuela D, Tanguay RM (2015) Molecular changes associated with chronic liver damage and neoplastic lesions in a murine model of hereditary tyrosinemia type 1. Biochim Biophys Acta. doi:10.1016/j.bbadis.2015.09.002

Banerjee P, Basu A, Datta D, Gasser M, Waaga-Gasser AM, Pal S (2011) The heme oxygenase-1 protein is overexpressed in human renal cancer cells following activation of the Ras-Raf-ERK pathway and mediates anti-apoptotic signal. J Biol Chem 286(38):33580–33590. doi:M111.248401 [pii]. doi:10.1074/jbc.M111.248401

Bergeron A, Jorquera R, Tanguay RM (2003) Hereditary tyrosinemia: an endoplasmic reticulum stress disorder? Med Sci (Paris) 19(10):976–980. doi:007170ar [pii]. doi:10.1051/medsci/20031910976

Bergeron A, Jorquera R, Orejuela D, Tanguay RM (2006) Involvement of endoplasmic reticulum stress in hereditary tyrosinemia type I. J Biol Chem 281(9):5329–5334. doi:M506804200 [pii]. doi:10.1074/jbc.M506804200

Bliksrud YT, Ellingsen A, Bjoras M (2013) Fumarylacetoacetate inhibits the initial step of the base excision repair pathway: implication for the pathogenesis of tyrosinemia type I. J Inherit Metab Dis 36(5):773–778. doi:10.1007/s10545-012-9556-0

Bruey JM, Paul C, Fromentin A, Hilpert S, Arrigo AP, Solary E, Garrido C (2000) Differential regulation of HSP27 oligomerization in tumor cells grown in vitro and in vivo. Oncogene 19(42):4855–4863. doi:10.1038/sj.onc.1203850

Budanov AV, Karin M (2008) p53 target genes sestrin1 and sestrin2 connect genotoxic stress and mTOR signaling. Cell 134(3):451–460. doi:10.1016/j.cell.2008.06.028

Buitrago-Molina LE, Pothiraju D, Lamle J, Marhenke S, Kossatz U, Breuhahn K, Manns MP, Malek N, Vogel A (2009) Rapamycin delays tumor development in murine livers by inhibiting proliferation of hepatocytes with DNA damage. Hepatology 50(2):500–509. doi:10.1002/hep.23014

Calderwood SK, Gong J (2016) Heat Shock proteins promote cancer: it's a protection racket. Trends Biochem Sci. doi:10.1016/j.tibs.2016.01.003

Choudhury AR, Ju Z, Djojosubroto MW, Schienke A, Lechel A, Schaetzlein S, Jiang H, Stepczynska A, Wang C, Buer J, Lee HW, von Zglinicki T, Ganser A, Schirmacher P, Nakauchi H, Rudolph KL (2007) Cdkn1a deletion improves stem cell function and lifespan of mice with dysfunctional telomeres without accelerating cancer formation. Nat Genet 39(1):99–105. doi:10.1038/ng1937

Ciocca DR, Arrigo AP, Calderwood SK (2013) Heat shock proteins and heat shock factor 1 in carcinogenesis and tumor development: an update. Arch Toxicol 87(1):19–48. doi:10.1007/s00204-012-0918-z

Dieter MZ, Freshwater SL, Miller ML, Shertzer HG, Dalton TP, Nebert DW (2003) Pharmacological rescue of the 14CoS/14CoS mouse: hepatocyte apoptosis is likely caused by endogenous oxidative stress. Free Radic Biol Med 35(4):351–367. doi:S0891584903002739 [pii]

Espeillac C, Mitchell C, Celton-Morizur S, Chauvin C, Koka V, Gillet C, Albrecht JH, Desdouets C, Pende M (2011) S6 kinase 1 is required for rapamycin-sensitive liver proliferation after mouse hepatectomy. J Clin Invest 121(7):2821–2832. doi:10.1172/JCI44203

Gao T, Newton AC (2002) The turn motif is a phosphorylation switch that regulates the binding of Hsp70 to protein kinase C. J Biol Chem 277(35):31585–31592. doi:10.1074/jbc.M204335200

Grompe M, Lindstedt S, Al-Dhalimy M, Kennaway NG, Papaconstantinou J, Torres-Ramos CA, Ou CN, Finegold M (1995) Pharmacological correction of neonatal lethal hepatic dysfunction in a murine model of hereditary tyrosinaemia type I. Nat Genet 10(4):453–460. doi:10.1038/ng0895-453

Hui L, Bakiri L, Mairhorfer A, Schweifer N, Haslinger C, Kenner L, Komnenovic V, Scheuch H, Beug H, Wagner EF (2007) p38alpha suppresses normal and cancer cell proliferation by antagonizing the JNK-c-Jun pathway. Nat Genet 39(6):741–749. doi:10.1038/ng2033

Hui L, Zatloukal K, Scheuch H, Stepniak E, Wagner EF (2008) Proliferation of human HCC cells and chemically induced mouse liver cancers requires JNK1-dependent p21 downregulation. J Clin Invest 118(12):3943–3953. doi:10.1172/JCI37156

Jaramillo MC, Zhang DD (2013) The emerging role of the Nrf2-Keap1 signaling pathway in cancer. Genes Dev 27(20):2179–2191. doi:10.1101/gad.225680.113

Jiang Y, Qian X, Shen J, Wang Y, Li X, Liu R, Xia Y, Chen Q, Peng G, Lin SY, Lu Z (2015) Local generation of fumarate promotes DNA repair through inhibition of histone H3 demethylation. Nat Cell Biol 17(9):1158–1168. doi:10.1038/ncb3209

Jorquera R, Tanguay RM (1997) The mutagenicity of the tyrosine metabolite, fumarylacetoacetate, is enhanced by glutathione depletion. Biochem Biophys Res Commun 232(1):42–48. (47). doi:http://dx.doi.org/10.1006/bbrc.1997.6220

Jorquera R, Tanguay RM (1999) Cyclin B-dependent kinase and caspase-1 activation precedes mitochondrial dysfunction in fumarylacetoacetate-induced apoptosis. FASEB J 13:2284–2298

Jorquera R, Tanguay RM (2001) Fumarylacetoacetate, the metabolite accumulating in hereditary tyrosinemia, activates the ERK pathway and induces mitotic abnormalities and genomic instability. Hum Mol Genet 10(17):1741–1752

Kampinga HH, Hageman J, Vos MJ, Kubota H, Tanguay RM, Bruford EA, Cheetham ME, Chen B, Hightower LE (2009) Guidelines for the nomenclature of the human heat shock proteins. Cell Stress Chaperones 14(1):105–111. doi:10.1007/s12192-008-0068-7

Kubo S, Sun M, Miyahara M, Umeyama K, Urakami K, Yamamoto T, Jakobs C, Matsuda I, Endo F (1998) Hepatocyte injury in tyrosinemia type 1 is induced by fumarylacetoacetate and is inhibited by caspase inhibitors. Proc Natl Acad Sci U S A 95(16):9552–9557

Labbadia J, Morimoto RI (2015) The biology of proteostasis in aging and disease. Annu Rev Biochem 84:435–464. doi:10.1146/annurev-biochem-060614-033955

Lamle J, Marhenke S, Borlak J, von Wasielewski R, Eriksson CJ, Geffers R, Manns MP, Yamamoto M, Vogel A (2008) Nuclear factor-eythroid 2-related factor 2 prevents alcohol-induced fulminant liver injury. Gastroenterology 134(4):1159–1168. doi:10.1053/j.gastro.2008.01.011

Lee JS, Lee JJ, Seo JS (2005) HSP70 deficiency results in activation of c-Jun N-terminal Kinase, extracellular signal-regulated kinase, and caspase-3 in hyperosmolarity-induced apoptosis. J Biol Chem 280(8):6634–6641. doi:M412393200 [pii]10.1074/jbc.M412393200

Lehmann K, Tschuor C, Rickenbacher A, Jang JH, Oberkofler CE, Tschopp O, Schultze SM, Raptis DA, Weber A, Graf R, Humar B, Clavien PA (2012) Liver failure after extended hepatectomy in mice is mediated by a p21-dependent barrier to liver regeneration. Gastroenterology 143(6):1609–1619. e1604. doi:10.1053/j.gastro.2012.08.043

Manabe S, Sassa S, Kappas A (1985) Hereditary tyrosinemia. Formation of succinylacetone-amino acid adducts. J Exp Med 162(3):1060–1074

Marhenke S, Lamle J, Buitrago-Molina LE, Canon JM, Geffers R, Finegold M, Sporn M, Yamamoto M, Manns MP, Grompe M, Vogel A (2008) Activation of nuclear factor E2-related factor 2 in hereditary tyrosinemia type 1 and its role in survival and tumor development. Hepatology 48(2):487–496. doi:10.1002/hep.22391

Morris SM, Baek JY, Koszarek A, Kanngurn S, Knoblaugh SE, Grady WM (2012) Transforming growth factor-beta signaling promotes hepatocarcinogenesis induced by p53 loss. Hepatology 55(1):121–131. doi:10.1002/hep.24653

Orejuela D, Jorquera R, Bergeron A, Finegold MJ, Tanguay RM (2008) Hepatic stress in hereditary tyrosinemia type 1 (HT1) activates the AKT survival pathway in the fah−/− knockout mice model. J Hepatol 48(2):308–317. doi:S0168-8278(07)00590-9 [pii]. doi:10.1016/j.jhep.2007.09.014

Oya-Ito T, Liu BF, Nagaraj RH (2006) Effect of methylglyoxal modification and phosphorylation on the chaperone and anti-apoptotic properties of heat shock protein 27. J Cell Biochem 99(1):279–291. doi:10.1002/jcb.20781

Plentz RR, Park YN, Lechel A, Kim H, Nellessen F, Langkopf BH, Wilkens L, Destro A, Fiamengo B, Manns MP, Roncalli M, Rudolph KL (2007) Telomere shortening and inactivation of cell cycle checkpoints characterize human hepatocarcinogenesis. Hepatology 45(4):968–976. doi:10.1002/hep.21552

Qiu W, Wang X, Leibowitz B, Yang W, Zhang L, Yu J (2011) PUMA-mediated apoptosis drives chemical hepatocarcinogenesis in mice. Hepatology 54(4):1249–1258. doi:10.1002/hep.24516

Saibil H (2013) Chaperone machines for protein folding, unfolding and disaggregation. Nat Rev Mol Cell Biol 14(10):630–642. doi:10.1038/nrm3658

Sakurai T, He G, Matsuzawa A, Yu GY, Maeda S, Hardiman G, Karin M (2008) Hepatocyte necrosis induced by oxidative stress and IL-1 alpha release mediate carcinogen-induced compensatory proliferation and liver tumorigenesis. Cancer Cell 14(2):156–165. doi:10.1016/j.ccr.2008.06.016

Sandhu IS, Maksim NJ, Amouzougan EA, Gallion BW, Raviele AL, Ooi A (2015) Sustained NRF2 activation in hereditary leiomyomatosis and renal cell cancer (HLRCC) and in hereditary tyrosinemia type 1 (HT1). Biochem Soc Trans 43(4):650–656. doi:10.1042/BST20150041

Schungel S, Buitrago-Molina LE, Nalapareddy P, Lebofsky M, Manns MP, Jaeschke H, Gross A, Vogel A (2009) The strength of the Fas ligand signal determines whether hepatocytes act as type 1 or type 2 cells in murine livers. Hepatology 50(5):1558–1566. doi:10.1002/hep.23176

Sporn MB, Liby KT (2012) NRF2 and cancer: the good, the bad and the importance of context. Nat Rev Cancer 12(8):564–571. doi:10.1038/nrc3278

Tanguay RM, Jorquera R, Poudrier J, St-Louis M (1996) Tyrosine and its catabolites: from disease to cancer. Acta Biochim Pol 43(1):209–216

Vihervaara A, Sistonen L (2014) HSF1 at a glance. J Cell Sci 127(Pt 2):261–266. doi:10.1242/jcs.132605

Vogel A, van Den Berg IE, Al-Dhalimy M, Groopman J, Ou CN, Ryabinina O, Iordanov MS, Finegold M, Grompe M (2004) Chronic liver disease in murine hereditary tyrosinemia type 1 induces resistance to cell death. Hepatology 39(2):433–443. doi:10.1002/hep.20077

Vogel A, Aslan JE, Willenbring H, Klein C, Finegold M, Mount H, Thomas G, Grompe M (2006) Sustained phosphorylation of Bid is a marker for resistance to Fas-induced apoptosis during chronic liver diseases. Gastroenterology 130(1):104–119. doi:10.1053/j.gastro.2005.10.012

Was H, Dulak J, Jozkowicz A (2010) Heme oxygenase-1 in tumor biology and therapy. Curr Drug Targets 11(12):1551–1570. doi:BSP/CDT/E-Pub/00150 [pii]

Willenbring H, Sharma AD, Vogel A, Lee AY, Rothfuss A, Wang Z, Finegold M, Grompe M (2008) Loss of p21 permits carcinogenesis from chronically damaged liver and kidney epithelial cells despite unchecked apoptosis. Cancer Cell 14(1):59–67. doi:S1535-6108(08)00159-1 [pii]. doi:10.1016/j.ccr.2008.05.004

Yamaji S, Zhang M, Zhang J, Endo Y, Bibikova E, Goff SP, Cang Y (2010) Hepatocyte-specific deletion of DDB1 induces liver regeneration and tumorigenesis. Proc Natl Acad Sci U S A 107(51):22237–22242. doi:10.1073/pnas.1015793108

Zhivotovsky B, Kroemer G (2004) Apoptosis and genomic instability. Nat Rev Mol Cell Biol 5(9):752–762. doi:10.1038/nrm1443

Part III

Pathology

Tyrosinemia and Liver Transplantation: Experience at CHU Sainte-Justine

5

Fernando Alvarez and Grant A. Mitchell

Abstract

Tyrosinemia is a disease of the tyrosine metabolism, affecting mainly liver, kidney and peripheral nerves. Two forms of liver disease caused by a deficiency of FAH are recognised: (1) acute liver failure; (2) chronic liver disease. Since the introduction of NTBC [2-(2-nitro-4-trifluoromethyl benzoyl)-1-3-cyclohexanedione] (nitisinone[R]) in the treatment of tyrosinemia, no liver disease has been observed when started in the first weeks of life. Liver transplantation is a good option for the treatment of tyrosinemic patients developing liver nodules, with high suspicion of hepatocarcinoma. In the long-term outcome of the liver transplant, survival was of 90% in tyrosinemic patients.

Keywords

Tyrosinemia • Hepatocarcinoma • Acute failure • Chronic liver disease • Neurologic crisis • Liver transplantation • Kidney disease

F. Alvarez (✉)
Service de gastroentérologie, hépatologie et nutrition, CHU Sainte-Justine, Université de Montréal, 3175 chemin Côte-Sainte-Catherine, Montréal, Québec H3T 1C5, Canada
e-mail: fernando.alvarez@umontreal.ca;

G.A. Mitchell
Departments of Nutrition (MB) and Radiology (JD), Divisions of Gastroenterology, Hepatology and Nutrition (FA, UH), Medical Genetics (CBG, MG, GAM) and Nephrology (AM, VP), Department of Pediatrics and Department of Pharmacy (SA, JFB), CHU Sainte-Justine and Université de Montréal, Montreal, Canada
e-mail: grant.mitchell@recherche-ste-justine.qc.ca

Abbreviations

AFP	Alpha-fœtoprotein
FAH	Fumarylacetoacetate hydrolase
GFR	Glomerular filtration rate
HCC	Hepatocarcinoma

© Springer International Publishing AG 2017
R.M. Tanguay (ed.), *Hereditary Tyrosinemia*, Advances in Experimental Medicine and Biology 959, DOI 10.1007/978-3-319-55780-9_5

5.1 Introduction

Tyrosinemia is a disease of the tyrosine metabolism, characterized by a defect in the fumarylacetoacetate hydrolase (FAH), the last enzyme of the tyrosine degradation pathway. The main organs affected in a tyrosinemia are the liver, kidney and peripheral nerves. Most probably fumarylacetoacetate and its metabolites, succinylacetoacetate and succinylacetone, as well as the precursor of fumarylacetoacetate, maleylacetoacetate, are responsible for the liver and kidney toxicity. Neurologic crises are produced by an accumulation of δ-amino-levulinic acid due to the succinylacetone inhibition of the δ-amino-levulinic acid dehydratase (Mitchell et al. 2014).

5.2 Clinical Findings

Two forms of liver disease caused by a deficiency of FAH are recognised: (1) acute liver failure; (2) chronic liver disease. Neonatal diagnosis allowed the prescription of a diet early in life preventing the development of an acute liver failure, in spite of it cirrhosis developed in those patients. Since the introduction of NTBC [2-(2-nitro-4- trifluoromethyl benzoyl)-1-3-cyclohexanedione] (nitisinone^{R}) in the treatment of tyrosinemia, no liver disease has been observed when started in the first weeks of life (Larochelle et al. 2012).

5.2.1 Acute Liver Disease

Acute liver disease presents before the age of 2 years, rarely before 2 months of life; a fulminant liver failure at onset can be observed. Patients treated by diet only can develop liver failure and cirrhosis. Treatment with NTBC can prevent the development of these complications and correction of liver failure (Larochelle et al. 2012; Lindstedt et al. 1992).

Symptoms and signs in non-NTBC-treated patients are: anorexia, vomiting and irritability. Infections and associated signs can be recorded, indicating that it has probably been the cause of the liver decompensation in the tyrosinemic patients. Interestingly, jaundice is not present or is mild at disease onset; marked jaundice can occur late in the course of the disease when liver failure develops. Patients may have an odor that has been likened to that of boiled cabbage.

Laboratory findings leading to a suspicion of tyrosinemia include the following: a profound decrease of clotting factors, associated with a mild or moderate increase of serum aminotransferase levels.

5.2.2 Chronic Liver Disease

The percentage of patients presenting as a chronic liver disease, without previous liver failure, is variable in different regions of the world. In Québec, before NTBC began to be used as a treatment, practically all patients developed cirrhosis in spite of neonatal screening followed by early dietary intervention.

Hepato-splenomegaly as signs of cirrhosis and portal hypertension are present in more than 70% of patients. Peripheral signs of chronic liver disease such as spider angioma and palmar erythema are equally seen (Mitchell et al. 2014). Half of these children show low clotting factors, associated with an increase of serum aminotransferase and of GGT. Alpha-foetoprotein (AFP) is usually elevated at onset (Mitchell et al. 2014).

5.2.3 Hepatocarcinoma

The differential diagnosis of liver nodules in tyrosinemia includes: (a) hepatocarcinoma, (b) dysplastic hepatocytes, and (c) revertant nodules. The latter are frequent in tyrosinemic livers, and are the consequence of spontaneous somatic mutations of one FAH gene allele, that restores normal activity (Kvittingen et al. 1993).

Liver nodules were a frequent finding in patients with tyrosinemia, before the NTBC era. In around one third of these cases, a high degree dysplasia or a hepatocarcinoma (HCC) was found at histologic examination (Mitchell et al. 2014). In most cases, HCC develops after 2 years of age; the youngest patient reported was 15 months old. AFP increases in patients with acute or chronic tyrosinemic liver diseases, although in those under NTBC treatment it should be normal or close to normal. Increase of AFP detected during the long-term follow-up should alert us on the possibility of HCC development. Conversely, some patients with HCC do not have an elevation of AFP, and a normal level of AFP is not sufficient to exclude the diagnosis of HCC.

5.2.4 Other Complications Related to Liver

Neurologic crises are produced by an accumulation of δ-amino-levulinic acid due to the succinylacetone inhibition of the enzyme δ-amino-levulinic acid dehydratase. Succinylacetone is mainly produced by the liver, and both liver transplantation and NTBC treatment stop the development of neurologic crises. The crises are episodes of acute polyneuropathy that usually cause pain, often severe, in the legs and can also cause paralysis requiring mechanical ventilation if it involves the diaphragm and months of intense rehabilitation if it involves the lower limbs. Only some tyrosinemic patients develop neurologic crises but this predisposition cannot be predicted in advance. Prior to the availability of NTBC treatment, the occurrence of neurologic crises was a major indication for liver transplantation in tyrosinemia.

5.3 Liver Transplantation

Thirty-one children with tyrosinemia were transplanted at CHU Sainte-Justine at the mean age of 5.2 years. In 26 of them, liver transplantation was indicated for the presence of liver nodules. In five patients, a variety of other complications led to a liver transplant. Among tyrosinemic patients with liver nodules, 4 had an HCC in the histologic examination of the explanted liver, and 22 dysplasia of various grades (in four cases of grade 3) (Fig. 5.1).

Since 1996, only five tyrosinemic patients needed a liver transplant. All of these patients started NTBC treatment after 1 month of age, either because NTBC became available after that time or because the diagnosis of tyrosinemia was delayed due to non-detection by neonatal screening or birth outside of Québec in places where newborn screening was not performed (Table 5.1). Three of them were transplanted because of cirrhosis and liver nodules; the fourth presented with a HCC in a cirrhotic liver at the age of 13 years (Fig. 5.2). The fifth patient developed an HCC after almost 16 years under NTBC and diet; he showed a progressive

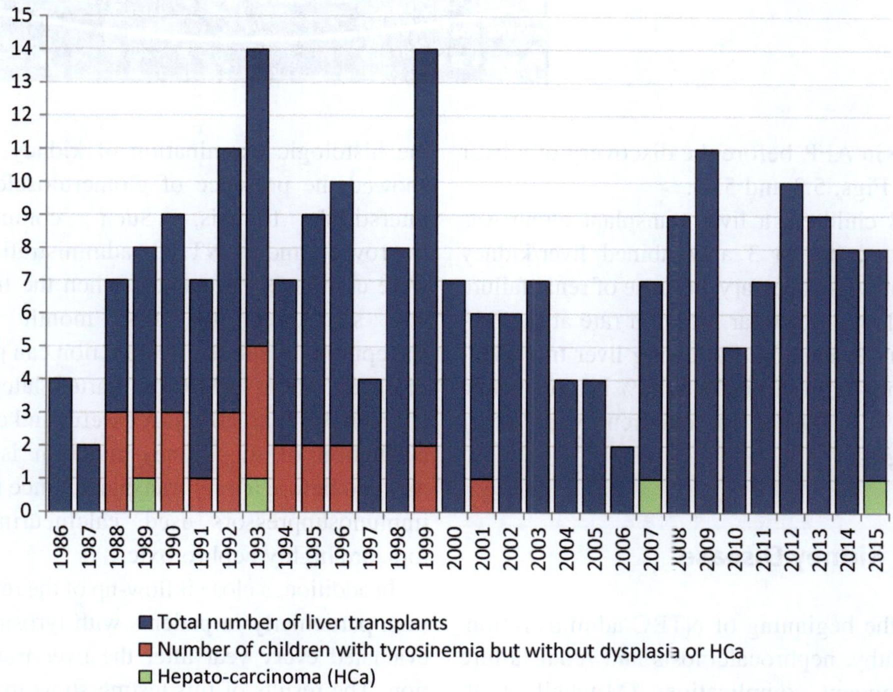

Fig. 5.1 Since 1986, 254 liver transplantations were carried on at CHU Sainte-Justine (*Blue*). In *red*, the number and year of liver transplantations for tyrosinemia without HCC. In *green*, patients with HCC

Table 5.1 Patients with failure in the neonatal diagnosis of tyrosinemia

Patients	Liver failure	NTBC – age	Follow-up nodules	Liver transplantation
1	Yes	3,5 months	Yes	Yes
2	Yes	4 months	Yes	Yes
3	Yes	7 months	Yes	Yes
4	No	2 months	No	No
5	No	No	Yes	Yes

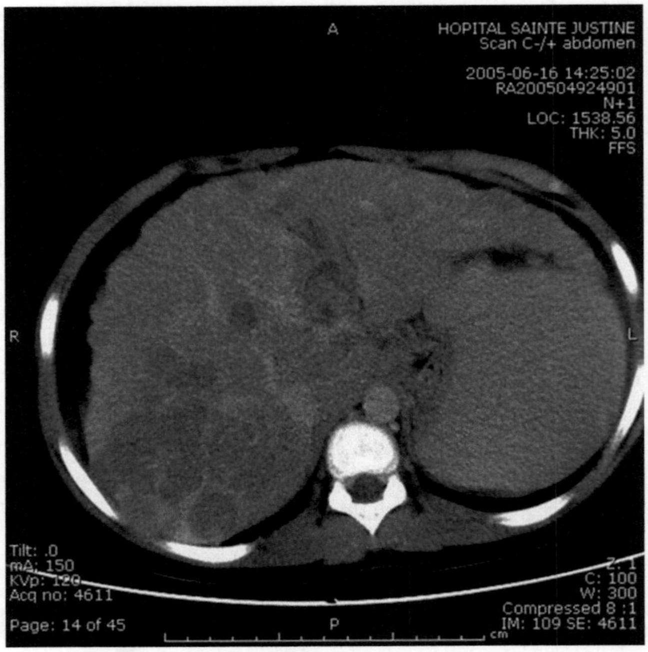

Fig. 5.2 CT-Scan of a 13 years old girl, with a Hepatocarcinoma in a nodular liver, leading to the diagnosis of tyrosinemia

increase in AFP, before the discovery of a liver nodule (Figs. 5.3 and 5.4).

In 28 children, a liver transplant alone was carried-out, but in 3 a combined liver/kidney transplant was necessary, because of renal failure or very low glomerular filtration rate at the pre-transplant evaluation. Following liver transplantation, long-term survival was 90% in tyrosinemic patients vs 80.5% in children with other pathologies.

5.4 Kidney Disease

Before the beginning of NTBC administration, tubulopathy, nephrocalcinosis and renal failure were frequent complications (Mitchell et al. 2014; Larochelle et al. 2012). In some patients, the histologic examination of kidney biopsies showed the presence of glomerulosclerosis or interstitial fibrosis. Such complications improved under NTBC administration, and were completely prevented when the treatment was started in the first month of life. Exceptionally, tubular dysfunction can persist in those in whom NTBC is started later in life (Larochelle et al. 2012). A careful and complete evaluation of the kidney function is recommended before a liver transplant, since the main immunosuppressors used, calcineurin inhibitors, are highly nephrotoxic.

In addition, a close follow-up of the renal function, particularly in patients with tyrosinemia, is evaluated every year after the liver transplantation. The results of this testing show that after a liver transplant, no significant deterioration of the

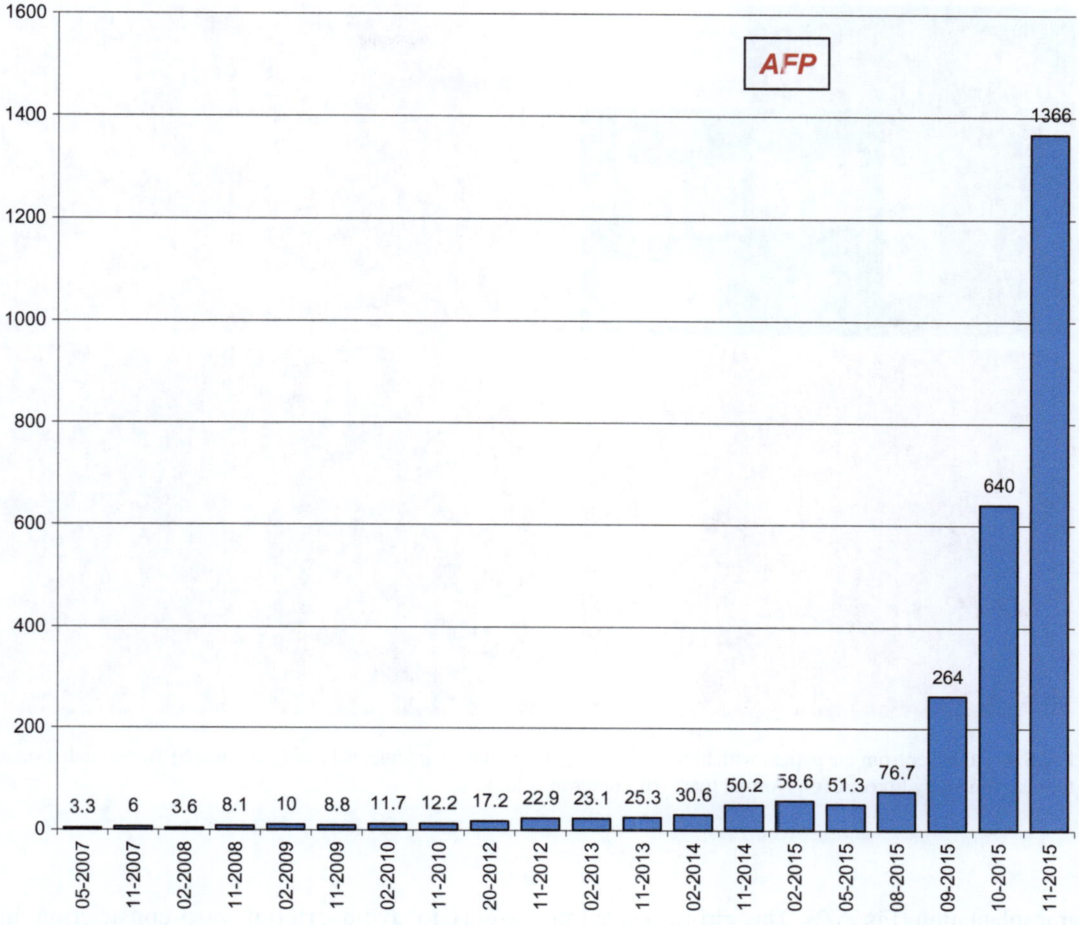

Fig. 5.3 Slow progressive increase of serum alpha-fetoprotein in a patient treated by NTBC. Treatment was started during infancy because of liver failure at onset

glomerular filtration rate (GFR) was recorded (Herzog et al. 2006). For immunosuppression, those patients received cyclosporine between 1986 and 2003, and tacrolimus thereafter. Cyclosporine total dosage was divided in three daily doses. This method allowed decrease by about 30% of total daily amount of cyclosporine, while maintaining a similar area under the curve. In addition, those with the lowest GFR also received diltiazem as a kidney protector.

5.5 Particular Comments

In our experience, the success of liver transplantation in tyrosinemia is higher than in children with other liver diseases. Potential explanations for this difference in tyrosinemic patients include: (1) no previous surgery; (2) normal nutritional status at the time of the liver transplant; and (3) more frequent use of entire livers. Post-transplant complications in tyrosinemic children were less frequent but not essentially different from the non-tyrosinemic group. One patient needed a liver re-transplant at the age of 20 years due to "chronic" hepatocellular rejection and cirrhosis. This child had received 15 years before, as a first graft, a liver from a 60 year-old donor showing a slow recovery of function during the immediate post-transplant period, probably influencing the final outcome.

One child had a late diagnosis of tyrosinemia, and presented with a non-resectable hepatocarcinoma that did not fulfill the criteria for liver

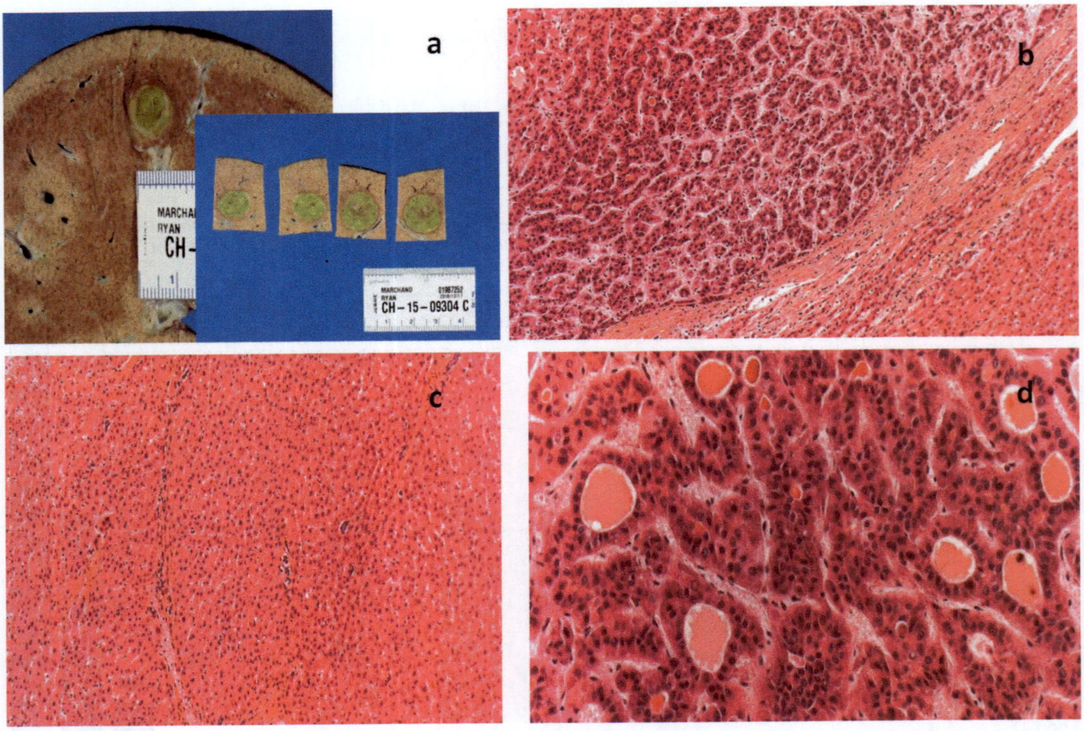

Fig. 5.4 (**a**) Liver from the patient with high AFP (Fig. 5.3), showing a nodule of 11 × 15 mmm. (**b**) Tumor and normal tissu separated by fibrosis. (**c**). Normal liver. (**d**) Tumor histology

transplantation (Fig. 5.2). This girl responded to chemotherapy, showing a remarkable decrease of the liver nodule. Thereafter, she was successfully transplanted. Currently, her follow-up time is 9 years, without relapse. The remarkably favorable response to chemotherapy in this patient has led to speculate that hepatocarcinoma in patients with HT1 may behave differently from those caused by other diseases.

Kidney function is a great concern in the long-term follow-up, since it could be affected before the liver transplant and could further deteriorate by the use of calcineurin inhibitors. Two measures were applied by our team to prevent such issue: (1) administration of calcineurin inhibitors in three daily doses (instead of two), leading to a decrease of the total amount administered (≅30%) and lower peaks without modification of the area under the curve (pharmacokinetics); and (2) prescription of calcium channel inhibitors (diltiazem) to avoid arterial vaso-constriction in the kidney. It was shown that calcium blockers inhibit calcineurin-inhibitor-induced hyperactivity of vascular smooth muscle cells (Grzesk et al. 2012). In addition, human and animal studies showed that through this mechanism, diltiazem attenuates nephrotoxicity of cyclosporine or tacrolimus (Gokce et al. 2012).

Succinylacetone in urine is found in patients transplanted for tyrosinemia; however, no changes in the GFR rate have been observed (Lindstedt et al. 1992; Pierik et al. 2005). These findings support the current protocol of treatment of children transplanted for tyrosinemia, consisting of an unrestricted diet without NTBC administration.

5.6 Conclusions

In conclusion, our experience shows that liver transplantation is a good option for the treatment of tyrosinemic patients developing liver nodules, with high suspicion of hepatocarcinoma. Patients under NTBC treatment and diet, without liver nodules are not considered to be candidates for transplantation because the advantages of transplantation (i.e., the convenience of an unrestricted diet, plus the withdrawal of NTBC treatment) are outweighed by the morbidity-mortality of the procedure and by the risk of long-term administration of toxic immunosuppressive medication.

References

Gokce M, Yuzbasioglu MF, Bulbuloglua E, Oksuz H, Yormaz S, Altınoren O et al (2012) Cilostazol and diltiazem attenuate cyclosporine-induced nephrotoxicity in rats. Transplant Proc 44:1738–1742

Grześk G, Wiciński M, Malinowski B, Grześk E, Manysiak S, Odrowąż-Sypniewska G et al (2012) Calcium blockers inhibit cyclosporine A-induced hyperreactivity of vascular smooth muscle cells. Mol Med Rep 5:1469–1474

Herzog D, Martin S, Turpin S, Alvarez F (2006) Normal glomerular filtration rate in long-term follow-up of children after orthotopic liver transplantation. Transplantation 81:672–677

Kvittingen EA, Rootwelt H, Brandtzaeg P, Bergan A, Berger R (1993) Hereditary tyrosinemia type I. Self-induced correction of the fumarylacetoacetase defect. J Clin Invest 91:1816–1821

Larochelle J, Alvarez F, Bussières JF, Chevalier I, Dallaire L, Dubois J et al (2012) Effect of nitisinone (NTBC) treatment on the clinical course of hepatorenal tyrosinemia in Québec. Mol Genet Metab 107:49–54

Lindstedt S, Holme E, Lock EA, Hjalmarson O, Strandvik B (1992) Treatment of hereditary tyrosinaemia type I by inhibition of 4-hydroxyphenylpyruvate dioxygenase. Lancet 340:813–817

Mitchell G, Russo PA, Dubois J, Alvarez F (2014) Tyrosinemia. In: Suchy FJ, Sokal RJ, Balistreri WF (eds) Liver disease in children. Cambridge, University Press, New York, pp 694–713

Pierik LJ, van Spronsen FJ, Bijleveld CM, van Dael CM (2005) Renal function in tyrosinaemia type I after liver transplantation: a long-term follow-up. J Inherit Metab Dis 28:871–876

The Liver in Tyrosinemia Type I: Clinical Management and Course in Quebec

6

Ugur Halac, Josée Dubois, and Grant A. Mitchell

Abstract

HT1 is a severe autosomal recessive disorder due to the deficiency of fumarylacetoacetate hydrolase (FAH), the final enzyme in the degradation of tyrosine. Before the era of treatment with 2-(2-N-4-trifluoromethylbenzoyl)-1,3-cyclohexanedione (NTBC), even with newborn screening and optimal diet therapy, HT1 patients often developed liver failure. Death was common in patients who did not undergo liver transplantation. For the last two decades, NTBC has revolutionized the management of HT1 patients. In screened newborns treated within the first month of life, we have not observed hepatocarcinoma. If patients are not detected at birth by neonatal screening, the diagnosis and treatment must be performed on an emergency basis, and patients are at risk for complications. Long term adhesion to treatment and reliable early detection of hepatocellular carcinoma (HCC) are two important challenges. In this chapter, we describe the clinical, biological, histo-pathological and imaging findings of HT1 in Québec before the era of NTBC. We also describe the hepatic status of nontransplanted tyrosinemic patients in Quebec and current management practices in the Quebec NTBC Study.

U. Halac (✉)
Division of Gastroenterology, Hepatology and Nutrition, Department of Pediatrics, CHU Sainte-Justine and Research Center, Université de Montréal, Montreal, QC, Canada
e-mail: ugur.halac@umontreal.ca

J. Dubois • G.A. Mitchell
Departments of Nutrition (MB) and Radiology (JD), Divisions of Gastroenterology, Hepatology and Nutrition (FA, UH), Medical Genetics (CBG, MG, GAM) and Nephrology (AM, VP), Department of Pediatrics and Department of Pharmacy (SA, JFB), CHU Sainte-Justine and Université de Montréal, Montreal, Canada

© Springer International Publishing AG 2017

R.M. Tanguay (ed.), *Hereditary Tyrosinemia*, Advances in Experimental Medicine and Biology 959,
DOI 10.1007/978-3-319-55780-9_6

Keywords
Hereditary tyrosinemia type 1 • Liver failure • Hepatocellular carcinoma • NTBC

Abbreviations

AFP	Alpha feto protein
FAA	Fumarylacetoacetate
FAH	Fumarylacetoacetate hydrolase
Gd-BOPTA	Gadobenate dimeglumine
Gd-EOB-DTPA	Gadolinium ethoxybenzyl diethylenetriamine pentaacetic acid (or gadoxetic acid)
HCC	Hepatocellular carcinoma
HT1	Hereditary tyrosinemia type 1
INR	International normalized ratio
MAA	Maleylacetoacetate
Min	Minutes
MR	Magnetic resonance
NTBC	2-(2-N-4-trifluoromethylbenzoyl)-1,3-cyclohexanedione
PT	Prolonged prothrombin
SAA	Succinylacetoacetate
SAC	Succinylacetone
Sec	Seconds
US	Ultrasound

6.1 Introduction

Hepatorenal tyrosinemia or tyrosinemia type I (HT1) is a severe autosomal recessive disorder caused by the deficiency of fumarylacetoacetate hydrolase (FAH), the final enzyme in the degradation of tyrosine. This leads to accumulation of proximal metabolites including maleylacetoacetate (MAA) and fumarylacetoacetate (FAA), which are converted to toxic byproducts, succinylacetoacetate (SAA) and succinylacetone (SA) (Mitchell et al. 2001).

The worldwide incidence of HT1 is estimated to be 1:100,000–1:125,000 and a high number of cases have been described in regions such as Quebec (Canada), Scandinavia and some parts of Turkey (see Aktuglu-Zeylek and Gigden, this book Chap. 15) and India (Mayorandan et al. 2014). In Quebec, the carrier rate is about 1/60, reaching 1/20 in the Saguenay-Lac Saint-Jean area (1/1,846 live births) where the founder mutation IVS12+5G>A is particularly prevalent (De Braekeleer and Larochelle 1990; Grompe et al. 1994). Universal newborn screening for HT1 has been performed in Quebec since 1970 (Grenier et al. 1982).

In this chapter, we summarize the liver features of HT1 patients before the era of 2-(2-N-4-trifluoromethylbenzoyl)-1,3-cyclohexanedione (NTBC), including clinical, biological, histopathological and imaging findings, focusing on the Quebec experience. Then we describe the current hepatic situation of patients treated in the Quebec NTBC Study and our recommendations for liver surveillance and management of the Quebec cohort. Early data about the course of HT1 patients before the availability of NTBC have been published elsewhere (Larochelle et al. 1973; Paradis et al. 1990). Data about liver transplantation in Quebec are reported in another chapter of this volume (Alvarez and Mitchell, Chap. 5).

6.2 Clinical Presentation

In children who are not detected by newborn screening and who do not receive treatment with NTBC, the clinical course of HT1 may be highly variable. The spectrum includes acute (or sub-acute) and chronic forms (Larochelle et al. 1967; van Spronsen et al. 1994). The acute form is associated with rapid disease progression, characterized by severe liver disease in the first 6 months of life. Liver synthetic function is particularly severely affected. Symptoms include bleeding abnormalities, hypoglycemia, ascites, oedema, vomiting, irritability and jaundice.

Laboratory findings show abnormally prolonged prothrombin (PT) and activated partial thromboplastin time, increased international normalized ratio (INR), and decrease of coagulation factors (factors II, V, VII, IX, X, XI and XII) as well as elevation of serum aminotransferases, gamma glutamyl transpeptidase. Since the defect in hepatocytes is primarily synthetic rather than cholestatic, hyperbilirubinemia tends to occur late, when liver failure is already established. Therefore, lack of hyperbilirubinemia is not infrequent and does not exclude HT1 in a patient with liver failure. Conversely, a patient who presents with marked hyperbilirubinemia and relative preservation of synthetic function is unlikely to have HT1.

The chronic form classically affects patients older than 1 year of age. Fibrosis and cirrhosis develop, but episodes of acute liver failure can occur in the context of infections and other stresses. Some patients have renal tubular dysfunction with a Fanconi syndrome of variable severity resulting in phosphaturia with hypophosphatemia and sometimes rickets, glycosuria, generalized aminoaciduria, hypercalciuria, hyperchloremic metabolic acidosis and hyperuricosuria/hypouricemia). Cardiomyopathy is described but in our experience is rarely of clinical importance. Episodes of acute hepatic porphyria-like neurologic crises can occur (Mitchell et al. 1990). Not all patients fall neatly into these two groups. In practice, there is a spectrum of disease severity. Some patients have an intermediate, "subacute" course. Rare patients have a prolonged course, in which cirrhosis and renal failure develop insidiously in early adulthood.

A key consideration is the high risk of hepatocellular carcinoma (HCC) as a mid- or long-term complication of HT1 (Paradis et al. 1994; van Spronsen et al. 2005; Koelink et al. 2006). The presence of a liver nodule in patients with tyrosinemia is an indication for urgent evaluation and is considered to be cancerous until proof to the contrary. Liver parenchymal changes of some HT1 patients, including fibrosis and steatosis, may make difficult the diagnosis of HCC on imaging techniques.

Serum alphafetoprotein (AFP) is increased in virtually all patients at the time of diagnosis. When NTBC treatment is started during the first month of life, AFP level typically normalizes over 12 to maximum 24 months of life. In non-NTBC treated patients AFP levels tend to improve with age and diet treatment, but usually do not normalize (Larochelle et al. 1967; van Spronsen et al. 1994).

HT1 patients are followed with repeated assay of serum AFP. This test should not be used alone as the only marker of HCC because it lacks specificity and sensitivity. However it is useful clinically when integrated with clinical and imaging considerations. Serum AFP can fluctuate, but a steady increase of AFP levels raises the suspicion of a HCC. Serum AFP levels are elevated in most cases of HCC but also can be high in patients with regenerative nodules (such as in the recuperation phase from severe liver crises in non-NTBC treated patients), fatty nodules and if NTBC treatment is stopped. Importantly, in some HT1 patients, HCC can occur with normal or only slightly elevated serum AFP levels (Paradis et al. 1990; van Ginkel et al. 2015).

In the Québec NTBC Study, we use an integrated clinical approach that includes (1) careful consideration of the clinical history (time of starting NTBC treatment, rapidity of AFP response, (2) adherence to treatment as seen by serial plasma NTBC and plasma and urine SA levels), (3) monthly surveillance of serum AFP levels and serial imaging with careful comparison to previous imaging studies. No single parameter suffices to diagnose or to exclude HCC.

Abdominal imaging is an essential element. Abdominal ultrasound (US) and magnetic resonance (MR) imaging are the modalities of choice for serial evaluation of liver, kidneys and pan-

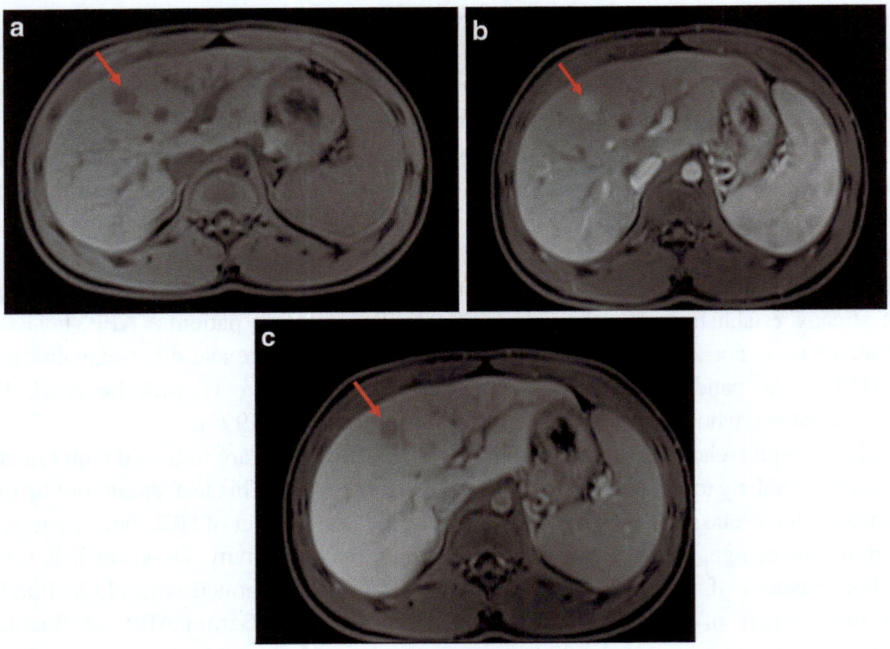

Fig. 6.1 MR imaging results suggesting HCC in a 16-year old boy with HT1presenting with AFP increase. Gd-EOB-DTPA-enhanced 3D T1–weighted images obtained at (**a**) precontrast (**b**) arterial phase 20 sec and (**c**) at 60 sec (portovenous phase). On segment VIII of the liver, an hypointense lesion was seen in pre-contrast image with a significant arterial phase enhancement with wash-out on the porto-venous phase compared to the liver parenchyma

creas in HT1 patients, preferably by the same radiologist using detailed comparison with previous studies. The liver screening is important to exclude liver nodules. US permit to detect small nodules sometimes difficult to see on MR. Nodules are hyper or hypoechoic in a heterogeneous (when macronodular cirrhosis is present) or normal parenchyma on US. It is impossible to make the difference between regenerative nodules, dysplasic nodules or HCC. However, MR imaging with the hepatocellular contrast allows a better characterization of lesions. Several publications suggest the Gd-BOPTA contrast (gadobenate dimeglumine, MultiHance®, Bracco Imaging SpA, Milano, Italy) and more recently Gd-EOB-DTPA (gadolinium ethoxybenzyl diethylenetriamine pentaacetic acid [or gadoxetic acid: *Primovist*®/ *Evovist*®]) as improved tool to detect small HCC nodules (Grazioli et al. 2010; Granito et al. 2013; Haradome et al. 2011). The MR dynamic acquisition is performed with the intravenous injection of 0.1 cc/kg = 0.025 mmol/ml of *Primovist*® with a pre-MR scanning followed by 25 seconds (sec), 70 sec, 3 minutes (min), 5 min, 20 min and 60 min post-injection scanning. It permits to increase the specificity of lesions. Most of the HCC nodules present with a portal washout followed by a hypointense lesion with a capsule and a hypointense lesion on delayed phase (Fig. 6.1).

In a recent meta-analysis, up to 400 articles related to agents contributing to the characterization of HCC nodules have been reviewed by Liu et al. (2013). Authors have also studied more than 200 publications focusing on agents to diagnose liver metastases from various origins. Altogether, with *Primovist*® as contrast agent in the diagnosis of HCC, the sensibility is around 91% (false negative rate may be related to a hyperintense aspect of HCC in ~7% of cases) and the specificity reaches 95% (false positive rate may be related to a hypo-intense aspect of other lesion like cholangiocarcinoma).

Possible extrahepatic findings on abdominal imaging include nephromegaly, nephrocalcinosis and renal cortical hyperechogenicity. The pancreas can occasionally be hyperechogenic but this is not typically associated with clinical signs.

6.3 Pathology

The histo-pathologic changes are variable depending on acute or chronic presentation mode (Dehner et al. 1989; Watanabe et al. 1983; Demers et al. 2003). In the early form, macroscopically, the liver will be enlarged and frequently pale. Usually, histo-pathologic changes include usually micronodular cirrhosis, fibrotic septa, bile duct proliferation within portal tracts and steatosis. The normal hepatic architecture is substituted by pseudoacinar or nodular formations around a central duct with bile plugging. Accumulation of iron in hepatocytes or Kuppfer cells can occur, as can giant cell transformation. In the chronic form of HT1, the liver is nodular, coarse and enlarged and histological analysis shows micro and macronodular cirrhosis, variable amount of steatosis, fibrotic septa and mild lymphoplasmatic infiltrate. Ductular proliferation, cholestasis or intralobular inflammation is less marked. Large or small type liver cell dysplasia may be present and is considered to be a premalignant lesion. Ultra-structural features include fatty inclusions into the hepatocytes with dilatation of endoplasmic reticulum and variable changes in the mitochondria.

HCC lesion size may be variable but, usually, small HCC measures less than 2 cm on diameter and macroscopically, it can be indistinguishable from macroregenerative nodules, in particular when liver cirrhosis occurs. A fibrous capsule and an approximately nodular aspect with irregular borders are usually present. Some green areas may alternate with yellow parts corresponding

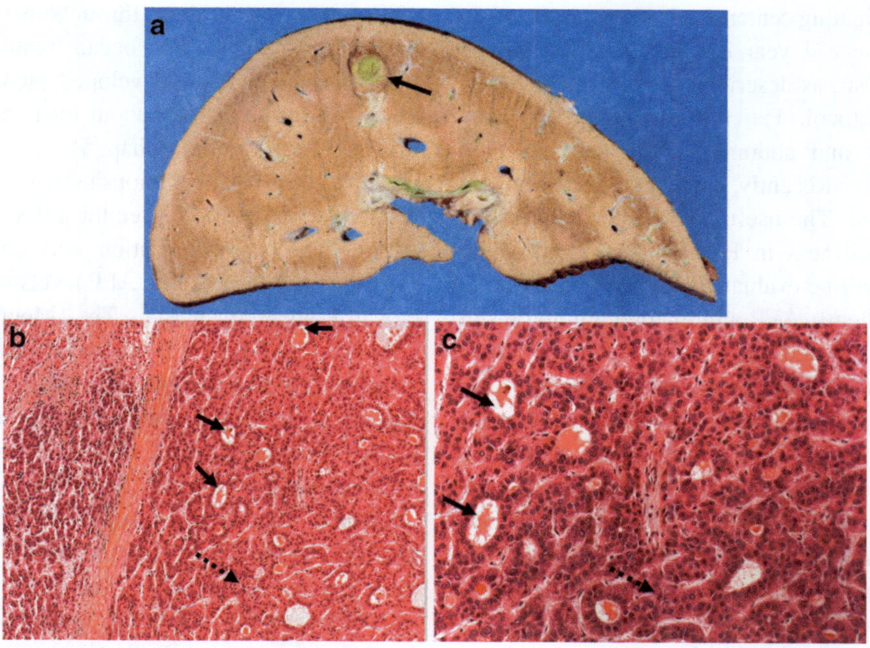

Fig. 6.2 (**a**) Macroscopic appearance of a liver from an adolescent who presented with hepatomegaly and renal Fanconi syndrome and received NTBC treatment from the age of 11 months. He had with a nodular lesion located in segment VIII, measuring around 15 × 15 mm. (**b, c**) Microscopically, the lesion is multinodular, well limited by a thin fibrotic capsule. The lesion is highly acinar (*continuous arrow*), with substantial cell proliferation. Multiple hepatocytes strands are present (*dotted arrow*) (Hematoxylin and eosin, ×200 (**b**) and ×400 (**c**))

respectively to bile staining and fat accumulation into tumor cells. The diagnosis is confirmed microscopically. HCC measuring less than 2 cm of diameter is well-differentiated and is composed of thin and multiple strands of hepatocytes. Some trabecular and acinar patterns are often mixed at the microscopic analyze. Tumor cells are assembled is cords of variable thickness separated by sinusoidal blood spaces. Initially, lesions have a thin trabecular pattern but with clonal expansion inducing the dedifferentiation, trabeculae become thicker. Mallory hyaline and intracytoplasmic proteinaceous accumulations are visible as well as loss of reticulin fibers (Washington and Harris 2010) (Fig. 6.2).

6.4 Quebec Cohort

In this section, the clinical and imaging features of the liver and the hepatological management of Quebec patients with HT1, is described.

Currently, 88 patients (43 females) are individually followed by a pediatric hepatologist at the coordinating center. Each patient is examined at least once a year during a interdisciplinary clinical visit, as described in the Quebec NTBC Study protocol. Patients have liver US every 6 months and abdominal MR imaging every 12 months. Recently, an annual fibroscan has been added. The usefulness of this examination of liver stiffness in HT1, and its optimal frequency, will be evaluated over the next years. In stable patients, AFP and alanine aminotransferase are measured every 3 months and albumin, INR and PT are assessed every 6 months.

Among entire cohort, 85 patients had a neonatal screening for HT1. Currently the median age is 13.9 years (range (r): 41 days (d) to 42.4 years (y)). The first use of NTBC occurred on February 1994 in a 5 year-old girl. NTBC treatment was started during the first month of life in 67 of 88 patients (median age: 25 days, range: 2 days to 21.7 years). Currently, the median duration is 13.8 years. At the evaluation in spring 2016, 81/88 patients (92%) had normal levels of alanine aminotransferase (<35 U/L). In the seven remaining patients (including five males in whom median BMI was 31.2, range, 26–50.1) with increased ALT (mean level 49.7 U/L, range 36–76 U/L), liver steatosis was confirmed on abdominal ultrasound. No other cause was found for the elevated level of AFP in these patients.

Serum AFP (N<10 ng/ml after the age of 12–18 months) is measured every 3 months. Excluding patients younger than 18 months, two patients had chronically increased AFP levels. The first is a 22-year-old man who started NTBC at 5 months of age in whom AFP levels decreased to about 50 ng/ml, and have remained at this level for almost 15 years. Extensive evaluations have been performed in this patient. No sign of HCC is evident. The second patient is a 23-year-old man who started NTBC at 8 months of age. His AFP levels are normal but liver imaging is potentially compatible with HCC. He is currently under evaluation for a possible liver transplantation.

In our experience, no patient who received NTBC treatment before the age of 1 month has developed HCC. Conversely, even early-treated patients are considered to be at risk, and current plans are to follow them throughout their lives. Among the patients who began treatment at an older age, some have developed HCC and are described in the chapter about liver transplantation in HT1 in Quebec (Chap. 5).

Altogether, 14 liver biopsies have been performed in 12 patients over the 20 y follow-up period. The main indication was unexplained cytolysis and/or increased AFP levels with suspicion of liver dysplasia. The Metavir score (Bedossa and Poynard 1996) was normal (F0-F1) for all except two patients who had an abnormal Metavir score of (F3). One of these patients eventually developed HCC and underwent liver transplantation.

Among the nontransplanted patients in the Quebec cohort, 21 patients began NTBC treatment after the first month of life due for example to birth before the NTBC era or birth outside of Quebec in a region without screening, with delayed diagnosis. To date, all patients except the two young men discussed above, show normal AFP and imaging without suspicion of hepatic nodules,. Typically, serum SA levels are near to or within the control reference range (<24 nmol/L), discussed in the chapter dealing with clinical challenges in tyrosinemia (Chap. 19).

6.5 Management of Liver Complications

The management of liver disease is challenging during the acute phase of HT1. In our experience, in recent years, due to the effectiveness of generalized neonatal screening combined with rapid NTBC and diet treatment, acute liver crisis have not occurred. Historically, during liver crises, non-NTBC treated HT1 patients were hospitalized. Caloric intake was maximized by enteral or parenteral routes, in order to limit catabolism and the enhanced breakdown of phenylalanine and tyrosine to the toxic metabolites that accumulate in HT1. Intake of phenylalanine and tyrosine was restricted and subsequently increased as permitted by the clinical course. Management of infections, which are important precipitants of acute liver crises, was a major goal. The clinical description and biological findings in HT1 patients prior the era of NTBC are described elsewhere (Mitchell et al. 2001).

In screened, NTBC-treated HT1 patients, our clinical approach for chronic liver disease focuses mainly in the detection and follow-up of nodules in order to assess the indication for liver transplantation. It is difficult to be certain of the malignant potential of nodules, in particular in those with imaging signs compatible with HCC. For this reason, we still believe that any patient metabolically well controlled with therapeutic levels of NTBC, who presents with increase of serum AFP and with repeated imaging confirming such nodules, should be assessed for liver transplantation. In our experience, the major risk factor for the development of liver nodules and of HCC, is the absence of early diagnosis leading to a delayed NTBC treatment (after the first month of life). Over the last 22 years, no patient who was diagnosed following a positive neonatal screen and who was treated before 1 month of age, has required liver transplantation.

6.6 Future Perspectives

Research into new therapeutic avenues, including stem cells and gene therapy, is providing interesting preliminary results (Hickey et al. 2014, 2016). The main challenge of these approaches is to substitute all hepatocytes without leaving a single FAH-deficient cell, which is not yet possible. Research in genetically modified stem cells is discussed elsewhere in this volume. Although it is not currently applicable clinically, these techniques may play a role in the future for the management of HT1.

6.7 Conclusion

Since the publication of Lindstedt and Holme (Lindstedt et al. 1992), NTBC therapy has revolutionized the medical management of HT1 patients. For two decades, we moved from the symptomatic treatment of life threatening hepatic emergencies to the elective regular surveillance of patients who have a high quality of life and who function normally in society. In Quebec, and increasingly elsewhere, identification of HT1 by newborn screening permits treatment to be started before the development of clinical symptoms or of detectable organ damage. Of course, this treatment requires adherence to a special diet and to the prescription of NTBC. Adherence has proven difficult for some patients, particularly during adolescence, as discussed in the chapter in this volume devoted to future clinical challenges. The clinical course to date is promising and continued follow-up of the Québec NTBC study cohort will continue to provide answers about the long-term hepatic outcome of the medical treatment of HT1 patients.

Acknowledgements We thank Dr. Dorothée Bouron – Dal Soglio and Dr. Nathalie Patey from the Division of Pathology and Cellular Biology (CHU Sainte-Justine, Université de Montréal, Quebec, Canada) for their helpful comments and valuable assistance with the histopathological analyses.

References

Bedossa P, Poynard T (1996) An algorithm for the grading of activity in chronic hepatitis C. The METAVIR Cooperative Study Group. Hepatology 24(2):289–293

De Braekeleer M, Larochelle J (1990) Genetic epidemiology of hereditary tyrosinemia in Quebec and in Saguenay-Lac-St-Jean. Am J Hum Genet 47:302–307

Dehner LP, Snover DC, Sharp HL, Ascher N, Nakhleh R, Day DL (1989) Hereditary tyrosinemia type I (chronic form): pathologic findings in the liver. Hum Pathol 20(2):149–158

Demers SI, Russo P, Lettre F, Tanguay RM (2003) Frequent mutation reversion inversely correlates with clinical severity in a genetic liver disease, hereditary tyrosinemia. Hum Pathol 34(12):1313–1320

Granito A, Galassi M, Piscaglia F, Romanini L, Lucidi V, Renzulli M, Borghi A, Grazioli L, Golfieri R, Bolondi L (2013) Impact of gadoxetic acid (Gd-EOB-DTPA)-enhanced magnetic resonance on the non-invasive diagnosis of small hepatocellular carcinoma: a prospective study. Aliment Pharmacol Ther 37(3):355–363

Grazioli L, Bondioni MP, Faccioli N, Gambarini S, Tinti R, Schneider G, Kirchin M (2010) Solid focal liver lesions: dynamic and late enhancement patterns with the dual phase contrast agent gadobenate dimeglumine. J Gastrointest Cancer 41(4):221–232

Grenier A, Lescault A, Laberge C, Gagné R, Mamer O (1982) Detection of succinylacetone and the use of its measurement in mass screening for hereditary tyrosinemia. Clin Chim Acta 123(1–2):93–99

Grompe M, St-Louis M, Demers SI, Al-Dhalimy M, Leclerc B, Tanguay RM (1994) A single mutation of the fumarylacetoacetate hydrolase gene in French Canadians with hereditary yrosinaemia type I. New Eng J Med 331:353–357

Haradome H, Grazioli L, Tinti R, Morone M, Motosugi U, Sano K, Ichikawa T, Kwee TC, Colagrande S (2011) Additional value of gadoxetic acid-DTPA-enhanced hepatobiliary phase MR imaging in the diagnosis of early-stage hepatocellular carcinoma: comparison with dynamic triple-phase multidetector CT imaging. J Magn Reson Imaging 34(1):69–78

Hickey RD, Mao SA, Glorioso J, Lillegard JB, Fisher JE, Amiot B, Rinaldo P, Harding CO, Marler R, Finegold MJ, Grompe M, Nyberg SL (2014) Fumarylacetoacetate hydrolase deficient pigs are a novel large animal model of metabolic liver disease. Stem Cell Res 13(1):144–153

Hickey RD, Mao SA, Glorioso J, Elgilani F, Amiot B, Chen H, Rinaldo P, Marler R, Jiang H, DeGrado TR, Suksanpaisan L, O'Connor MK, Freeman BL, Ibrahim SH, Peng KW, Harding CO, Ho CS, Grompe M, Ikeda Y, Lillegard JB, Russell SJ, Nyberg SL (2016) Curative ex vivo liver-directed gene therapy in a pig model of hereditary tyrosinemia type 1. Sci Transl Med 8(349):349ra99

Koelink CJ, van Hasselt P, van der Ploeg A et al (2006) Tyrosinemia type I treated by NTBC: how does AFP predict liver cancer? Mol Genet Metab 89(4):310–315

Larochelle J, Mortezai A, Belanger M, Tremblay M, Claveau JC, Aubin G (1967) Experience with 37 infants with tyrosinemia. Can Med Assoc J 97(18):1051–1054

Larochelle J, Privé L, Bélanger M, Bélanger L, Tremblay M, Claveau JC, Aubin G, Paradis D (1973) Hereditary tyrosinemia 1. Clinical and biological study of 62 cases. Pediatrie 28(1):5–18

Lindstedt S, Holme E, Lock EA, Hjalmarson O, Strandvik B (1992) Treatment of hereditary tyrosinaemia type I by inhibition of 4-hydroxyphenylpyruvate dioxygenase. Lancet 340:813–817

Liu X, Zou L, Liu F, Zhou Y, Song B (2013) Gadoxetic acid disodium-enhanced magnetic resonance imaging for the detection of hepatocellular carcinoma: a meta-analysis. PLoS ONE 8(8):e70896

Mayorandan S, Meyer U, Gokcay G, Segarra NG, de Baulny HO, van Spronsen F, Zeman J, de Laet C, Spiekerkoetter U, Thimm E, Maiorana A, Dionisi-Vici C, Moeslinger D, Brunner-Krainz M, Lotz-Havla AS, Cocho de Juan JA, Couce Pico ML, Santer R, Scholl-Bürgi S, Mandel H, Bliksrud YT, Freisinger P, Aldamiz-Echevarria LJ, Hochuli M, Gautschi M, Endig J, Jordan J, McKiernan P, Ernst S, Morlot S, Vogel A, Sander J, Das AM (2014) Cross-sectional study of 168 patients with hepatorenal tyrosinaemia and implications for clinical practice. Orphanet J Rare Dis 9:107

Mitchell G, Larochelle J, Lambert M, Michaud J, Grenier A, Ogier H, Gauthier M, Lacroix J, Vanasse M, Larbrisseau A et al (1990) Neurologic crises in hereditary tyrosinemia. N Engl J Med 322(7):432–437

Mitchell GA, Grompe M, Lambert M, Tanguay RM (2001) Hypertyrosinemia. In: Scriver CS, Beaudet AL, Sly WS, Valle D (eds) The metabolic and molecular bases of inherited disease. McGraw-Hill, New York, pp 1777–1806

Paradis K, Weber A, Seidman EG, Larochelle J, Garel L, Lenaerts C, Roy CC (1990) Liver transplantation for hereditary tyrosinemia: the Quebec experience. Am J Hum Genet 47(2):338–342

Paradis K, Mitchell GA, Russo P (1994) Tyrosinemia. In: Suchy FJ (ed) Liver disease in children. Mosby, St-Louis, pp 803–818

van Ginkel WG, Gouw AS, van der Jagt EJ, de Jong KP, Verkade HJ, van Spronsen FJ (2015) Hepatocellular carcinoma in tyrosinemia type 1 without clear increase of AFP. Pediatrics 135(3):e749–e752

van Spronsen FJ, Thomasse Y, Smit GP, Leonard JV, Clayton PT, Fidler V, Berger R, Heymans HS (1994) Hereditary tyrosinemia type I: a new clinical classification with difference in prognosis on dietary treatment. Hepatology 20:1187–1191

van Spronsen FJ, Bijleveld CM, van Maldegem BT, Wijburg FA (2005) Hepatocellular carcinoma in hereditary tyrosinemia type I despite 2-(2 nitro-4-3

trifluoromethylbenzoyl)-1, 3-cyclohexanedione treatment. J Pediatr Gastroenterol Nutr 40(1):90–93
Washington K, Harris E (2010) Masses of the liver. In: Mills SE, Carter D, Greenson JK, Reuter VE, Stoler MH (eds) Sternberg's diagnostic surgical pathology, 5th edn. Lippincott Williams & Wilkins, Philadelphia, pp 1553–1568
Watanabe S, Okita K, Harada T, Kodama T, Numa Y, Takemoto T, Takahashi T (1983) Morphologic studies of the liver cell dysplasia. Cancer 51(12):2197–2205

7 Liver Transplantation for Hereditary Tyrosinaemia Type 1 in the United Kingdom

Patrick McKiernan

Abstract

Fourteen children have undergone liver transplantation for hereditary tyrosinaemia type 1 (HT1) at Birmingham Children's hospital (BCH) since 1989; six were treated prior to the availability of Nitisinone in 1993 and eight in the post Nitisinone era. Prior to 1993 essentially all children with HT1 were referred for transplantation. In the Nitisinone era only those with unresponsive liver failure or suspected malignancy were considered for transplantation. Those who were treated pre-emptively following newborn screening have no evidence of liver disease and none have required transplantation.

Absolute patient survival is 86% for the whole group and 100% in the Nitisinone era. There has been a functional correction of the metabolic defect in all cases allowing a normal diet. Persistent renal succinylacetone production was universal but did not appear to have any clinical consequence. Renal function appeared better, and hypertension less common in those treated in the Nitisinone era.

Outcome was poorer for those four children with established malignancy; one was unfit for transplantation and another developed a pulmonary metastasis, which was successfully resected.

Keywords

Tyrosinemia type I • Liver transplantation • Inherited metabolic disease • Hepatocellular carcinoma • Nitisinone treatment

P. McKiernan (✉)
Division of Gastroenterology/Hepatology/Nutrition, Children's Hospital of Pittsburgh of UPMC, University of Pitttsburgh,
4401 Penn Avenue, Pittsburgh, PA 15224, USA
e-mail: patrick.mckiernan@chp.edu

Abbreviations

AFP Alpha-fetoprotein
cGFR Calculated glomerular filtration rate

© Springer International Publishing AG 2017
R.M. Tanguay (ed.), *Hereditary Tyrosinemia*, Advances in Experimental Medicine and Biology 959,
DOI 10.1007/978-3-319-55780-9_7

HB	Hepatoblastoma
HCC	Hepatocellular carcinoma
HT1	Hereditary tyrosinaemia type 1
PCR	Protein:creatinine ratio
SA	Succinylacetone
TRP	Tubular reabsorption of phosphate

7.1 Introduction

Liver transplantation for hereditary tyrosinaemia type 1 (HT1) was first undertaken in 1976 by Professor Starzl in a 9-year-old girl who had developed hepatocellular carcinoma (HCC) (Fisch et al. 1978). Although she died 3 months later from infective complications, her metabolic defect was corrected. Since that time more than 150 liver transplants have been undertaken for HT1 with greater than 80% long term survival (Bartlett et al. 2014; Mayorandan et al. 2014). The metabolic effect has been consistent with an immediate functional correction of the systemic defect allowing a normal diet in all cases. Fourteen children have undergone Liver transplantation for HT1 at Birmingham Children's hospital (BCH) since 1989.

The outcome for children with HT1 continues to improve era on era (Mayorandan et al. 2014). This is multifactorial and encompasses progress in medical management, the possibility of preemptive treatment following newborn screening, the availability of more palatable dietary products and in the management of liver transplantation and immunosuppression.

Liver transplantation in children for metabolic disease is a highly successful treatment with an expected 1 year survival following elective transplantation of greater than 95% in experienced centres (Mazariegos et al. 2014). Children surviving more than 1 year can confidently expect long-term survival (McKiernan 2010). Follow up studies are consistent in showing that correction of the metabolic defect is lifelong, and that quality-of-life for children and their families are excellent for the great majority.

7.1.1 Indications for Transplantation

As with any hepatic based inborn error of metabolism whether and not liver transplant is appropriate will depend on the interaction of current clinical status, success and availability of liver transplantation and the effectiveness and tolerability of alternative treatments. Traditionally the major indication for transplantation in HT1 has been to prevent the development of hepatocellular carcinoma (HCC). This was based on a "not if but when"philosophy. In practice all children with HT1 were candidates for transplantation and the exact timing of listing was determined on local criterion. These ranged from either arbitrary size or age criterion to some individual biological marker suggesting increased risk of HCC (Mohan et al. 1999; Sokal et al. 1992). These latter included rising alpha-fetoprotein (AFP), histological evidence of hepatocyte dysplasia or radiological features such as the development of new nodules or progression of established ones.

The availability of Nitisinone from 1993 transformed this approach. From this time all were treated with Nitisinone and a more expectant approach taken (McKiernan 2013). Indications for transplantation became failure of Nitisinone and suspected (or proven) HCC. As shown in Table 7.1, referral for transplantation became the exception rather than the rule. Only one child was transplanted because of liver failure

Table 7.1 Referral for transplantation at Birmingham Children's Hospital in two eras

	Pre-nitisinone (1989–1992)	Post-nitisinone (1993–2015)	P value
No. referred for transplantation	6/7	9[a]/34	<0.05
Age at transplant months (range)	61 (19–126)	53 (5–173)	0.94
Time on nitisinone prior to transplantation (months)		39 (2–161)	
Median age starting nitisinone treatment (days)		OLT 428 No OLT 52	0.03

[a]Only eight underwent transplantation

Table 7.2 Radiological findings and alpha-fetoprotein (AFP) evolution in eight children treated with Nitisinone referred for liver transplantation for suspected malignancy at Birmingham Children's Hospital 1993–2015

	AFP evolution	Radiological appearance
Proven malignancy (3)	Persistent elevation 1	Dominant nodule 3
	Initial fall and secondary decrease 2	Extrahepatic spread (1)
Hepatic adenoma (1)	Fell to normal	Dominant nodule
Suspected malignancy (4)	Initial fall but never normalised	Non dominant nodules 3
		Dominant nodule 1

despite Nitisinone. Eight children were referred for transplantation because of proven or suspected malignancy. Interestingly the age at transplantation was similar in both eras.

Of the eight children referred with suspected malignancy following Nitisinone treatment this was confirmed in three. Two children had developed HCC and one developed hepatoblastoma (HB). One other child proved to have an adenoma, which has not been reported in HT1 before, but in the context was probably a premalignant lesion. The AFP patterns and radiological appearances in the referred children are summarised in Table 7.2. All children had abnormal cross-sectional radiology, which showed a multinodular appearance. In the cases with proven malignancy there was a dominant nodule in combination with an abnormal evolution of AFP. In contrast, the four children without proven malignancy showed a logarithmic decease in AFP, but which failed to completely normalise. Only one of these four had a dominant nodule on imaging.

This demonstrates how difficult management decisions can be in individual cases. Given the risk of irreversible consequences from established malignancy and the excellent results of transplantation; where there is genuine diagnostic doubt, transplantation will usually be indicated.

Unfortunately the one child with HB had evidence of extrahepatic disease with extensive splanchnic vascular thrombosis when the tumour was recognised. There was no meaningful response to chemotherapy and the tumour proved rapidly fatal. There have been two previous reports of hepatoblastoma in HT1 (Buyukpamukcu et al. 2006; Nobili et al. 2010), which suggests that it is not a coincidental association. However the very different aetiological spectrums of HCC and HB make a mechanistic link difficult to explain at this time.

It is also important to highlight the impact of age when Nitisinone was commenced. Of the 34 children treated at BCH since 1993, 12 were treated pre-emptively following detection by newborn screening. This cohort remains well with normal AFP and hepatic imaging at median age of 9 years (McKiernan et al. 2015) and other centres have reported similar experience (Larochelle et al., 2012). None of this cohort have been considered for transplantation and it seems unlikely they ever will be. Going forward, the indications for liver transplantation in HT1 will continue to be unresponsive liver failure and where hepatic malignancy is suspected in patients who initially presented symptomatically.

7.1.2 Management of Transplantation

All children received orthotopic transplants from cadaveric donors. Immunosuppression management was according to the contemporaneous protocol, with some minor modifications. Prior to 2000, immunosuppression consisted of lifelong Cyclosporin in combination with Prednisolone for 3 months and Azathioprine in the first year. Following 2000, Tacrolimus was substituted for Cyclosporin and Azathioprine. Since 2004 anti IL-2 induction and mycophenolate mofetil were added in tandem with lower Tacrolimus target levels with the aim of preventing nephrotoxicity.

7.1.3 Post Transplant Monitoring

Conventional monitoring of renal function, liver function and immunosuppression was as according to the contemporaneous protocol. Renal function (calculated glomerular filtration rate, cGFR) was evaluated using the Schwartz formula for height:creatinine ratio expressed in ml/min/1.73m^2 and calculated as (height in cm × 40)/plasma creatinine (Schwartz et al. 1987). cGFR was classified using the National Kidney

Foundation stages of chronic kidney disease classification (National Kidney Foundation 2002). Tubular function was assessed by tubular reabsorption of phosphate (TRP) and urinary protein:creatinine ratio (PCR). Normal TRP was considered to be >80% and normal PCR <20 mg/mmol. Formal measures of glomerular filtration rate were undertaken at least every 5 years. In those transplanted for malignancy, or where malignancy was discovered in the explanted liver, AFP was repeated every 3 months for 1 year and subsequently annually. The extent of the residual metabolic defect was assessed by annual measures of urinary and plasma succinylacetone (SA) and plasma amino acids.

7.1.4 Outcome of Liver Transplantation for HT1

Fourteen children underwent transplantation at a median age of 5 years (4 months–13 years) of whom 12 survive, currently aged 23 (2–34) at a median of 17 years (2–26) post transplantation. Six underwent transplantation prior to the availability of Nitisinone and eight had received Nitisinone prior to transplantation (Fig. 7.1). There were two deaths in the early cohort giving an absolute patient survival of 86%. One patient developed hepatic artery thrombosis with chronic rejection and had a re-graft 11 months later. He developed the same complications in the second graft and died 5 months later. A second patient from the early era developed primary non-function and died 10 days later despite urgent re-transplantation.

All eight patients undergoing transplant in the second cohort are alive and well. Two have required repeat transplantation giving an absolute graft survival of 71%. One patient underwent transplant for proven HCC aged 13 and received a second transplant 8 months later for chronic rejection. Another patient who underwent first transplant age 4 months required repeat transplantation due to a combination of vascular outflow obstruction and biliary fibrosis 12 years later.

7.1.5 Outcome of Transplantation for Established HCC

Three children had established HCC at the time of transplant. In two cases this had been detected preoperatively and was the primary indication for transplantation while in the other it was an incidental finding in the explanted liver. One child had presented with chronic liver disease aged 20 months, which responded well to Nitisinone. At age 13 routine monitoring simultaneously detected a new hepatic nodule and a rapidly rising AFP. He underwent transplant 2 months later and by then there was evidence of vascular invasion. Due to the high risk of recurrence his

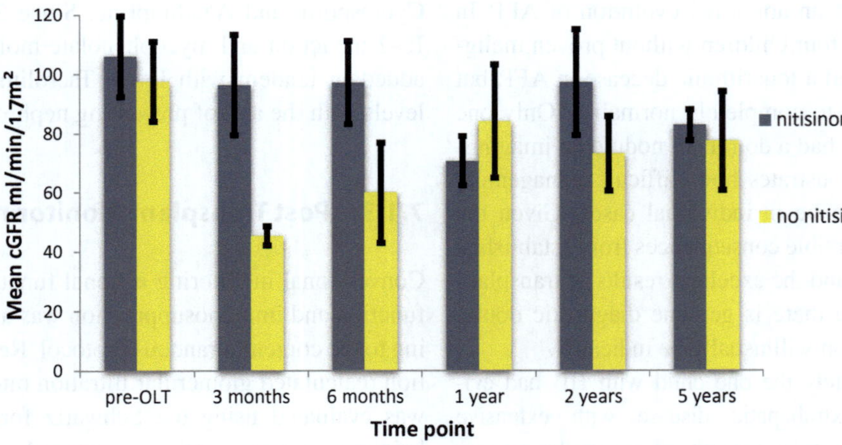

Fig. 7.1 Change in cGFR following liver transplantation in pre and post Nitisinone eras

immunosuppression was minimised but he developed chronic rejection requiring repeat transplant 8 months later. Two years after the original transplant routine monitoring showed an acute rise in AFP and chest radiology revealed a pulmonary metastasis. Abdominal imaging showed no recurrence and the pulmonary metastasis was fully resected. He remains well without evidence of further recurrence 9 years later.

The second case was asymptomatic until presentation with abdominal distension aged 4 months. A large hepatic mass was found with grossly raised AFP. Histology was suggestive of a hepatoblastoma and she received appropriate chemotherapy. Although the mass showed some response it was never resectable. Further investigation of persistent coagulopathy confirmed she had HT1 and Nitisinone was commenced. She underwent liver transplant aged 7 months and remains well 3 years later. Histology of the explant confirmed the tumour to be HCC. She did not receive any further chemotherapy.

The final case was treated in the pre-Nitisinone era. He underwent liver transplantation aged 5 because of established hepatocyte dysplasia. The explant was found to contain an established HCC without vascular invasion. He developed chronic rejection, which recurred following repeat transplantation and he died 16 months after original transplantation. There was no evidence of tumour recurrence at any time.

This experience demonstrates that liver transplantation in children with HT1 and established malignancy has an acceptable outcome, even when vascular invasion has been shown. However the morbidity is higher compared to those where malignancy was suspected rather than proven prior to transplantation. An additional child with malignancy had already developed extrahepatic metastases by the time malignancy was recognised and hence was never suitable for transplantation. There is an ongoing need for close monitoring for malignancy in children with HT1 who present clinically and a need for improved methods for early detection (Baumann et al. 2006).

7.1.6 Glomerular Function

Changes in cGFR pre- and post-OLT are shown in Fig. 7.1. Median pre-operative cGFR values for patients treated with Nitisinone (104 ml/min/1.73m^2, range 54–152) were similar to those of the early cohort (100 ml/min/1.73m^2, range 58–146). By 3 months after OLT, cGFR had significantly decreased in the early cohort (46 ml/min/1.73m^2 range 40–51, $p = 0.02$) but not in the Nitisinone treated cohort (90 ml/min/1.73m^2 range 51–172 $p = 0.5$). At later time points, median cGFR remained slightly below normal in those treated with Nitisinone equivalent to stage 1 or stage 2 (60–89 ml/min/1.73m^2) chronic kidney disease. In patients who had not received Nitisinone, median cGFR remained lower than those who had received Nitisinone. These were equivalent to stage 3A chronic kidney disease (45–59 ml/min/1.73m^2) for up to 6 months but later they improved to within the stage 2 category. However, there was no statistically significant difference between the two groups after 3 months. One patient who did not receive Nitisinone and had stage 2 chronic kidney disease prior to transplant (cGFR 64 ml/min/1.73 m^2) was the only patient whose renal function failed to improve following OLT. He developed renal failure and he underwent successful renal transplant 21 years post liver transplant

7.1.7 Tubular Function

There was a trend towards higher TRP pre-OLT in the Nitisinone-treated group compared to those not treated with Nitisinone (93% range 91–98% vs. 82% range 50–83%, $p = 0.05$) although values were within the normal range for all except one patient in the non-nitisinone group who had a pre-OLT TRP of 50%. Following OLT, TRP remained normal in all patients in both groups up to 5 years with no significant difference between groups.

Urinary PCR was raised in about 50% of children pre-OLT with no significant difference between those who received Nitisinone and those who did not (median 25.7 range 14.0–32.2 vs.

median 19.0 range 13.5–32.0, $p = 0.6$). Following OLT urinary PCR remained elevated in the majority of children who did not receive Nitisinone and normal in the majority of children who did receive Nitisinone up to 5 years although these differences did not reach significance.

7.1.8 Hypertension

Following OLT, all four surviving patients (75 %) who did not receive Nitisinone are currently on antihypertensive medication with three patients requiring two or more agents and the third patient on a single agent. In the Nitisinone-treated group, only one of the seven patients (14.3%) is currently requiring antihypertensive treatment with a single agent.

7.1.9 Correction of the Metabolic Defect

All were allowed an unrestricted diet following transplantation. In the early era plasma amino acids normalised within 48 h of transplantation and remain normal during subsequent follow up. In those treated with Nitisinone prior to transplant, initial tyrosine levels were modestly elevated for the first month, which we attributed to the long half-life of Nitisinone. In all cases these subsequently became persistently normal.

An opposite pattern was seen with urinary and plasma SA. Those in the early era had high urinary SA prior to transplant, which fell rapidly to <5% of baseline, but still well above the normal range. Plasma SA levels were only available from >10 years post transplant and these were persistently raised at levels <5 % of that expected in an untreated patient with HT1 (Bartlett et al. 2013).

In those treated with Nitisinone prior to transplant, urinary and plasma SA were undetectable at the time of transplantation. Over the first 5 years following transplant plasma and urinary SA recurred and gradually rose to the persistent levels seen in the early cohort.

In both groups PBG synthase activity mirrored the plasma SA levels. PBG synthase levels were normal at the time of transplant in those treated with Nitisinone. In both groups PBG synthase fell to the low normal range by 5 years post transplantation.

The source of persistent SA is thought to be the kidney where FAH is also active (Tuchman et al. 1987) but the functional significance of this is still not fully understood. We did not detect any clinical consequence of these abnormalities and in particular there were no features of porphyria. There was no correlation between post transplant SA levels and any index of renal function.

It is known that tubular function normalises within 1 year in patients treated with Nitisinone and subsequently remains normal for more than 10 years (Santra et al. 2008). Continued Nitisinone treatment post-OLT has been suggested as a means of controlling the renal SA production and hence improving long term renal function (Pierik et al. 2005). We have not taken this approach for a number of reasons; there was no correlation of renal dysfunction with the extent of the SA production, renal dysfunction does not appear to be progressive; options for preventing and ameliorating immunosuppression associated nephropathy have increased. An additional factor to consider is that even low dose Nitisinone causes significantly raised tyrosine levels (Introne et al. 2011), raising the issue of whether reintroduction of dietary restriction would be necessary. This would likely have a significant negative impact on the quality of life for these transplant recipients.

References

Bartlett DC, Lloyd C, McKiernan PJ, Newsome PN (2014) Early nitisinone treatment reduces the need for liver transplantation in children with tyrosinaemia type 1 and improves post-transplant renal function. J Inherit Metab Dis 37(5):745–752

Bartlett DC, Preece MA, Holme E, Lloyd C, Newsome PN, McKiernan PJ (2013) Plasma succinylacetone is persistently raised after liver transplantation in tyrosinaemia type 1. J Inherit Metab Dis 36(1):15–20

Baumann U, Duhme V, Auth MK, McKiernan PJ, Holme E (2006) Lectin-reactive alpha-fetoprotein in patients with tyrosinemia type I and hepatocellular carcinoma. J Pediatr Gastroenterol Nutr 43(1):77–82

Buyukpamukcu M, Varan A, Haberal M, Buyukpamukcu N, Koksal Y, Coskun T, Yuce A, Kale G, Akyuz C, Kutluk T (2006) The efficacy of liver transplantation in malignant liver tumors associated with tyrosinemia: clinical and laboratory findings of five cases. Pediatr Transplant 10(4):517–520

Fisch RO, McCabe ER, Doeden D, Koep LJ, Kohlhoff JG, Silverman A, Starzl TE (1978) Homotransplantation of the liver in a patient with hepatoma and hereditary tyrosinemia. J Pediatr 93(4):592–596

Introne WJ, Perry MB, Troendle J, Tsilou E, Kayser MA, Suwannarat P, O'Brien KE, Bryant J, Sachdev V, Reynolds JC, Moylan E, Bernardini I, Gahl WA (2011) A 3-year randomized therapeutic trial of nitisinone in alkaptonuria. Mol Genet Metab 103(4):307–314

Larochelle J, Alvarez F, Bussieres JF, Chevalier I, Dallaire L, Dubois J, Faucher F, Fenyves D, Goodyer P, Grenier A, Holme E, Laframboise R, Lambert M, Lindstedt S, Maranda B, Melancon S, Merouani A, Mitchell J, Parizeault G, Pelletier L, Phan V, Rinaldo P, Scott CR, Scriver C, Mitchell GA (2012) Effect of nitisinone (NTBC) treatment on the clinical course of hepatorenal tyrosinemia in Quebec. Mol Genet Metab 107(1–2):49–54

Mayorandan S, Meyer U, Gokcay G, Segarra N, de BH, van SF, Zeman J, de LC, Spiekerkoetter U, Thimm E, Maiorana A, Dionisi-Vici C, Moeslinger D, Brunner-Krainz M, Lotz-Havla A, Cocho de JJ, Couce PM, Santer R, Scholl-Burgi S, Mandel H, Bliksrud Y, Freisinger P, Aldamiz-Echevarria L, Hochuli M, Gautschi M, Endig J, Jordan J, McKiernan P, Ernst S, Morlot S, Vogel A, Sander J, Das A (2014) Cross-sectional study of 168 patients with hepatorenal tyrosinaemia and implications for clinical practice. Orphanet J Rare Dis 9(1):107

Mazariegos G, Shneider B, Burton B, Fox IJ, Hadzic N, Kishnani P, Morton DH, McIntire S, Sokol RJ, Summar M, White D, Chavanon V, Vockley J (2014) Liver transplantation for pediatric metabolic disease. Mol Genet Metab 111(4):418–427

McKiernan PJ (2010) Long-term care following paediatric liver transplantation. Arch Dis Child Educ Pract Ed

McKiernan PJ (2013) Nitisinone for the treatment of hereditary tyrosinemia type I. Expert Opin Orphan Drugs 1(6):491–497

McKiernan PJ, Preece MA, Chakrapani A (2015) Outcome of children with hereditary tyrosinaemia following newborn screening. Arch Dis Child 100(8):738–741

Mohan N, McKiernan P, Preece MA, Green A, Buckels J, Mayer AD, Kelly DA (1999) Indications and outcome of liver transplantation in tyrosinaemia type 1. Eur J Pediatr 158(Suppl 2):S49–S54

National Kidney Foundation (2002) K/DOQI clinical practice guidelines for chronic kidney disease: evaluation, classification, and stratification. Am J Kidney Dis 39(2 Suppl 1):S1–266

Nobili V, Jenkner A, Francalanci P, Castellano A, Holme E, Callea F, Dionisi-Vici C (2010) Tyrosinemia type 1: metastatic hepatoblastoma with a favorable outcome. Pediatrics 126(1):e235–e238

Pierik LJ, van Spronsen FJ, Bijleveld CM, van Dael CM (2005) Renal function in tyrosinaemia type I after liver transplantation: a long-term follow-up. J Inherit Metab Dis 28(6):871–876

Santra S, Preece MA, Hulton SA, Mckiernan PJ (2008) Renal tubular function in children with tyrosinaemia type I treated with nitisinone. J Inherit Metab Dis 31(3):399–402

Schwartz GJ, Brion LP, Spitzer A (1987) The use of plasma creatinine concentration for estimating glomerular filtration rate in infants, children, and adolescents. Pediatr Clin N Am 34(3):571–590

Sokal EM, Bustos R, Van Hoof F, Otte JB (1992) Liver transplantation for hereditary tyrosinemia – early transplantation following the patient's stabilization. Transplantation 54(5):937–939

Tuchman M, Freese DK, Sharp HL, Ramnaraine ML, Ascher N, Bloomer JR (1987) Contribution of extrahepatic tissues to biochemical abnormalities in hereditary tyrosinemia type I: study of three patients after liver transplantation. J Pediatr 110(3):399–403

NTBC and Correction of Renal Dysfunction

8

Arianna Maiorana and Carlo Dionisi-Vici

Abstract

Hereditary tyrosinemia type 1 (HT1) is characterized by severe progressive liver disease and renal tubular dysfunction. Kidney involvement is characterized by hypophosphatemic rickets and Fanconi syndrome. Different animal models were useful to investigate the pathophysiology of the disease and the effects of NTBC therapy on liver and kidney function. NTBC has revolutionized the prognosis of HT1 and its acute and chronic effects on renal tubular function have been proved, with normalization of tubular function within a few weeks, particularly hypophosphatemia and proteinuria. NTBC therapy is highly effective in improving renal function both at short and long-term. However, its efficacy critically depends on the age at start of treatment with normal outcome in patients diagnosed at birth by newborn screening.

Keywords

Renal tubular dysfunction • Rickets • Fanconi syndrome • NTBC therapy

Abbreviations

cGFR	Calculated glomerular filtration rate
FAA	Fumarylacetoacetate
Fah	Fumarylacetoacetate hydrolase
HCC	Hepatocellular carcinoma
HGA	Homogentisic acid
Hpd	Hydroxyphenylpyruvate dioxygenase
HT1	Hereditary tyrosinemia type 1
MAA	Maleylacetoacetate
OLT	Orthotopic liver transplantation
SA	Succinylacetone
SAA	Succinyl acetoacetate
TRP	Tubular reabsorption of phosphate

A. Maiorana (✉) • C. Dionisi-Vici
Metabolic Unit, Department of Pediatric Specialties, Bambino Gesù Children's Research Hospital, Rome, Italy
e-mail: arianna.maiorana@opbg.net; carlo.dionisivici@opbg.net

© Springer International Publishing AG 2017
R.M. Tanguay (ed.), *Hereditary Tyrosinemia*, Advances in Experimental Medicine and Biology 959, DOI 10.1007/978-3-319-55780-9_8

8.1 Renal Disease in HT1

Hereditary tyrosinaemia type 1 (HT1) is caused by deficient activity of fumarylacetoacetase (Fah), the final enzyme of tyrosine degradation pathway. As a result of the metabolic block, toxic metabolites succinylacetone (SA), maleylaceto-acetate (MAA) and fumarylacetoacetate (FAA) are formed and these are responsible for severe disruption of intracellular metabolism of the liver and kidney. Patients may present in infancy, childhood and as adults, and the age at onset broadly correlates with disease severity (Chakrapani et al. 2012). Liver disease is a major cause of morbidity and mortality and includes cirrhosis, liver failure and development of HCC (Arnon et al. 2011). Renal disease may be the predominant feature in patients with subacute and chronic presentation, however there is always some co-existing liver disease. The characteristic renal disease is a Fanconi syndrome, with variable severity of tubular dysfunction. Patients may exhibit hypophosphataemic rickets, which can be severe. In untreated patients renal signs may progress during disease course with development of nephrocalcinosis, glomerulosclerosis and chronic renal failure (de Laet et al. 2013).

The typical features of Fanconi syndrome include generalized aminoaciduria, glycosuria, phosphaturia and renal tubular acidosis. However, not all may be necessarily present. In a series of 45 patients, tubulopathy was observed at diagnosis in 80% of patients and rickets in 40% (Masurel-Paulet et al. 2008), whereas other studies reported nephrocalcinosis in 16–33% of patients (Paradis et al 1990, Forget et al. 1999, Couce et al. 2011). Interestingly, patients homozygous for the IVS6-1 (G>T) mutation had a lower frequency of nephrocalcinosis in comparison with those harboring other mutations (Couce et al. 2011). A recent multicenter study reported renal dysfunction in about 50% of 114 patients presenting at diagnosis with combined liver and renal dysfunction signs (Mayorandan et al. 2014). Renal manifestations in some cases may appear very early, as seen in one patient diagnosed by newborn screening who already suffered from renal dysfunction with nephromegaly, renal tubulopathy, and nephrocalcinosis at the age of 12 days (Mayorandan et al. 2014).

8.2 Pathophysiology and Renal Damage in HT1: Cellular and Animal Models

Lack of activity of Fah causes the accumulation of MAA and FAA, and of their derivatives, SA and succinyl acetoacetate (SAA), which have important pathogenic effects. MAA, FAA and SA disrupt sulfhydryl metabolism by forming glutathione adducts, thereby rendering cells susceptible to free radical damage (Jorquera and Tanguay 1997). Moreover, FAA and MAA are alkylating agents, capable of binding DNA, promoting mutagenesis and apoptosis (Manabe et al. 1985). As a result of these multiple effects, hepatic and renal cells exposed to high levels of these compounds undergo either apoptotic cell death or an adaptive alteration of gene expression, leading patients at risk of future liver malignant transformation (Jorquera and Tanguay 2001).

Two strains of mutant mice carrying Fah deficiency have been useful to study the pathophysiology of HT1 and the effect of NTBC therapy. The first mouse model is an albino lethal c14CoS mouse, which is neonatally lethal (Gluecksohn-Waelsch 1979). These mice have a large deletion on chromosome 7, including the albino locus and the *Fah* gene. The lethal phenotype has been related to impaired liver expression of hepatocyte-specific genes during the perinatal period (Nakamura et al. 2007). The other Fah-deficient mouse was generated by targeted disruption of the *Fah* gene, which is mutated in the exon 5 ($Fah^{\Delta exon\,5}$), and also this knock-out mouse is neonatally lethal, likely because of hypoglycemia (Grompe et al. 1993). No gross abnormalities in the histology of the liver of both Fah-deficient mice have been found, however ultrastructural investigation revealed altered membranous components. The kidneys of $Fah^{\Delta exon\,5}$ mice appeared enlarged and pale, histology showed cystic dilatation of the proximal tubules (Grompe et al. 1998), whereas dilatation and vesiculation of the rough endoplasmic reticulum and Golgi apparatus in the albino lethal c14CoS mice were evident at electron microscopy (Trigg and Gluecksohn-Waelsch 1973). Fah-deficient mice treated with NTBC survived, however after several months HCC developed. In the treated mice, kidneys showed signs of tubular apoptosis and regeneration (Grompe et al. 1995). Withdrawal of NTBC therapy in the Fah-deficient mice caused marked alterations of renal and liver architecture, with tubular disarray and regeneration, focal vacuolization, glomerular inflammatory cell infiltration, nuclear size and chromatin variation, induction of apoptosis, along with appearance of dysplastic cells in the liver (Orejuela et al. 2008). Rescue of Fah-deficient mice has been achieved by the introduction of a mutant hydroxyphenylpyruvate dioxygenase (*Hpd*) gene into the homozygous c14CoS mice: the double mutant mice (*Fah*$^{-/-}$/*Hpd*$^{-/-}$) are viable, and their liver and kidney histology is normal (Endo et al. 1997). The recovery of *Hpd* gene, obtained by a recombinant adenovirus, caused early death for massive hepatocyte apoptosis. Similarly, the injection of homogentisic acid (HGA) in the double mutant mice (*Fah*$^{-/-}$/*Hdp*$^{-/-}$) caused death after a few hours for a sudden rise of its oxidative products (Kubo et al. 1997). Remarkably, apoptosis and severe cellular damage were seen in proximal renal tubular epithelial cells, with vast bleeding areas, vacuolation, mitochondrial swelling, lysosomal enlargement at brush borders, along with compaction and degradation of chromatin (Sun et al. 2000). Mechanism of cellular damage has been linked to the accumulation of FAA (the downstream product of HGA), a potent oxidizing agent capable to bind DNA through electrophile response elements and to induce phase II dioxine-inducible genes, resulting in genomic instability (Grompe 2001). Furthermore, induction of apoptosis, through activation of caspase 1 and 3, with consequent release of cytochrome C oxidase and decrease of reduced glutathione, was also attributed to the effect of FAA (Grompe 2001). Remarkably, apoptotic death of renal tubular epithelial cells was prevented when the caspase inhibitor YVAD was given 2 h prior the administration of HGA (Sun et al. 2000). The double mutant mice (*Fah*$^{-/-}$/*Hdp*$^{-/-}$) was also used to evaluate the renal tubular function. The urinary glucose/creatinine ratio was markedly increased after HGA administration, indicating its pathogenic role on renal glucose reabsorption. The caspase inhibitor YVAD, which effectively prevented apoptosis of renal tubular epithelial cells, did not change the urinary excretion of glucose and phosphate, indicating that tubular dysfunction was not directly related to apoptosis. Furthermore, levels of SA (the reduction product of FAA) were markedly increased following the injection of HGA. The authors of this study concluded that the accumulation of FAA can induce apoptosis and cellular death, whereas SA can act more likely as a reversible inhibitor of cellular metabolism, causing renal Fanconi syndrome (Sun et al. 2000). Consistent with this hypothesis, injection of SA to healthy adult rats produced urinary abnormalities similar to those seen in human renal Fanconi syndrome (Roth et al. 1985). Further in vitro studies in isolated renal brush border vesicles of tubule fragments obtained from healthy adult rats demonstrated that SA administration reduced sugar and amino acids uptake (Spencer et al. 1988). Inhibition of Na-H antiport and glycine-Na cotransport, rise of membrane fluidity, and reduction of O_2 consumption by tubular mitochondria were thought to be the mechanisms causing disruption of glucose and amino acids uptake (Roth et al. 1985, Spencer and Roth 1987, Spencer et al. 1988). A disrupted gene expression, induced by the accumulation of toxic intermediates, could also contribute to cellular damage. In the hepatocytes of albino lethal mice and in HT1 patients, the expression of several specific hepatocyte transcription factors (i.e. G6Pase, HNF1, HNF4 and C/EBP) is reduced, and similar changes were hypothesized to occur also in the kidney (Haber et al. 1996). Indeed, targeted disruption of HNF1 in mice results in renal Fanconi syndrome, suggesting that the abnormal expression of transcription factors in HT1 tubular cells of double mutant mice (*Fah*$^{-/-}$/*Hdp*$^{-/-}$) could be involved in the tubular dysfunction (Sun et al. 2000). There is also evidence that changes in the expression of AKT and Bad-BCl survival pathways may contribute to liver and kidney abnormalities (Orejuela et al. 2008).

In summary, studies on animal models of HT1 show that SA mainly impairs tubular solute reabsorption, whereas the oxidizing action of FAA causes tubular cells apoptosis. Furthermore, both intermediates might induce tubular dysfunction affecting gene expression in the kidney.

8.3 NTBC Effects on Renal Dysfunction: Early Effect of NTBC Therapy in HT1

In the original study reporting the efficacy of NTBC in HT1, one patient with renal involvement normalized phosphatemia, hyperaminoaciduria and α1-microglobulin excretion within the first month of treatment (Lindstedt et al. 1992).

Maiorana et al. (2014) investigated the early effect of NTBC on renal tubular function in 5 patients, evaluated before and during the first 2 weeks of therapy with NTBC. At diagnosis, all children manifested signs of renal dysfunction, which included Fanconi syndrome, renal tubular acidosis, and variable degree of proteinuria and hypercalciuria, and most of them received phosphate and/or vitamin D and/or calcium supplementation. Start of NTBC therapy resulted in the normalization of plasma phosphate levels within 1–2 weeks. The tubular reabsorption of phosphate corrected for the glomerular filtration rate (TmP/GFR) became near normal in the first 48 h, while other markers of renal dysfunction showed an improving trend over 2 weeks. At a longer term follow-up, phosphate and vitamin D were withdrawn in all patients.

In one patient, not receiving vitamin or mineral supplementations and closely monitored, NTBC reduced the urinary SA of the 28% the first day after its initiation, and most of abnormal values of renal function normalized within the first days. All biomarkers of renal tubular dysfunction normalized in the first 2 weeks of NTBC therapy. Remarkably, improvement of tubular dysfunction correlated with fall of urinary SA, confirming the strong renal toxicity of SA and the efficacy of NTBC in suppressing its generation (Maiorana et al. 2014). A further patient, diagnosed with a chronic form of HT1, presented severe rickets with bone fractures, loss of walking, overt Fanconi syndrome and dyselectrolytemia (Maiorana and Dionisi-Vici, personal observation). Despite a very rapid disappearance of urinary SA 24 hours after the start of NTBC therapy, her clinical and biochemical response was not as rapid as seen in the previous one (Maiorana et al. 2014). She presented a longer persistence of hypocalcemia, hypophosphatemia and acidosis, which required calcium, phosphate, vitamin D and alkali therapy and an increased dose of NTBC, up to 1.5 mg/kg/day. After 1 month, she restarted walking, dyselectrolytemia disappeared and TRP normalized, with no further need of supplementations. The slow response to NTBC therapy in this patient can be attributed to the combination of at least two events. A severe osteopenia – that determined the so called "hungry bone syndrome", with great and prolonged demand of calcium and phosphate – and a renal loss of NTBC, sustained by a major impairment of tubular function, as shown by detectable amounts of NTBC in the urine. Despite some temporal differences in the response to therapy, 7 days after the start of NTBC both patients showed a positive correlation between plasma NTBC concentrations and TmP/GFR, and between plasma phosphate levels and TmP/GFR (Fig. 8.1). Furthermore, the levels of SA in plasma and in urine were negatively correlated with plasma NTBC concentration and with TmP/GFR.

In summary, all these findings showed that the effect of NTBC therapy on renal tubular cells is very rapid, allowing quick normalization of plasma phosphate levels and TmP/GFR, the latter resulting as the most reliable index of tubular function.

8.4 Long-Term Effect of NTBC Therapy in HT1

In the pre-NTBC era, renal architecture and tubular function have been reported to be abnormal in the majority of HT1 patients, with nephrocalcinosis in about one third (Paradis et al. 1990, Forget et al. 1999). Following the advent of

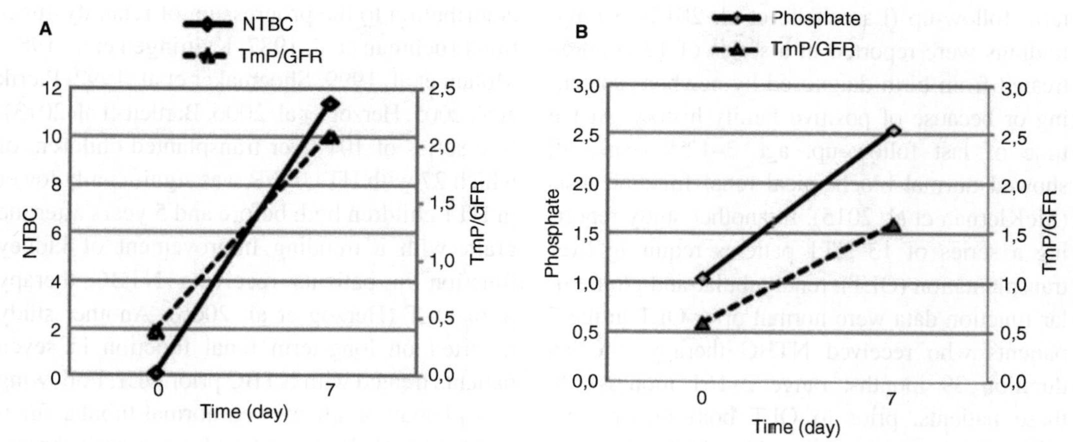

Fig. 8.1 (**a**) Correlation between plasma NTBC concentrations (μmol/l) and TmP/GFR (mg/dl); (**b**) correlation between plasma phosphate levels (mg/dl) and TmP/GFR (mg/dl). Data are expressed as the mean in the two patients at baseline and after 7 days of treatment with NTBC

NTBC, the long-term outcome of renal function in HT1 patients changed substantially, with persistence of tubulopathy in some cases. Pronicka et al. (1996) reported on two patients, with different ages at diagnosis, presenting sub-acute/chronic HT1 with rickets as dominant clinical feature. Both had hypophosphatemia, hypocalcemia, high alkaline phosphatase levels, reduced TRP, hyperaminoaciduria, high fractional urate excretion, bicarbonate loss and elevated α1-microglobulin in urine. Initial treatment with protein-restricted diet, supplemented with phosphate, vitamin D and alkali therapy, did not improve tubular function. When NTBC therapy became available, both patients showed normalization of plasma alkaline phosphatase and fractional urate excretion, and increase of TRP. Aminoaciduria gradually improved, but remained slightly elevated in one patient. Similarly, urinary α1-microglobulin normalized in 18 months in one patient but remained elevated in the other, indicating that some signs of tubulopathy may persist in patients receiving NTBC therapy late in the disease course.

A more recent multicentre study analyzed a series of 45 HT1 patients, all treated from diagnosis with NTBC (initial dose 0.5–2.5 mg/kg/day, maintenance dose 0.5–1.7 mg/kg/day) combined with tyrosine- and phenylalanine-restricted diet. At diagnosis over 80% of total manifested tubulopathy and rickets was the initial symptom in 40% of patients. On NTBC therapy, one third of patients had long-term persistence of biochemical signs of tubulopathy, however without apparent clinical consequences. No glomerular filtration abnormality was observed in this series (Masurel-Paulet et al. 2008). Another study described 21 HT1 children treated with NTBC for 10 years at a dose ranging from 0.6 to 1 mg/kg/day. In this cohort, proteinuria, plasma phosphate and TRP normalized within the first year of NTBC treatment and remained normal in the following 10 years. Some children required temporary phosphate supplementation and all patients remained on vitamin D therapy for a median of 3 years. In one patient, pre-existent nephrocalcinosis did not reverse on NTBC therapy, despite normalization of biochemical indices of tubular function (Santra et al. 2008). In a large retrospective study from Quebec, comparing three genetically homogenous cohorts (those who never received NTBC, those who were treated after 1 month of age and those treated before 1 month), renal failure, hypophosphatemia, glycosuria and generalized aminoaciduria was never recorded at long-term follow-up in patients receiving NTBC therapy. One patient with hypophosphatemic rickets not identified by newborn screening and starting NTBC at the age of 11 months, showed rapid improvement of renal tubular function with complete resolution at a long-term follow-up (Larochelle et al. 2012). Similar findings were reported in a series of 12 children treated from birth diagnosed by newborn screening or because of positive family history. At the time of last follow–up, age 3–12,5 years, all showed normal biochemical renal function tests (McKiernan et al. 2015). In another study reporting a series of 13 HT1 patients requiring liver transplantation (OLT), renal tubular and glomerular function data were normal prior OLT in the 7 patients who received NTBC therapy (median duration 39 months, range 2–161 months). In these patients, prior to OLT both urinary and plasma SA became undetectable with NTBC treatment (Bartlett et al. 2013). El-Karaksy et al. reported on 3 patients with a subacute/chronic presentation with severe rickets who normalized renal tubular function with low dose (0.55 mg/kg) NTBC therapy (El-Karaksy et al. 2010). As clearly shown by a large multicenter study, the rate of clinical complications, including those related to renal function, critically depends on the age at start of NTBC treatment (Mayorandan et al. 2014). The odds ratios for renal tubular dysfunction were 1.2 (95% CI 0.1–7.9) for treatment started at age 1–6 months, 1.9 (95% CI 0.2–14.9) for treatment started at age 7–12 months, and 4.3 (95% CI 0.8–21.6) for treatment started over 12 months, compared to the start of treatment in the neonatal period.

8.5 The Effect of Liver Transplantation on Renal Function. Is NTBC Still Necessary?

The effect of OLT on renal function has been evaluated in several studies. Before the advent of NTBC therapy, resolution of renal dysfunction was seen in nearly 25% of patients undergoing OLT, and persistence of renal abnormalities after OLT was attributed to a pre-existent irreversible renal damage, to severe perioperative alteration in renal perfusion, or to long-term toxicity of immunosuppression (Shoemaker et al. 1992). Residual urinary SA excretion after OLT has been frequently reported, suggesting a potential contribution to the progression of renal dysfunction (Tuchman et al. 1987, Kvittingen et al. 1986, Mohan et al. 1999, Shoemaker et al. 1992, Pierik et al. 2005, Herzog et al. 2006, Bartlett et al. 2013). In a series of 101 liver transplanted children, of which 27 with HT1, GFR was significantly lower in HT1 children both before and 5 years after the graft, with a trending improvement of kidney function in patients receiving NTBC therapy prior OLT (Herzog et al. 2006). Another study reported on long-term renal function in seven patients treated with NTBC prior OLT. Following transplantation, all showed normal tubular function but the GFR was reduced at a stage 2 chronic kidney disease (cGFR 60–89 ml/min/1.73m^2) with urinary and plasma SA still detectable, however to a lesser extent (15–20×) than pre-NTBC therapy (Bartlett et al. 2013). A more recent study by the same group retrospectively analyzed kidney function on 13 HT1 patients undergone OLT, 6 before and 7 after the advent of NTBC (Bartlett et al. 2014). The prior use of NTBC therapy resulted in a preservation of subsequent renal function. Although before OLT the GFR was similar in both groups, NTBC prevented its early decline after OLT. Indeed, proteinuria normalized in NTBC treated patients, and remained elevated in the untreated ones. Similarly, hypertension was more common and severe in those not treated with NTBC, whereas TRP was normal or near normal in both groups pre-OLT and post-OLT (Bartlett et al. 2014).

Although it is evident that NTBC therapy prior to OLT has a long-term protective effect on renal function, there are insufficient data to establish whether the persistence of detectable SA after OLT may lead to a progressive renal damage and if NTBC therapy after OLT should be maintained.

8.6 Conclusions

NTBC therapy is highly effective in improving renal function both at short and long-term. However, its efficacy critically depends on the age at start of treatment with normal outcome in patients diagnosed at birth by newborn screening.

References

Arnon R, Annunziato R, Miloh T, Wasserstein M, Sogawa H, Wilson M, Suchy F, Kerkar N (2011) Liver transplantation for hereditary tyrosinemia type I: analysis of the UNOS database. Pediatr Transplant 15:400–405

Bartlett DC, Preece MA, Holme E, Lloyd C, Newsome PN, McKiernan PJ (2013) Plasma succinylacetone is persistently raised after liver transplantation in tyrosinaemia type 1. J Inherit Metab Dis 36:15–20

Bartlett DC, Lloyd C, McKiernan PJ, Newsome PN (2014) Early nitisinone treatment reduces the need for liver transplantation in children with tyrosinaemia type 1 and improves post-transplant renal function. J Inherit Metab Dis 37:745–752

Chakrapani A, Gissen P, McKiernan P (2012) Disorders of tyrosine metabolism. In: Saudubray J-M, van den Berghe G, Walter JH (eds) Inborn metabolic diseases, 5th edn. Springer, Heidelberg, pp 275–276. Chapter 18

Couce ML, Dalmau J, del Toro M, Pintos-Morell G, Aldámiz-Echevarría L, Spanish Working Group on Tyrosinemia type 1 (2011) Tyrosinemia type 1 in Spain: mutational analysis, treatment and long-term outcome. Ped Intern 53:985–989

de Laet C, Dionisi-Vici C, Leonard J, McKiernan P, Mitchell G, Monti L, de Baulny H, Pintos-Morell G, Spiekerkötter U (2013) Recommendations for the management of tyrosinaemia type 1. Orphanet J Rare Dis 8:8

El-Karaksy H, Rashed M, El-Sayed R, El-Raziky M, El-Koofy N, El-Hawary M, Al-Dirbashi O (2010) Clinical practice. NTBC therapy for tyrosinemia type 1: how much is enough? Eur J Pediatr 169(6):689–693

Endo F, Kubo S, Awata H, Kiwaki K, Katoh H, Kanegae Y, Saito I, Miyazaki J, Yamamoto T, Jakobs C, Hattori S, Matsuda I (1997) Complete rescue of lethal albino c14CoS mice by null mutation of 4-hydroxyphenylpyruvate dioxygense and induction of apoptosis of hepatocytes in these mice by an in vivo retrieval of the tyrosine catabolic pathway. J Biol Chem 272:24426–24432

Forget S, Patriquin HB, Dubois J, Lafortune M, Merouani A, Paradis K, Russo P (1999) The kidney in children with tyrosniemia: sonographic, CT and biochemical findings. Pediatr Radiol 29:104–108

Gluecksohn-Waelsch S (1979) Genetic control of morphogenetic and biochemical differentiation: lethal albino deletion in the mouse. Cell 16:225–237

Grompe M (2001) The pathophysiology and treatment of hereditary Tyrosinemia type 1. Semin Liver Dis 21:563–571

Grompe M, Al Dhalimy M, Finegold M, Ou CN, Burlingame T, Kennaway NG, Soriano P (1993) Loss of fumarylacetoacetate hydrolase is responsible for the neonatal hepatic dysfunction phenotype of lethal albino mice. Genes Dev 7:2298–2307

Grompe M, Lindstedt S, Al Dhalimy M, Kennaway NG, Papaconstantinou J, Torres-Ramos CA, Ou CA, Finegold M (1995) Pharmacological correction of neonatal lethal hepatic dysfunction in a murine model of hereditary tyrosinemia type I. Nat Genet 10:453–460

Grompe M, Overturf K, Al-Dhalimy M, Finegold M (1998) Therapeutic trials in the murine model of hereditary tyrosinemia type I: a progress report. J Inherit Metab Dis 21:518–531

Haber BA, Chuang E, Lee V, Taub R (1996) Variable gene expression with human tyrosinemia type 1 liver may reflect region-specific dysplasia. Hepatology 24:65–71

Herzog D, Martin S, Turpin S, Alvarez F (2006) Normal glomerular filtration rate in long-term follow-up of children after orthotopic liver transplantation. Transplantation 8:672–677

Jorquera R, Tanguay RM (1997) The mutagenicity of the tyrosine metabolite, fumarylacetoacetate, is enhanced by glutathione depletion. Biochem Biophys Res Commun 232:42–48

Jorquera R, Tanguay RM (2001) Fumarylacetoacetate, the metabolite accumulating in hereditary tyrosinemia, activates the ERK pathway and induces mitotic abnormalities and genomic instability. Hum Mol Genet 10:1741–1752

Kubo S, Kiwaki K, Awata H, Katoh H, Kanegae Y, Saito I, Yamamoto T, Miyazaki J, Matsuda I, Endo F (1997) In vivo correction with recominant adenovirus of 4-hydroxyphenylpyruvate acid dioxygenase deficiencies in train III mice. Hum Gene Ther 8:65–71

Kvittingen EA, Jellum E, Stokke O, Flatmark A, Bergan A, Sødal G, Halvorsen S, Schrumpf E, Gjone E (1986) Liver transplantation in a 23-year-old tyrosinaemia patient: effects on the renal tubular dysfunction. J Inherit Metab Dis 9:216–224

Larochelle J, Alvarez F, Bussières JF, Chevalier I, Dallaire L, Dubois J, Faucher F, Fenyves D, Goodyer P, Grenier A, Holme E, Laframboise R, Lambert M, Lindstedt S, Maranda B, Melançon S, Merouani A, Mitchell J, Parizeault G, Pelletier L, Phan V, Rinaldo P, Scott CR, Scriver C, Mitchell GA (2012) Effect of nitisinone (NITISINONE) treatment on the clinical course of hepatorenal tyrosinemia in Québec. Mol Genet Metab 107:49–54

Lindstedt S, Holme E, Lock EA, Hjalmarson O, Strandvik B (1992) Treatment of hereditary tyrosinaemia type I by inhibition of 4-hydroxyphenylpyruvate dioxygenase. Lancet 340:813–817

Maiorana A, Malamisura M, Emma F, Boenzi S, Di Ciommo VM, Dionisi-Vici C (2014) Early effect of NTBC on renal tubular dysfunction in hereditar tyrosinemia type 1. Mol Genet Metab 113:188–193

Manabe S, Sassa S, Kappas A (1985) Hereditary tyrosinemia. Formation of succinylacetone-amino acid adducts. J Exp Med 162:1060–1074

Masurel-Paulet A, Poggi-Bach J, Rolland MO, Bernard O, Guffon N, Dobbelaere D, Sarles J, de Baulny HO, Touati G (2008) NTBC treatment in tyrosinaemia type I: long-term outcome in French patients. J Inherit Metab Dis 31:81–87

Mayorandan S, Meyer U, Gokcay G, Segarra NG, de Baulny HO, van Spronsen F, Zeman J, de Laet C,

Spiekerkoetter U, Thimm E, Maiorana A, Dionisi-Vici C, Moeslinger D, Brunner-Krainz M, Lotz-Havla AS, Cocho de Juan JA, Couce Pico ML, Santer R, Scholl-Bürgi S, Mandel H, Bliksrud YT, Freisinger P, Aldamiz-Echevarria LJ, Hochuli M, Gautschi M, Endig J, Jordan J, McKiernan P, Ernst S, Morlot S, Vogel A, Sander J, Das AM (2014) Cross-sectional study of 168 patients with hepatorenal tyrosinaemia and implications for clinical practice. Orphanet J Rare Dis 9:107

McKiernan PJ, Preece MA, Chakrapani A (2015) Outcome of children with hereditary tyrosinaemia following newborn screening. Arch Dis Child 100:738–741

Mohan N, McKiernan P, Preece MA, Green A, Buckels J, Mayer AD, Kelly DA (1999) Indications and outcome of liver transplantation in tyrosinaemia type 1. Eur J Pediatr 158(Suppl 2):S49–S54

Nakamura K, Tanaka Y, Mitsubuchi H, Endo F (2007) Animal models of Tyrosinemia. J Nutr 137:1556S–1560S

Orejuela D, Jorquera R, Bergeron A, Finegold MJ, Tanguay RM (2008) Hepatic stress in hereditary tyrosinemia type 1 (HT1) activates the AKT survival pathway in the fah−/− knockout mice model. J Hepatol 48(2):308–317

Paradis K, Weber A, Seidman EG, Larochelle J, Garel L, Lenaerts C, Roy CC (1990) Liver transplantation for he reditary tyrosinemia: the Quebec experience. Am J Hum Genet 47:338–342

Pierik LJ, van Spronsen FJ, Bijleveld CM, van Dael CM (2005) Renal function in tyrosinaemia type I after liver transplantation: a long-term follow-up. J Inherit Metab Dis 28(6):871–876

Pronicka E, Rowinska E, Bentkowski Z, Zawadzki J, Holme E, Lindstedt S (1996) Treatment of two children with hereditary tyrosinaemia type I and long-standing renal disease with a 4 hydroxyphenylpyruvate dioxygenase inhibitor (NTBC). J Inherit Metab Dis 19:234–238

Roth K, Spencer PD, Higgins ES, Spencer RF (1985) Effects of succinylacetone on methyl α-D-glucoside uptake by the rat renal tubule. Biochim Biophys Acta 820:140–146

Santra S, Preece MA, Hulton SA, McKiernan PJ (2008) Renal tubular function in children with tyrosinaemia type I treated with nitisinone. J Inherit Metab Dis 31:399–402

Shoemaker LR, Strife CF, Balistreri WF, Ryckman FC (1992) Rapid improvement in the renal tubular dysfunction associated with tyrosinemia following hepatic replacement. Pediatrics 89(2):251–255

Spencer PD, Roth KS (1987) Effects of succinylacetone on amino acid uptake in the rat kidney. Biochem Med Metab Biol 37:101–109

Spencer PD, Medow MS, Moses LC, Roth KS (1988) Effects of succinylacetone on the uptake of sugars and amino acids by brush border vesicles. Kidney Int 34(5):671–677

Sun MS, Hattori S, Kubo S, Awata H, Matsuda I, Endo F (2000) A mouse model of renal tubular Injury of tyrosinemia type 1: development of de Toni Fanconi syndrome and apoptosis of renal tubular cells in Fah/Hpd double mutant mice. J Am Soc Nephrol 11:291–300

Trigg MJ, Gluecksohn-Waelsch S (1973) Ultrastructural basis of biochemical effects in a series of lethal alleles in the mouse. J Cell Biol 58:549–563

Tuchman M, Freese DK, Sharp HL, Ramnaraine ML, Ascher N, Bloomer JR (1987) Contribution of extrahepatic tissues to biochemical abnormalities in hereditary tyrosinemia type I: study of three patients after liver transplantation. J Pediatr 110(3):399–403

Liver Cancer in Tyrosinemia Type 1

9

Willem G. van Ginkel, Jan P. Pennings, and Francjan J. van Spronsen

Abstract

Hereditary Tyrosinemia type I (HT1) is clinically mainly characterised by severe liver disease. Most patients present in their first months of life with liver failure, but others can present later with issues of compensated cirrhosis, renal tubulopathy or acute intermittent porphyria. If patients survive the acute phase with liver failure or if they present later with compensated cirrhosis, they often develop hepatocellular carcinoma early but also later in life. The course of the disease changed after the introduction of 2-(2 nitro-4-3 trifluoro-methylbenzoyl)-1, 3-cyclohexanedione (NTBC), which blocks the tyrosine degradation pathway at an earlier step. Therefore, the toxic products did not accumulate anymore and all clinical problems resolved. However, the risk (although clearly decreased) for developing liver cancer remained, especially if NTBC treatment is initiated late, a slow decrease of the tumor marker α-fetoprotein is seen or if the α-fetoprotein concentrations remain just above the normal range. A rise of α-fetoprotein in these HT1 patients is more or less pathognomonic for liver cancer. Although hepatoblastoma development occurs in HT1 patients, most HT1 patients develop hepatocellular carcinoma (HCC) or a mixed type of carcinoma consisting of HCC and hepatoblastoma. Due to the small risk of liver cancer development, screening for liver cancer (especially HCC) is still recommended in HT1 patients using regular measures of α-fetoprotein and imaging. Ultrasound is mostly the modality of choice for surveillance, because it is widely available, it does not use radiation and is noninvasive. When a suspicious lesion is present, the higher sensitivity of MRI could be used for characterization and staging of

W.G. van Ginkel • F.J. van Spronsen (✉)
Beatrix Children's Hospital, Division of Metabolic Diseases, University of Groningen, University Medical Center of Groningen, HPC CA33, Antwoordnummer 333, 9700 VB Groningen, The Netherlands
e-mail: f.j.van.spronsen@umcg.nl

J.P. Pennings
Department of Radiology, University of Groningen, University Medical Center of Groningen, Groningen, The Netherlands

© Springer International Publishing AG 2017

R.M. Tanguay (ed.), *Hereditary Tyrosinemia*, Advances in Experimental Medicine and Biology 959,
DOI 10.1007/978-3-319-55780-9_9

lesions. At this moment, no HCC development in pre-symptomatically treated patients is reported. These different situations could possibly indicate that NTBC can prevent the start of the development of HCC when initiated early, but can't stop the development of HCC if it is prescribed at a later stage, stressing the importance of early diagnosis.

Keywords
Tyrosinemia • Liver cancer • Hepatocellular carcinoma • HCC • AFP • Ultrasound • MRI

Abbreviations

AFP	Alpha feto protein
CT	Computed tomography
FAA	Fumarylacetoacetate
FAH	Fumarylacetoacetate hydrolase
HCC	Hepatocellular carcinoma
HT1	Hereditary tyrosinemia type 1
MRI	Magnetic resonance imaging
NTBC	2-(2 nitro-4-3 trifluoro-methylbenzoyl)-1, 3-cyclohexanedione

9.1 Introduction

Hereditary Tyrosinmia type 1 (HT1) is caused by an enzymatic defect in the tyrosine degradation pathway. Due to a genetic defect in the enzyme fumarylacetoacetate hydrolase (FAH), its substrate fumarylacetoacetate (FAA) could not be metabolized into fumarate and acetoacetate. In this way, FAA and associated metabolites such as its precursor maleylacetoacetate and alternative degradation products succinylacetoacetate and succinylacetone are accumulating. The accumulation of these toxic products especially causes severe liver dysfunction. This liver dysfuntion secondarily causes a decrease in various enzyme systems including 4-hydroxy phenylpyruvate dioxygenase, phenylalanine hydroxylase and methionine adenosyltransferase leading to high blood concentrations of phenylalanine, tyrosine and methionine in clinically presenting and dietary treated HT1 patients. Treatment with 2-(2 nitro-4-3 trifluoromethylbenzoyl)-1, 3-cyclohexanedione (NTBC) resolved most of the clinical problems by placing a new metabolic block upstream of the primary enzymatic defect. However, treatment with NTBC leads to increased blood tyrosine and phenylalanine concentrations as these products are just above the new metabolic block, making dietary restriction of both large neutral amino acids necessary.

9.2 Epidemiology of Hepatocellular Carcinoma

In various countries, HT1 is diagnosed presymptomatically with newborn screening by using succinylacetone in blood spot. In countries without a screening program or with a program using tyrosine as the biomarker in newborn screening, HT1 patients can present clinically with heterogeneous symptoms mainly affecting the liver, kidney and peripheral nerves or otherwise pre-symptomatically due to affected siblings (Hadzic and Vara 2015; van Spronsen et al. 1994). Three different groups of HT1 patients are identified, based on the timing of diagnosis: very early (onset of symptoms <2 months), early (2–6 months) and late presenting form (> 6 months) (van Spronsen et al. 1994). The first group that presents at the hospital within 2 months of life is mostly suffering from acute liver failure. The prognosis of these patients is rather poor, with a 2 year survival probability of

29%. In the patients presenting between 2 and 6 months and after 6 months of life, renal tubulopathy and porphyria like syndrome are also seen next to the liver problems. The later the patients present, the better their survival probability is as they mostly survive the acute initial phase. However, the liver in all these surviving patients may show cirrhosis already or proceed towards cirrhosis and therefore these patients have a high risk for developing hepatocellular carcinoma (HCC) later on (Russo et al. 2001; van Spronsen et al. 1989, 1994, 2005).

Worldwide, HCC is the most common primary hepatic malignancy, mostly occurring after the onset chronic liver disease. The world leading causes of HCC in adults are chronic Hepatitis B and C infection and alcohol abuses. These risk factors lead to the formation and progression of cirrhosis, which is present in 80–90% of patients with HCC (El-Serag 2011; Forner et al. 2012). Overall, the 5 year cumulative risk for the development of HCC in patients with cirrhosis ranges between 5% and 30% depending on the cause, region or ethnic group and stage of cirrhosis (El-Serag 2012).

When only treated with a tyrosine and phenylalanine restricted diet (before introduction of NTBC) (almost) all HT1 patients who survived the first critical period of life developed cirrhosis and in this group of patients, the incidence of HCC was really high, probably up to 37% (van Spronsen et al. 1994; Weinberg et al. 1976). This incidence of HCC in HT1 patients is higher than in other cirrhotic liver diseases and considerably above the incidence of HCC in cirrhotic liver of adults (El-Serag 2012; Weinberg et al. 1976). Due to the high incidence of HCC and its severity, HCC development is the main cause of death of HT1 patients who survived the critical first period (Weinberg et al. 1976). The peak incidence for tumor development in HT1 patients is between 4 and 5 years old, however, the variation is large with malignant transformation after long periods of time (Weinberg et al. 1976).

Most of the malignancies in HT1 patiënts are HCC, however, development of hepatoblastoma is reported at least once in a HT1 patient (Nobili et al. 2010), and next to this one of our patients had a mixed type of HCC and hepatoblastoma. Hepatoblastoma is common among young children and has a favourable outcome compared to HCC. Therefore, a neoplastic mass in HT1 patients is likely, but not definitely HCC. It is important to distinguish between both, because of the difference in treatment and outcome.

The drug or NTBC was released in 1992 (Lindstedt et al. 1992). By placing a metabolic block upstream from the primary enzymatic defect, accumulation of the toxic metabolites was prevented and clinical symptoms, such as liver failure and renal involvement, resolved. Next to this, the incidence of HCC decreased tremendously (Holme and Lindstedt 1998; Larochelle et al. 2012). However, patients receiving NTBC are still at risk for developing HCC, especially when initiation of treatment with NTBC was late due to delayed diagnosis or unavailability of NTBC, a slow decrease of the tumor marker α-fetoprotein (AFP) or an AFP level that remains just above the normal range (<10 ug/L) (Koelink et al. 2006; van Ginkel et al. 2015; van Spronsen et al. 2005). So far, no HCC development in presymptomatically treated patients is reported. These different situations could possibly indicate that NTBC can prevent the start of the development of HCC if initiated early, but can't stop the development of HCC when it is prescribed at later stage, stressing the importance of early diagnosis.

9.3 Etiology of Hepatocellular Carcinoma

Due to the enzyme deficiency in HT1 (FAH deficiency), the substrate for FAA, accumulates. This FAA is partly converted into succinylacetoacetate and afterwards in succinylacetone. The accumulating FAA in the cells where it is generated (hepatocytes and cells in the proximal tubule of the kidneys) causes oxidative stress by reacting with glutathione and sulfhydryl groups of proteins. Especially this FAA, but maybe also its precursor, maleylacetoacetate, has shown to be mutagenic and to cause chromosomal instability as well as lead to cell cycle arrest and apoptosis, leading first to cirrhosis in the liver and to HCC formation later on (Bergeron et al. 2006; Jorquera and Tanguay 2001).

Microscopic features in the liver of HT1 patients will reveal micronodulair cirrhosis that will develop to macronodulair cirrhosis (Dehner et al. 1989). Therefore, the chronic hepatic phase usually shows mixed micro- and macro-nodular cirrhosis with minimal ductular proliferation and mild lymphoplasmacytic infiltrates within the fibrous septa. Varying degrees of steatosis are also seen and may show variation within a nodule and/or between nodules. The most significant histological feature is again the foci of dysplasia (large and small cell types) and/or HCC (Schady et al. 2015; Weinberg et al. 1976).

9.4 Diagnosis of Hepatocellular Carcinoma

9.4.1 Clinical Presentation

In theory, HCC can give complaints ranging from fatigue due to anemia to abdominal distress due to the increasing mass or a bleeding in the liver due to tumour rupture. However, in general, patients are detected before clinical symptoms arise due to several screening mechanisms such as laboratory investigation and imaging, which are described below.

9.4.2 Laboratory Parameters

The main laboratory parameter suggestive of HCC development is AFP. AFP is a glycoprotein that typically is very high just after birth. However, AFP concentrations in HT1 patients are at birth already much higher than the AFP concentrations of healthy infants (Hostetter et al. 1983). These AFP concentrations stay higher than normal during infancy in untreated HT1 patients due to the liver regeneration that already starts during fetal life (Hostetter et al. 1983). So far, high AFP levels at diagnosis have not been related to any prognostic value for initial recovery of acute physical complaints or development of HCC later on.

After initiation of adequate treatment, consisting of NTBC and tyrosine and restricted diet, hepatotoxic metabolites should not accumulate anymore and AFP levels should decrease more or less exponentially to normal levels (<10 μg/L) within 1 year like it is shown in the figure published by Koelink et al. in 1996 (de Laet et al. 2013; Koelink et al. 2006) (Fig. 9.1).

The tumormarker AFP has been used as a marker for HCC for many years. AFP concentrations could rise in various circumstances, for example in chronic hepatitis with reactivation, but without HCC. On the other hand it could be normal when there is only a small HCC (Paul et al. 2007). Furthermoref, in other diseases with a high risk of developing HCC (e.g., hepatitis B and C), up to 44% of patients diagnosed with HCC show no clear increase in AFP levels (Beale et al. 2008; Giannelli et al. 2007). Due to this, the reported sensitivity of AFP is variable across different studies.

However, in HT1 patients, AFP was considered to be a reliable marker for liver cancer development. Until recently, all reported HT1 patients with proven HCC and hepatoblastoma had exhibited a well-defined rise in AFP levels and a clear lesion at imaging (de Laet et al. 2013; Koelink et al. 2006; Nobili et al. 2010). But recently, a case study showed that HCC in HT1 patients is not always accomplished with the rise in AFP, like it is shown in Fig. 9.2 published by van Ginkel et al. in 2015 (van Ginkel et al. 2015) (Fig. 9.2). Therefore the search for other reliable markers is necessary. At this moment, no other markers have been investigated extensively, although promising results were seen with Lectin-reactive alfa-fetoprotein (Baumann et al. 2006). Unfortunately, this did not get follow-up so far.

9.4.3 Imaging Modalities

Many patients with tyrosinaemia show signs of cirrhosis on imaging. In severe cirrhosis the size of the liver can change with (1) hypertrophy of the caudate lobe and lateral segments of the left lobe (segment 2 and 3) and (2) atrophy of the posterior segments of the right lobe (segment 6 and 7). Other findings are nodularity of the surface and heterogeneity or nodular aspect of the

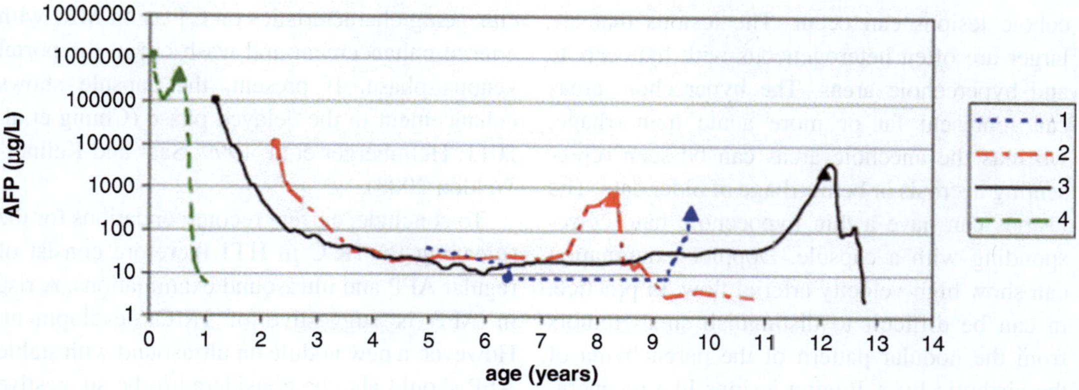

Fig. 9.1 Tyrosinemia type 1 patients with proven liver cancer, all showing a clear increase in AFP concentrations. Year 0 represents birth, ● start NTBC treatment; patient 4 started with NTBC treatment before the first AFP concentration was measured. ▲ diagnosis of liver cancer.

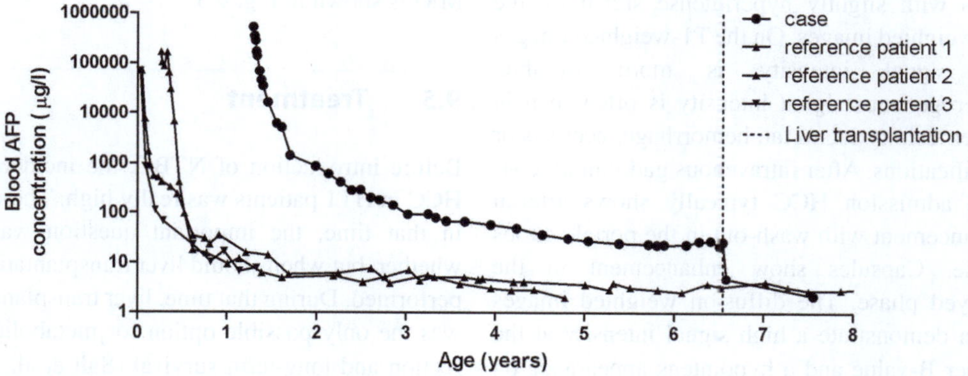

Fig. 9.2 Course of AFP from diagnosis until 1 year after liver transplantation (histologically proven HCC). After an initial decline in AFP, AFP levels stabilised around 25 μg/L. After liver transplantation AFP further decreased to normal values. This is in contrast with the reference patients who almost immediately reached normal values (0–10 μg/L)

parenchyma. Signs of portal hypertension like splenomegaly, portosystemic collaterals and ascites can be present (Dubois et al. 1996).

The three major goals of imaging in patients with Tyrosinaemia is the screening for lesions; the characterization of the lesion, and a change in the characterization or the volume. Ultrasound is mostly the modality of choice for surveillance, because it is widely available, it does not use radiation and is noninvasive. Ultrasound has a good sensitivity and specificity for the detection of HCC in cirrhotic patients with a sensitivity of 60–80% and a specificity of 45–94 % particularly depending on the experience of the performer. For the purpose of surveillance, Computed Tomography (CT) is less useful due to the radiation exposure. Magnetic Resonance Imaging (MRI) has a high sensitivity in the detection of HCC (89–100%), especially for smaller lesions between 1 and 2 cm. However, the higher sensitivity of MRI is mainly used for characterization and staging of lesions that are suspicious for HCC, caused by the higher costs and lower availability of MRI (Ayuso et al. 2012; Bolondi 2003; Burrel et al. 2003).

The sonographic appearance of HCC is nonspecific. Small lesions are often hypoechoic to the liver parenchyma, but isoechoic or hyperechoic lesions can occur. The lesions that are larger are often heterogeneous with hypoechoic and hyperechoic areas. The hyperechoic areas can represent fat or more acute hemorrhage, whereas the anechoic areas can be seen representing necrosis or hemorrhage of older date. The lesions can have a thin hypoechoic halo corresponding with a capsule. Doppler examination can show high-velocity arterial flow. In practice, in can be difficult to distinguish small tumors from the nodular pattern of the parenchyma of the cirrhotic liver. Benign lesions like regenerative and dysplastic nodules are frequently present in cirrhotic livers and can be indistinguishable from small HCC (Ayuso et al. 2012; Chung et al. 2011; Helmberger et al. 1999).

On MRI the typical appearance of HCC is a mass with slightly hyperintense signal on the T2-weighted images. On the T1-weighted images the signal intensity is more variable. Heterogeneous signal intensity is often seen in larger lesions due to fat, hemorrhage, necrosis or calcifications. After intravenous gadolinium contrast admission HCC typically shows arterial enhancement with wash-out in the portal venous phase. Capsules show enhancement in the delayed phase. The diffusion weighted images often demonstrate a high signal intensity at the higher B-value and a hypointens appearance on the apparent diffusion coefficient corresponding to diffusion restriction. Regenerative nodules are isointense or slightly hyperintense on the T1 weighted images and isointense or hypointense on the T2 weighted images. A distinction between HCC and regenerative nodules can be made, as regenerative nodules do not show arterial enhancement and portal venous wash-out. Because dysplastic nodules can be low or high grade they a show a variable appearance. The low grade resembles regenerative nodules while the high grade resembles HCC (Chung et al. 2011; Helmberger et al. 1999; Saar and Kellner-Weldon 2008).

If using CT, on unenhanced CT HCC is a hypoattenuating mass, which can show areas with different attenuation as a result of fat, hemorrhage, necrosis and calcification. After contrast administration HCC typically shows the same enhancing characteristics on CT as on MRI with arterial enhancement and wash-out in the portal venous phase. If present, the capsule shows enhancement in the delayed phase (Chung et al. 2011; Helmberger et al. 1999; Saar and Kellner-Weldon 2008).

To conclude, current recommendations for the screening for HCC in HT1 therefore consist of regular AFP and ultrasound examinations. A rise in AFP is suggestive of HCC development. However, a new nodule on ultrasound with stable AFP should also be considered to be suggestive of HCC (van Ginkel et al. 2015). If nodules are present in the liver, further imaging preferably with MRI should be done. An example of this surveillance and staging protocol with the radiological appearance of HCC on ultrasound and MRI is shown in Fig. 9.3.

9.5 Treatment

Before introduction of NTBC, the incidence of HCC in HT1 patients was really high. Therefore, in that time, the important question was not whether, but when should liver transplantation be performed. During that time, liver transplantation was the only possible option for metabolic correction and long-term survival (Salt et al. 1992; van Spronsen et al. 1989, 1995; Weinberg et al. 1976). This changed after the introduction of NTBC. Due to NTBC, the incidence of HCC diminished and, therefore, the need for transplantation in HT1 patients decreased as well. Due to the required major abdominal surgery, postoperative complications and life long immunosuppression with risks for secondary malignant diseases, liver transplantation is currently in most centers only an option in patients not responding on NTBC or who develop HCC despite treatment with NTBC. However, as NTBC is costly, some countries seem to prefer liver transplantation that in the long run may be less expensive (personal communication). Therefore, studies that compare the outcome of patients treated with NTBC versus liver transplantation are needed.

Pediatric liver transplantation is done under various circumstances, with biliary atresia still

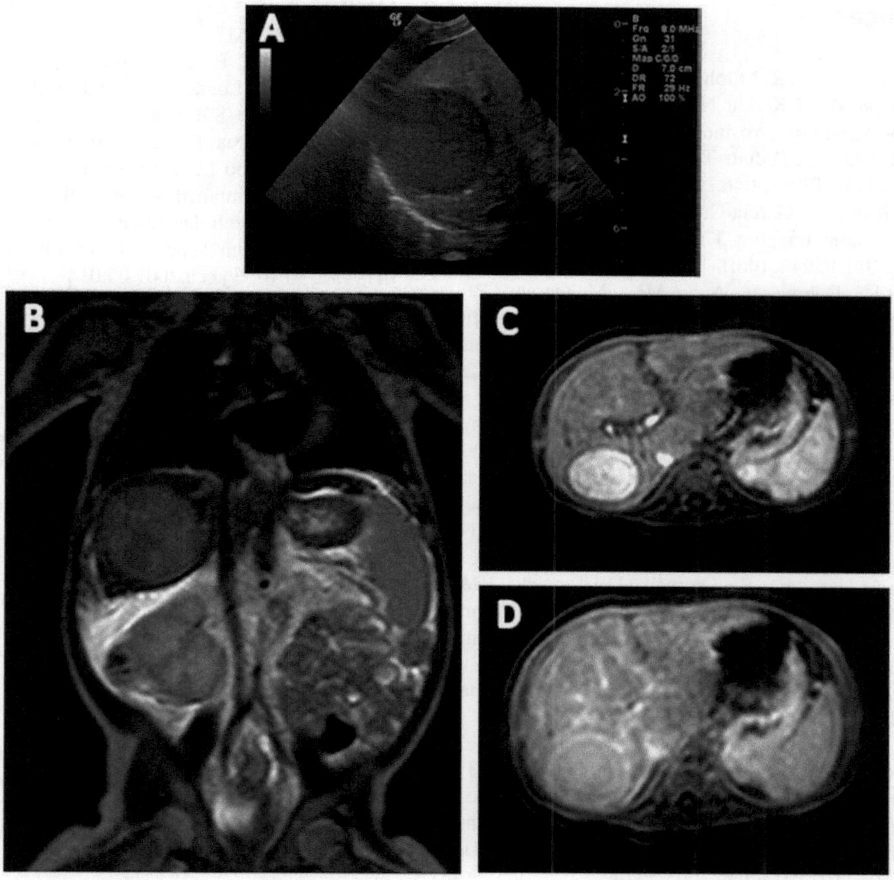

Fig. 9.3 Imaging features from a case presenting with HCC, images provided by courtesy of dr. S. McGuirk from the Birmingham Children's Hospital, UK. (**a**) US demonstrates a hypoechoic lesion in the right liver lobe (histologically proven HCC). (**b**) On the coronal T2 weighted the lesions demonstrates a hyperintens signal. On the T1weighted images the lesion shows arterial enhancement (**c**) with wash-out in the portal venous phase (**d**)

being the main indication for liver transplantation in children. Since the first liver transplantation in 1953 this treatment option has changed to one with an excellent long term outcome, having high short and long term survival rates (Hackl et al. 2015). Liver graft survival has improved as well to 80% at 5 years and an estimated half-life of 13 years (Yazigi 2013).

In HT1 patients, recent articles about the outcome after liver transplantation are limited, as most of the manuscripts are published before 2000, mostly describing patients before introduction of NTBC. At that time, the survival after transplantation was relatively good with survival rates around 90% (Paradis 1996; Weinberg et al. 1976), although post-operative complications such as non-functioning of the graft due to artery thrombosis and graft rejection were seen (Weinberg et al. 1976). More recently, Arnon et al. showed in a large cohort of patients that HT1 patients have an excellent outcome after transplantation with a 5 year survival rate over 90% (Arnon et al. 2011). However, a clear and continuous decrease in the need for liver transplantation has been seen after introduction of NTBC (Arnon et al. 2011).

References

Arnon R, Annunziato R, Miloh T, Wasserstein M, Sogawa H, Wilson M,…(Kerkar N (2011) Liver transplantation for hereditary tyrosinemia type I: analysis of the UNOS database. Pediatr Transplant 15(4): 400–405. doi:10.1111/j.1399-3046.2011.01497.x [doi]

Ayuso C, Rimola J, Garcia-Criado A (2012) Imaging of HCC. Abdom Imaging 37(2):215–230. doi:10.1007/s00261-011-9794-x [doi]

Baumann U, Duhme V, Auth MK, McKiernan PJ, Holme E (2006) Lectin-reactive alpha-fetoprotein in patients with tyrosinemia type I and hepatocellular carcinoma. J Pediatr Gastroenterol Nutr 43(1):77–82. doi:10.1097/01.mpg.0000228112.29359.f8

Beale G, Chattopadhyay D, Gray J, Stewart S, Hudson M, Day C,… Reeves H (2008) AFP, PIVKAII, GP3, SCCA-1 and follisatin as surveillance biomarkers for hepatocellular cancer in non-alcoholic and alcoholic fatty liver disease. BMC Cancer 8: 200-2407-8-200. doi:10.1186/1471-2407-8-200; 10.1186/1471-2407-8-200

Bergeron A, Jorquera R, Orejuela D, Tanguay RM (2006) Involvement of endoplasmic reticulum stress in hereditary tyrosinemia type I. J Biol Chem 281(9):5329–5334

Bolondi L (2003) Screening for hepatocellular carcinoma in cirrhosis. J Hepatol 39(6):1076–1084

Burrel M, Llovet JM, Ayuso C, Iglesias C, Sala M, Miquel R,… Barcelona Clinic Liver Cancer Group (2003) MRI angiography is superior to helical CT for detection of HCC prior to liver transplantation: an explant correlation. Hepatology (Baltimore, Md), 38(4), 1034–1042. 10.1053/jhep.2003.50409

Chung EM, Lattin GE Jr, Cube R, Lewis RB, Marichal-Hernandez C, Shawhan R, Conran RM (2011) From the archives of the AFIP: pediatric liver masses: radiologic-pathologic correlation. part 2. malignant tumors. Radiographics: A Rev Publ Radiol Soc N Am Inc 31(2):483–507. doi:10.1148/rg.312105201

de Laet C, Dionisi-Vici C, Leonard JV, McKiernan P, Mitchell G, Monti L,…, Spiekerkotter U (2013) Recommendations for the management of tyrosinaemia type 1. Orphanet J Rare Dis 8: 8–1172–8-8. doi:10.1186/1750-1172-8-8

Dehner LP, Snover DC, Sharp HL, Ascher N, Nakhleh R, Day DL (1989) Hereditary tyrosinemia type I (chronic form): pathologic findings in the liver. Hum Pathol 20(2):149–158

Dubois J, Garel L, Patriquin H, Paradis K, Forget S, Filiatrault D et al (1996) Imaging features of type 1 hereditary tyrosinemia: a review of 30 patients. Pediatr Radiol 26(12):845–851

El-Serag HB (2011) Hepatocellular carcinoma. N Engl J Med 365(12):1118–1127. doi:10.1056/NEJMra1001683 [doi]

El-Serag HB (2012) Epidemiology of viral hepatitis and hepatocellular carcinoma. Gastroenterology 142(6):1264–1273.e1. doi:10.1053/j.gastro.2011.12.061

Forner A, Llovet JM, Bruix J (2012) Hepatocellular carcinoma. Lancet (London, England) 379(9822):1245–1255. doi:10.1016/S0140-6736(11)61347-0

Giannelli G, Fransvea E, Trerotoli P, Beaugrand M, Marinosci F, Lupo L,… Antonaci S (2007) Clinical validation of combined serological biomarkers for improved hepatocellular carcinoma diagnosis in 961 patients. Clin Chim Acta; Int J Clin Chem, 383(1–2): 147–152. 10.1016/j.cca.2007.05.014

Hackl C, Schlitt HJ, Melter M, Knoppke B, Loss M (2015) Current developments in pediatric liver transplantation. World J Hepatol 7(11):1509–1520. doi:10.4254/wjh.v7.i11.1509

Hadzic N, Vara R (2015) Neonatal screening for hereditary tyrosinaemia: are we there yet? Arch Dis Child 100(8):720–721. doi:10.1136/archdischild-2015-308265

Helmberger TK, Ros PR, Mergo PJ, Tomczak R, Reiser MF (1999) Pediatric liver neoplasms: a radiologic-pathologic correlation. Eur Radiol 9(7):1339–1347. 90091339.330 [pii]

Holme E, Lindstedt S (1998) Tyrosinaemia type I and NTBC (2-(2-nitro-4-trifluoromethylbenzoyl)-1,3-cyclohexanedione). J Inherit Metab Dis 21(5):507–517

Hostetter MK, Levy HL, Winter HS, Knight GJ, Haddow JE (1983) Evidence for liver disease preceding amino acid abnormalities in hereditary tyrosinemia. N Engl J Med 308(21):1265–1267. doi:10.1056/NEJM198305263082105

Jorquera R, Tanguay RM (2001) Fumarylacetoacetate, the metabolite accumulating in hereditary tyrosinemia, activates the ERK pathway and induces mitotic abnormalities and genomic instability. Hum Mol Genet 10(17):1741–1752

Koelink CJ, van Hasselt P, van der Ploeg A, van den Heuvel-Eibrink MM, Wijburg FA, Bijleveld CM, van Spronsen FJ (2006) Tyrosinemia type I treated by NTBC: how does AFP predict liver cancer? Mol Genet Metab 89(4):310–315. doi:10.1016/j.ymgme.2006.07.009

Larochelle J, Alvarez F, Bussieres JF, Chevalier I, Dallaire L, Dubois J,… Mitchell GA (2012) Effect of nitisinone (NTBC) treatment on the clinical course of hepatorenal tyrosinemia in quebec. Mol Genet Metab 107(1–2):49–54. doi:10.1016/j.ymgme.2012.05.022; 10.1016/j.ymgme.2012.05.022

Lindstedt S, Holme E, Lock EA, Hjalmarson O, Strandvik B (1992) Treatment of hereditary tyrosinaemia type I by inhibition of 4-hydroxyphenylpyruvate dioxygenase. Lancet 340(8823):813–817. doi:0140-6736(92)92685-9 [pii]

Nobili V, Jenkner A, Francalanci P, Castellano A, Holme E, Callea F, Dionisi-Vici C (2010) Tyrosinemia type 1: metastatic hepatoblastoma with a favorable outcome. Pediatrics 126(1):e235–e238. doi:10.1542/peds.2009-1639; 10.1542/peds.2009-1639

Paradis K (1996) Tyrosinemia: the quebec experience. Clinical and investigative medicine. Med Clin Exp 19(5):311–316

Paul SB, Gulati MS, Sreenivas V, Madan K, Gupta AK, Mukhopadhyay S, Acharya SK (2007) Evaluating patients with cirrhosis for hepatocellular carcinoma: Value of clinical symptomatology, imaging and alpha-fetoprotein. Oncology 72(Suppl 1):117–123

Russo PA, Mitchell GA, Tanguay RM (2001) Tyrosinemia: a review. Pediatr Dev Pathol: Off J Soc Pediatr Pathol Paediat Pathol Soc 4(3):212–221

Saar B, Kellner-Weldon F (2008) Radiological diagnosis of hepatocellular carcinoma. Liver Int: Off J Int Assoc Study Liver 28(2):189–199. doi:10.1111/j.1478-3231.2007.01655.x

Salt A, Barnes ND, Rolles K, Calne RY, Clayton PT, Leonard JV (1992) Liver transplantation in tyrosinaemia type 1: the dilemma of timing the operation. Acta Paediatr (Oslo, Norway : 1992) 81(5):449–452

Schady DA, Roy A, Finegold MJ (2015) Liver tumors in children with metabolic disorders. Transl Pediatr 4(4):290–303. doi:10.3978/j.issn.2224-4336.2015.10.08

van Ginkel WG, Gouw AS, van der Jagt EJ, de Jong KP, Verkade HJ, van Spronsen FJ (2015) Hepatocellular carcinoma in tyrosinemia type 1 without clear increase of AFP. Pediatrics 135(3):e749–e752. doi:10.1542/peds.2014-1913

van Spronsen FJ, Berger R, Smit GP, de Klerk JB, Duran M, Bijleveld CM et al (1989) Tyrosinaemia type I: orthotopic liver transplantation as the only definitive answer to a metabolic as well as an oncological problem. J Inherit Metab Dis 12(Suppl 2): 339–342

van Spronsen FJ, Bijleveld CM, van Maldegem BT, Wijburg FA (2005) Hepatocellular carcinoma in hereditary tyrosinemia type I despite 2-(2 nitro-4-3 trifluoro- methylbenzoyl)-1, 3-cyclohexanedione treatment. J Pediatr Gastroenterol Nutr 40(1):90–93

van Spronsen FJ, Smit GP, Wijburg FA, Thomasse Y, Visser G, Heymans HS (1995) Tyrosinaemia type I: considerations of treatment strategy and experiences with risk assessment, diet and transplantation. J Inherit Metab Dis 18(2):111–114

van Spronsen FJ, Thomasse Y, Smit GP, Leonard JV, Clayton PT, Fidler V, … Heymans HS (1994) Hereditary tyrosinemia type I: a new clinical classification with difference in prognosis on dietary treatment. Hepatology (Baltimore, Md.) 20(5):1187–1191. doi:S027091399400340X [pii]

Weinberg AG, Mize CE, Worthen HG (1976) The occurrence of hepatoma in the chronic form of hereditary tyrosinemia. J Pediatr 88(3):434–438

Yazigi NA (2013) Long term outcomes after pediatric liver transplantation. Pediatr Gastroenterol Hepatol Nutr 16(4):207–218. doi:10.5223/pghn.2013.16.4.207

Neurological and Neuropsychological Problems in Tyrosinemia Type I Patients

10

Willem G. van Ginkel, Rianne Jahja, Stephan C.J. Huijbregts, and Francjan J. van Spronsen

Abstract

Clinically, Hereditary Tyrosinemia type I (HTI) is especially characterized by severe liver dysfunction in early life. However, recurrent neurological crises are another main finding in these patients when they are treated with a tyrosine and phenylalanine restricted diet only. This is caused by the accumulation of δ-aminolevulinic acid due to the inhibitory effect of succinylacetone on the enzyme that metabolizes δ-aminolevulinic acid. Due to the biochemical and clinical resemblance of these neurological crises and acute intermittent porphyria, this group of symptoms in HTI patients is mostly called porphyria-like-syndrome. The neurological crises in HTI patients disappeared after the introduction of treatment with 2-(2 nitro-4-3 trifluoro-methylbenzoyl)-1, 3-cyclohexanedione (NTBC). However, if NTBC treatment is stopped for a while, severe neurological dysfunction will reappear.

If NTBC treatment is started early and given continuously, all clinical problems seem to be solved. However, recent research findings indicate that HTI patients have a non-optimal neurocognitive outcome, showing (among others) a lower IQ and impaired executive functioning and social cognition. Unfortunately the exact neuropsychological profile of these HTI patients is not known yet, neither are the exact pathophysiological

W.G. van Ginkel • R. Jahja • F.J. van Spronsen (✉)
Beatrix Children's Hospital, Division of Metabolic Diseases, University of Groningen, University Medical Center of Groningen, HPC CA33, Antwoordnummer 333,
9700 VB Groningen, The Netherlands
e-mail: f.j.van.spronsen@umcg.nl

S.C.J. Huijbregts
Department of Clinical Neuropsychology, University of Leiden, Leiden, The Netherlands

© Springer International Publishing AG 2017

R.M. Tanguay (ed.), *Hereditary Tyrosinemia*, Advances in Experimental Medicine and Biology 959,
DOI 10.1007/978-3-319-55780-9_10

mechanisms underlying these impairments. It may be hypothesized that the biochemical changes such as high blood tyrosine or low blood phenylalanine concentrations are important in this respect, but an direct toxic effect of NTBC or production of toxic metabolites (that previously characterized the disease before introduction of NTBC) cannot be excluded either. This chapter discusses the neurological and neuropsychological symptoms associated with HTI in detail. An extended section on possible underlying pathophysiological mechanisms of such symptoms is also included.

Keywords
Tyrosinemia • Porphyria • Porphyria-like-syndrome • Neurological • Neurocognitive • Neuropsychological • IQ • Executive functioning

Abbreviations

5-ALA	δ-aminolevulinic acid
CSU	Cerebral spinal fluid
HTI	Hereditary tyrosinemia type I
HTII	Hereditary tyrosinemia type II
HTIII	Hereditary tyrosinemia type III
LAT-1	Late neutral aminoacid-1 transporter
LNAA	Large neutral amino acid
NTBC	2-(2 nitro-4-3 trifluoro-methylbenzoyl)-1, 3-cyclohexanedione
PKU	Phenylketonuria

10.1 Introduction

Hereditary Tyrosinemia type 1 (HTI) is caused by a defect of the enzyme fumarylacetoacetate hydrolase in the tyrosine degradation pathway. Due to this defect toxic products such as maleylacetoacetate, fumarylacetoacetate, succinylacetoacetate and succinylacetone accumulate that especially affect the liver. Therefore, before the introduction of neonatal screening, most HTI patients presented with symptoms of acute liver failure during the first 6 months of life, while a minority of the patients presented after 6 months of age, mainly with symptoms of chronic liver disease and/or renal tubulopathy (van Spronsen et al. 1994). Although less frequent than liver and renal problems, neurological symptoms (specifically: paralysis) were also observed (van Spronsen et al. 1994).

These clinical problems mainly occurred before the introduction of 2-(2 nitro-4-3 trifluoromethylbenzoyl)-1, 3-cyclohexanedione (NTBC). At that time, the survival probability of the HTI patients was rather low, with only about 30% of the very early presenting patients (i.e. <2 months of age) still alive 2 years after diagnosis (van Spronsen et al. 1994). In the patients presenting early (2–6 months of age) or very early (<2 months of age), acute liver failure with bleeding and ascites were the main causes of death (van Spronsen et al. 1994), if not transplanted in time (van Spronsen et al. 1995). If these (very) early presenting patients survived this initial period, they sometimes died later on due to the development of liver cancer, or respiratory failure. This respiratory failure was due to the neurological crises that could not be prevented in these dietary treated HTI patients, accounting for approximately 10% of deaths (van Spronsen et al. 1994). Recurrent neurological crises were observed in almost half of the HTI patients during multiple episodes when they were treated with a tyrosine and phenylalanine restricted diet alone (Mitchell et al. 1990), showing that the dietary treatment did not clearly positively influence the frequency or severity of the bouts of neurological crises.

10.1.1 Porphyria Like Syndrome: Giving Rise to Neurological Problems

Neurological crises usually arose after a minor infection and were characterized by severely painful parenthesis, hypertonia and weakness, mainly occurring in the legs or in the abdomen (Mitchell et al. 1990), resembling an acute abdominal syndrome for which sometimes – unnecessarily- surgical intervention took place. However, the symptoms could progress to paralysis and even respiratory failure and death. Less frequently occurring, but also present, were symptoms such as arterial hypertension, seizures and even auto-mutilation (Mitchell et al. 1990). The peripheral nerves were mainly affected in these neurological crises. After some days axonal degeneration, with undetectable or low-amplitude potentials, could be found and secondary demyelination took place (Mitchell et al. 1990). No epileptic activity was found in the brain of HTI patients facing a neurological crisis, but hyperreflexia in the lower extremities has been described during a neurological crisis at least once (Rank et al. 1991). Therefore, it was hypothesized that the brain or more specifically the upper motor neurons could be affected to some extent as well. The spectrum of neurological problems in dietary treated HTI patients ranged from normal function to the occurrence neurological crises with complete extinction of peripheral nerve-function (Mitchell et al. 1990).

In 1969 the first article that reported about increased δ-aminolevulinic acid (5-ALA) concentrations in HTI patients was published (Gentz et al. 1969). The 5-ALA is the first compound in the heme biosynthesis pathway and is formed out of succinyl-CoA and Glycine by ALA synthase. Normally, 5-ALA would be metabolized into porphobilinogen by 5-ALA dehydratase (also called porphobilinogen synthase). However, some compounds such as lead and succinylacetone are found to largely inhibit this 5-ALA dehydratase (Ebert et al. 1979; Warren et al. 1998). When HTI patients are treated with a diet only, succinylacetone concentrations are clearly elevated, leading to an inhibition of the 5-ALA dehydratase and, consequently, to increased concentrations of its substrate 5-ALA (Sassa and Kappas 1983). Although no clear correlation in individual patients between excretion of 5-ALA and the occurrence of neurological crises was established, HTI patients with these crises have, in general, higher levels of 5-ALA than patients without neurological crises (Mitchell et al. 1990). Disorders in this heme synthesis pathway are so-called porphyria's and thus the neurological crises in HTI may have a pathophysiological basis essentially identical to acute intermittent porphyria, and lead toxicity: therefore, the neurological crises in HTI have also been called porphyria-like-syndrome (Mitchell et al. 1990; Russo et al. 2001).

The clinical picture of these neurological crises also resembles Guillain-Barré syndrome. Guillain-Barré syndrome is an acute peripheral neuropathy that progresses over time. The progressive weakness in Guillain-Barré is due to demyelination and usually occurs after a minor infection (Kuwabara 2004). Despite the similarities, the pathophysiological mechanisms of both neurological diseases strongly differ, and therefore, treatment differs as well. Whereas Guillain-Barré is considered to be an auto-immune disease, the neurological crises in HTI patients are caused by the increased concentrations of 5-ALA. The clinical resemblance indicates that in patients presenting with symptoms associated with Guillain-Barré syndrome, acute intermittent porphyria due to HTI or lead toxicity have to be considered.

10.1.2 Porphyria Like Syndrome: Treatment by Liver Transplantation

Treatment of these crises of acute intermittent porphyria is mainly based on maintaining adequate calorie intake, respiratory support when needed and control of pain. Hematin and heme arginate inhibit the enzyme ALA-synthase and therefore the synthesis of 5-ALA. This can sometimes decrease the symptoms if given early enough during a crisis (Rank et al. 1991), although a side effect of hematin administration may be an exacerbating coagulopathy, especially when liver dysfunction is already present in these patients (Pierach 1982). Before the introduction of NTBC, the only definitive treatment of HTI, not only to prevent the neurological crises, but also to prevent of other consequences of the disease, including liver cancer, was performing a liver transplantation. After liver transplantation almost all the neurological problems resolved (van Spronsen et al. 1989; Mitchell et al. 1990; Rank et al. 1991). Therefore, before the introduction of NTBC the question was when, rather than whether, to perform liver transplantation in HTI patients (van Spronsen et al. 1994). The timing of the transplantation was based on the risk profile of the patient, which, in turn, largely depended on the patient's age at presentation (van Spronsen et al. 1994, 1995).

10.1.3 Porphyria Like Syndrome: Treatment by NTBC

In 1992, the course of the disease completely changed with the introduction of a new treatment option called NTBC. NTBC is a herbicide that blocks the tyrosine degradation pathway at an earlier step (i.e. before the enzyme fumarylacetoacetate hydrolase comes into play). Therefore, when given at adequate doses, the toxic metabolites such as succinylacetone are not formed anymore and correction of the almost complete inhibition of 5-ALA dehydratase in erythrocytes can be observed. 5-ALA formation and accumulation is in this way prevented and therefore NTBC protects against porphyric crises in HTI patients (Lindstedt et al. 1992). Due to NTBC, the neurological crises disappeared (Holme and Lindstedt 1998; Larochelle et al. 2012).

However, when NTBC treatment is interrupted, severe neurological crises may reappear (Schlump et al. 2008; Ucar et al. 2016). Until now, two case reports about severe neurological crises after discontinuation of NTBC are known. In both reports treatment with NTBC was interrupted and after a few weeks patients were complaining of a fever, vomiting, and diarrhea followed by a progressive ascending polyneuropathy, diaphragmatic paralysis and arterial hypertension, quite similar to the most severe neurological crises before introduction of NTBC (Schlump et al. 2008; Ucar et al. 2016).

10.2 Neuropsychological Problems in Tyrosinemia Type 1 Patients: A New Issue?

Before introduction of NTBC no neurocognitive or behavioural problems had been reported at all in HTI. In contrast, IQ was considered to be normal in all HTI patients including the patients facing neurological problems (Mitchell et al. 1990). However, it should be taken into account that treatment at that time was especially focusing on how to keep the patients alive, preventing death from liver failure, renal tubulopathy, liver cancer and neurological crises. The introduction of treatment with NTBC resolved most of the clinical problems (including liver, renal, and neurological problems) in HTI patients, especially when initiated early (Holme and Lindstedt 1998; Larochelle et al. 2012). Treatment with NTBC and a tyrosine and phenylalanine restricted diet greatly increased the life expectancy of HTI patients. However, some years ago, the first reports arose indicating non-optimal neurocognitive outcome in HTI patients. A remarkably high frequency of school problems and cognitive disturbances (8 out of 23) were found in French HTI patients (Masurel-Paulet et al. 2008). Some years later, the first reports on lower IQ-scores in HTI patients appeared. In a Belgian HTI patient group, nine out of ten patients scored below 100 and three patients even had an IQ below 85 (De Laet et al. 2011). With respect to the IQ scores, verbal IQ seems to be affected to a larger extent than performal IQ and these sub-optimal IQ scores may present together with attentional problems (Pohorecka et al. 2012). Recently, a study even suggested a regression of IQ in a small number of patients (Bendadi et al. 2014). However, IQ is not the only construct that seems to be affected in these HTI patients. Suboptimal motor abilities are found and higher-order brain functions such as executive functions and social cognition seem to be affected as well (Thimm et al. 2012; van Ginkel et al. 2016).

Executive functions are a set of inter-related high level skills across several cognitive domains, which develop during childhood and adolescence and include at least the following core abilities: inhibition, working memory and cognitive flexibility. In HTI patients, especially the working memory component seems to be affected (van Ginkel et al. 2016). Social cognitive abilities can be described as the mental operations underlying social interactions such as the perception, interpretation and generation of responses to the intentions, dispositions, and behaviors of others. In one specific article, HTI patients for example show problems in the ability to recognize emotions based on facial expressions (van Ginkel et al. 2016).

Thus, some recent articles are pointing towards neurocognitive problems in HTI patients that were not recognized before the introduction of NTBC, although the existence of such problems may have been masked by the low life expectancy of the HTI patients at that time. It is possible that this non-optimal neurocognitive outcome underlies problems that have been observed in daily life of HTI patients, such as the behavioural problems and school problems.

This more or less resembles the story of another metabolic disorder in the tyrosine metabolic pathway, namely Phenylketonuria (PKU). PKU is characterized by a genetic defect in the enzyme phenylalanine hydroxylase, leading to an opposite biochemical profile compared to HTI patients with high phenylalanine concentrations and somewhat low tyrosine concentrations (Blau et al. 2010). Untreated, both diseases do not appear similar. Whereas untreated HTI patients especially show liver and renal related problems in early life, untreated PKU is characterized by severe mental retardation, seizures and behavioural difficulties. However, if both diseases are treated adequately, both diseases do show some similarities regarding neurocognitive functioning and its sequelae. When PKU patients started with their phenylalanine-restricted diet (early 1970s), which is still the main treatment option in PKU, cognitive development seemed to be normal at first sight. However, subsequent studies showed that PKU patients still have somewhat lower IQ, impaired motor control and sub optimal executive functions and social cognition (Huijbregts et al. 2002a, b; Moyle et al. 2007; Albrecht et al. 2009; Jahja et al. 2016). This, in a way, resembles the present history on neurocognition in HTI as both groups of patients are showing deficiencies in IQ, executive functions and social cognition. So clearly, both diseases are associated with a quite similar sub-optimal neurocognitive outcome when they are treated. However, comparison of results of studies performed in our group showed that on many executive functions, HTI patients perform less optimal than PKU patients who already show some deficits if compared to healthy age and gender matched individuals (van Ginkel, in preparation).

10.2.1 Hypotheses for Neuropsychological Dysfunction

Several hypotheses have been proposed in order to explain the neuropsychological deficits observed in HTI, however, until now the mechanism has remained unknown. In the next few paragraphs, we will review some possible hypotheses and will explain them further using two schematic figures of the metabolites in the blood and the relation to the blood brain barrier (Figs. 10.1 and 10.2).

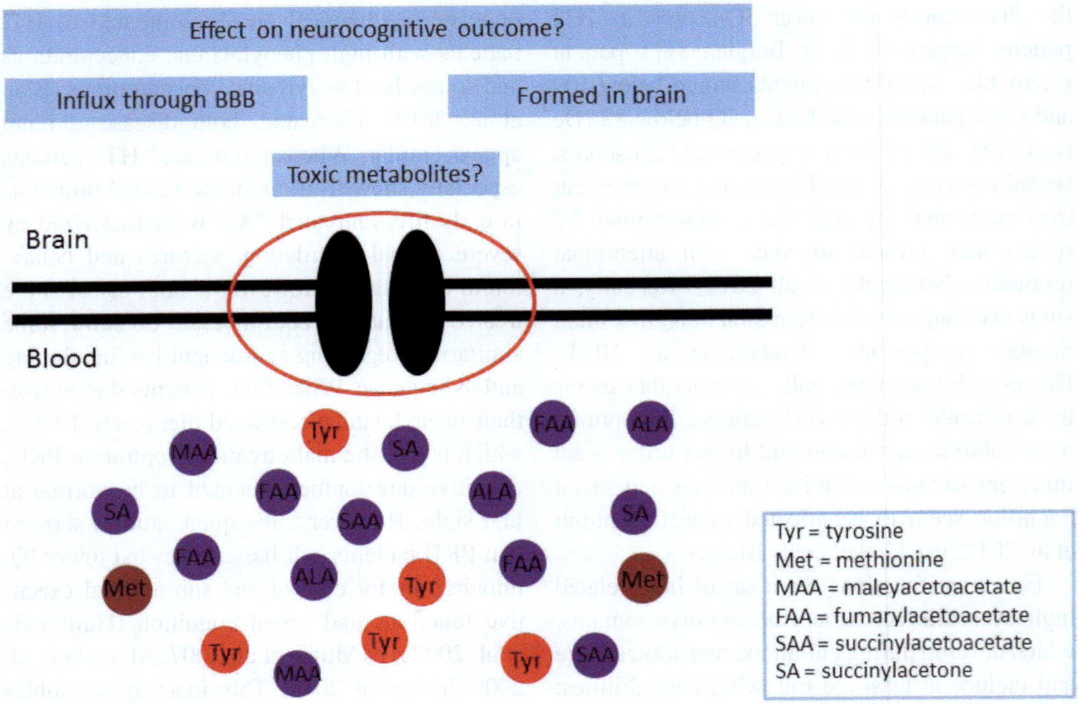

Fig. 10.1 Before the introduction of NTBC, tyrosine and methionine were slightly increased (for more details see Table 10.1), but several toxic metabolites were especially accumulating in the blood but possibly in the brain as well after transport across the blood brain barrier or after formation in the brain itself due to tyrosine degradation. However, at this time no neurocognitive deficiencies were seen

10.2.1.1 Toxic Products and Liver Disease Before Introduction of NTBC

First of all, developing severe liver disease of any cause, especially in infancy, is associated with both a low life expectancy and a risk of long term developmental sequelae and behavioural consequences (Alonso 2008). In the past, almost all HTI patients were diagnosed after a period of severe liver dysfunction. However, their low life expectancy, especially before initiation of treatment with NTBC, may have masked any impaired neurocognitive outcome associated with liver dysfunction in HTI. But not only liver failure in general could have had an effect on the neurocognitive outcome. When patients were treated with a diet only, tyrosine and methionine concentrations were increased (Table 10.1), but many toxic substances were accumulating, such as maleylacetoacetate, fumarylacetoacetate, succinylacetoacetate, succinylacetone and 5-ALA. It could be hypothesized that all these individual metabolites – or a combination of them- could be toxic for the brain, possibly depending on how easily they are transported across the blood brain barrier or in which concentrations it is formed in the brain itself (Fig. 10.1).

The deficient enzyme in HTI patients, fumarylacetoacetate hydrolase, is highly expressed in oligodendrocytes in the brain of rats (Labelle et al. 1993). If the toxic products accumulate in these oligodendrocytes, myelin formation and axonal integrity could be impaired (Nave 2010). However, as noted before, it should be taken into account that no reports of cognitive problems had appeared before the introduction of NTBC (Mitchell et al. 1990). Moreover, current cognitive deficiencies have not only been found in

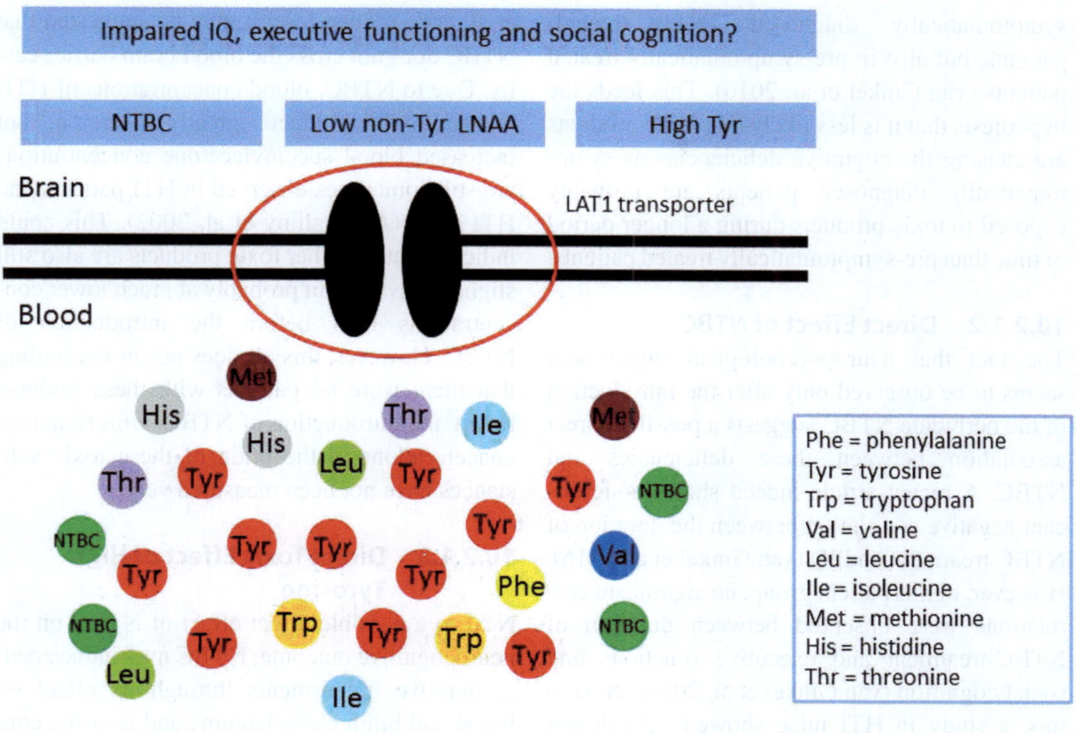

Fig. 10.2 Several mechanisms could underlie neurocognitive deficiencies when patients are treated with NTBC and diet. All amino acids are transported in a competitive way by the same transporter (LAT-1) transporter. Treatment with NTBC results in high blood tyrosine concentrations and therefore possibly also high brain tyrosine concentrations. Next to this the brain influx of other amino acids (especially phenylalanine as concentrations in blood are decreased, see Table 10.1) could be impaired because they are outcompeted by tyrosine. On the other hand NTBC could be toxic for the brain as it was known as herbicide before it was introduced as treatment option for HTI patients

Table 10.1 Estimated blood amino acid concentrations (μmol/l) in HTI patients treated with dietary treatment only compared to the situation that they used NTBC and dietary treatment (also compared to normal concentrations), based on two children treated in the University Medical Center Groningen, the Netherlands

Amino acids	Mean concentrations on diet only (n = 2)	Mean concentrations on NTBC and diet (n = 2)	Normal values children 10–18 years (Blau et al. 1996)
Phenylalanine	33	16	38–78
Tyrosine	135	318	40–92
Tryptophan	93	67	27–75
Valine	250	252	142–278
Leucine	116	133	76–168
Isoleucine	67	72	38–94
Methionine	45	23	16–36
Histidine	78	75	58–106
Threonine	169	152	72–192

symptomatically diagnosed (NTBC-treated) patients, but also in pre-symptomatically treated patients (van Ginkel et al. 2016). This feeds the hypothesis that it is less likely that toxic products are causing the cognitive deficiencies as symptomatically diagnosed patients are probably exposed to toxic products during a longer period of time than pre-symptomatically treated patients.

10.2.1.2 Direct Effect of NTBC

The fact that neuropsychological impairment seems to be observed only after the introduction of the herbicide NTBC suggests a possible direct association between these deficiencies and NTBC. A recent article indeed shows a significant negative correlation between the duration of NTBC treatment and IQ (van Ginkel et al. 2016). However, in this patient group, no significant correlations were observed between duration of NTBC-treatment and executive functions and social cognition (van Ginkel et al. 2016). Next to this, a study in HTI mice showed behavioural problems in the HTI mice that are hypothesized not to be caused by NTBC, but more likely by the toxic metabolites that are associated with the disease (Hillgartner et al. 2016). In this study, Hillgartner et al. compared NTBC treated HTI mice to NTBC treated wildtype mice and normal wildtype mice who did not receive NTBC. In the different tasks, especially the NTBC treated HTI mice showed memory problems, whereas the NTBC treated wildtype mice did not show differences compared to the normal wildtype mice. Based on these results, it was suggested that it was unlikely that the cognitive problems were a consequence of NTBC-use (Hillgartner et al. 2016). In contrast, it was deemed more likely that the problems were caused by the toxic products that accumulate when NTBC concentrations are too low, being the same toxic products as before the introduction of NTBC (Fig. 10.1).

Indirect support for the hypothesis by Hillgartner (2016) might stem from the fact that in rats, concentrations of NTBC are much lower in the brain than in the liver (Lock et al. 1996), although the enzyme that NTBC binds to (4-Hydroxyphenylpyruvate dioxygenase) is highly expressed in the brain of rats as well (Neve et al. 2003). Therefore, it may be suggested that NTBC does not cross the blood brain barrier easily. Due to NTBC, blood concentrations of HTI-related toxic products greatly decrease, but increased blood succinylacetone concentrations are still sometimes observed in HTI patients and HTI mice (Al-Dhalimy et al. 2002). This could indicate that the other toxic products are also still slightly elevated but probably at much lower concentrations than before the introduction of NTBC. However, this all does not fit the finding that there were no patients with these findings before the introduction of NTBC. Unfortunately, concentrations in the brain of these toxic substances have not been measured yet.

10.2.1.3 Direct Toxic Effect of High Tyrosine

Next to a possible direct effect of NTBC on the neurocognitive outcome, NTBC may induce neurocognitive impairments through its effect on blood and brain phenylalanine and tyrosine concentrations as well. NTBC blocks the enzyme 4-hydroxyphenylpyruvate dioxygenase and in this way creates a new metabolic defect upstream of the primary defect. Due to this new block, tyrosine concentrations increase significantly so that dietary restriction of tyrosine and its precursor phenylalanine becomes necessary. Despite these dietary restrictions, tyrosine concentrations are still higher than normal. In HTI children, recommended concentrations are 200–400 umol/L (de Laet et al. 2013), while healthy children have concentrations around 40–92 umol/L (Table 10.1) (Blau et al. 1996). Tyrosine concentrations during NTBC treatment are mostly higher than tyrosine concentrations before the introduction of NTBC. As tyrosine is transported in a competitive way across the blood brain barrier, high blood tyrosine concentrations possibly cause high concentrations of tyrosine in the brain as well (Pardridge 1998) (Fig. 10.2).

And indeed, in the cerebral spinal fluid (CSF) of (NTBC-treated) HTI patients, an increased concentration of tyrosine has been found. Furthermore, a linear relationship between blood tyrosine concentrations and tyrosine concentrations in the CSF was reported (Thimm et al. 2011).

Increased brain tyrosine concentrations can, in turn, lead to a direct toxic effect of tyrosine in the brain like phenylalanine has in PKU (van Spronsen et al. 2009; Surtees and Blau 2000). Research on the toxicity of tyrosine showed that increased tyrosine concentrations can induce all kinds of processes including oxidative stress and DNA repair issues, with, as a potential consequence, brain dysfunction (De Pra et al. 2014; Macedo et al. 2013; Ramos et al. 2013), but the exact tyrosine concentration that harms development still needs to be established.

High tyrosine concentrations may not only be associated with a sub-optimal neurocognitive outcome in HTI patients, but also in other defects affecting tyrosine metabolism such as Hereditary Tyrosinemia type II (HTII, Richner-Hanhart syndrome) and Hereditary Tyrosinemia type III (HTIII). In HTII, biochemically characterized by high blood tyrosine (sometimes above 1000 μmol/L) and normal blood phenylalanine concentrations, neurocognitive deficiencies have been shown as well. Especially when dietary treatment is started late, these patients are prone to cognitive deficiencies possibly caused by high tyrosine concentrations (Scott 2006). However, tyrosine concentrations are much higher in these HTII patients than in HTI patients. Nevertheless, it should be taken into account that high tyrosine could play an important role in HTI as well, as developmental problems have also been observed in HTIII patients, while this disease is characterized by much lower tyrosine concentrations than HTII. Actually, tyrosine concentrations in HTIII are quite similar to tyrosine concentrations in NTBC treated HTI patients, probably due to the fact that NTBC induces the same metabolic block as is present in HTIII.

On the other hand, as tyrosine is the precursor of dopamine (and noradrenalin), increased brain tyrosine concentrations could not only have a direct toxic effect on the brain, but also increase the synthesis of the neurotransmitter dopamine in the brain. Dopamine is important for various functions in the brain and is involved in emotional regulation, memory and attention, and executive functioning (Nieoullon 2002; Salgado-Pineda et al. 2005). For dopamine, there is clear evidence that both low and high levels are related to poor cognitive outcomes (Aarts et al. 2014; McGough et al. 2009; Vaillancourt et al. 2013). This supposed inverted U-shape for dopamine-outcome associations should be investigated further in HTI. Thimm et al. already found that in 2/3 patients an increased concentration of homovanillic acid, a degradation product of dopamine was found, indicating that dopamine concentrations are probably too high in these patients (Thimm et al. 2011).

10.2.1.4 Impaired Influx of Other Amino Acids

In PKU, research has suggested the presence of additional pathophysiological mechanisms, next to phenylalanine toxicity, that might affect brain functioning. All large neutral amino acids, such as tyrosine and phenylalanine are transported in a competitive way across the blood brain barrier by the Large Neutral Amino Acid -1 (LAT-1) transporter. High concentrations of one particular amino acid could in this way inhibit the transport of the others (Fig. 10.2). All amino acids have a different affinity for the LAT-1 transporter. Phenylalanine is supposed to have the highest affinity for the LAT-1 transporter in rats (Smith 2000) and therefore a small increase in plasma phenylalanine concentrations can already affect the influx of the other amino acids (de Groot et al. 2010; van Spronsen et al. 2009). Increased concentrations of tyrosine may have a similar negative effect on the influx of other amino acids. Impaired transport of those LNAA could have extensive effects on neurotransmitter synthesis and/or protein synthesis in the brain in general. Tryptophan is the metabolic precursor for the neurotransmitter serotonin. If tryptophan is outcompeted by tyrosine at the blood brain barrier, lower tryptophan concentrations in the brain will be the result. It may be hypothesized that this will lead to decreased serotonin synthesis (leading to cognitive and behavioural problems) comparable to the issues seen in PKU (van Spronsen et al. 2009; de Groot et al. 2010). Evidence for an impaired synthesis of serotonin in HTI patients stems from one study showing decreased levels of a degradation product of serotonin,

5-hydroxyindoleacetic acid, in the CSF (Thimm et al. 2011). Clinically, mood disorders caused by the depletion of serotonin have not yet been found in HTI patients, although impaired executive functioning and social cognition, which have been observed in HTI, could also be related to lower serotonin levels.

10.2.1.5 Low Blood Phenylalanine Concentrations

Since the introduction of NTBC low phenylalanine concentrations have been observed occasionally in HTI patients (Wilson et al. 2000; Daly et al. 2012; van Vliet et al. 2014). Most likely this is caused by a diet that is too strict, although NTBC seems to lower blood phenylalanine concentrations in phenylalanine hydroxylase deficient mice. Unfortunately, the mechanisms behind the decreases of blood phenylalanine due to NTBC are not known yet (Harding et al. 2014). Low phenylalanine concentrations could have peripheral implications such as eczema and faltering growth in inadequately treated PKU patients (Pode-Shakked et al. 2013), but also in HTI (van Vliet et al. 2014). Next to this, low phenylalanine concentrations in a human HTI patient have been associated with developmental delays (van Vliet et al. 2014). In HTI patients, it is possible that the influx of phenylalanine is already lower than normal due to competition with an excess of tyrosine at the blood brain barrier. When blood phenylalanine concentrations are also lower than normal, brain influx of phenylalanine could be even more impaired. Theoretically, this could for example result in lower protein synthesis in the brain. Therefore, phenylalanine supplementation is sometimes given to HTI patients (Wilson et al. 2000; van Vliet et al. 2014) although professionals have not yet agreed on the amount that should be given.

10.3 Conclusion

Although the exact neurocognitive profile of the HTI patients is not known yet, it is concluded that HTI patients not only show neurological issues but also neurocognitive problems in several different domains. Interestingly, the neurological issues have not been observed on treatment with NTBC and diet, whilst on the other hand neurocognitive problems have been reported for the first time after the introduction of NTBC and diet. Therefore, future studies have to relate this non-optimal neurocognitive outcome to treatment parameters such as tyrosine and phenylalanine concentrations throughout life. In addition to this, direct and indirect effects of NTBC or the toxic metabolites associated with the disease on the neurocognitive outcome need further investigation.

References

Aarts E, Wallace DL, Dang LC, Jagust WJ, Cools R, D'Esposito M (2014) Dopamine and the cognitive downside of a promised bonus. Psychol Sci 25(4):1003–1009

Albrecht J, Garbade SF, Burgard P (2009) Neuropsychological speed tests and blood phenylalanine levels in patients with phenylketonuria: a meta-analysis. Neurosci Biobehav Rev 33(3):414–421

Al-Dhalimy M, Overturf K, Finegold M, Grompe M (2002) Long-term therapy with NTBC and tyrosine-restricted diet in a murine model of hereditary tyrosinemia type I. Mol Genet Metab 75(1):38–45

Alonso EM (2008) Growth and developmental considerations in pediatric liver transplantation. Liver Transplant Off Publ Am Assoc Stud Liver Dis Int Liver Transplant Soc 14(5):585–591

Bendadi F, De Koning TJ, Visser G, Prinsen HC, De Sain MG, Verhoeven-Duif N, Sinnema G, Van Spronsen FJ, Van Hasselt PM (2014) Impaired cognitive functioning in patients with tyrosinemia type I receiving nitisinone. J Pediatr 164(2):398–401

Blau N, Duran M, Blaskovics ME (1996) Physician's guide to the laboratory diagnosis of metabolic diseases. Chapman & Hall, London

Blau N, Van Spronsen FJ, Levy HL (2010) Phenylketonuria. Lancet (London, England) 376(9750):1417–1427

Daly A, Gokmen-Ozel H, MacDonald A, Preece MA, Davies P, Chakrapani A, McKiernan P (2012) Diurnal variation of phenylalanine concentrations in tyrosinaemia type 1: should we be concerned? J Hum Nutr Diet Off J British Diet Assoc 25(2):111–116

De Groot MJ, Hoeksma M, Blau N, Reijngoud DJ, Van Spronsen FJ (2010) Pathogenesis of cognitive dysfunction in phenylketonuria: review of hypotheses. Mol Genet Metab 99(Suppl 1):S86–S89

De Laet C, Munoz VT, Jaeken J, Francois B, Carton D, Sokal EM, Dan B, Goyens PJ (2011) Neuropsychological outcome of NTBC-treated patients with tyrosinaemia type 1. Dev Med Child Neurol 53(10):962–964

De Laet C, Dionisi-Vici C, Leonard JV, Mckiernan P, Mitchell G, Monti L, De Baulny HO, Pintos-Morell G, Spiekerkotter U (2013) Recommendations for the management of tyrosinaemia type 1. Orphanet J Rare Dis 8:8-1172-8-8

De Pra SD, Ferreira GK, Carvalho-Silva M, Vieira JS, Scaini G, Leffa DD, Fagundes GE, Bristot BN, Borges GD, Ferreira GC, Schuck PF, Andrade VM, Streck EL (2014) L-tyrosine induces DNA damage in brain and blood of rats. Neurochem Res 39(1):202–207

Ebert PS, Hess RA, Frykholm BC, Tschudy DP (1979) Succinylacetone, a potent inhibitor of heme biosynthesis: effect on cell growth, heme content and delta-aminolevulinic acid dehydratase activity of malignant murine erythroleukemia cells. Biochem Biophys Res Commun 88(4):1382–1390

Gentz J, Johansson S, Lindblad B, Lindstedt S, Zetterstrom R (1969) Exertion of delta-aminolevulinic acid in hereditary tyrosinemia. Clinica Chim Acta Int J Clin Chem 23(2):257–263

Harding CO, Winn SR, Gibson KM, Arning E, Bottiglieri T, Grompe M (2014) Pharmacologic inhibition of L-tyrosine degradation ameliorates cerebral dopamine deficiency in murine phenylketonuria (PKU). J Inherit Metab Dis 37(5):735–743

Hillgartner MA, Coker SB, Koenig AE, Moore ME, Barnby E, Macgregor GG (2016) Tyrosinemia type I and not treatment with NTBC causes slower learning and altered behavior in mice. J Inherit Metab Dis 39(5):673–682. doi:10.1007/s10545-016-9949-6. Epub 2016 Jun 6

Holme E, Lindstedt S (1998) Tyrosinaemia type I and NTBC (2-(2-nitro-4-trifluoromethylbenzoyl)-1,3-cyclohexanedione). J Inherit Metab Dis 21(5):507–517

Huijbregts S, De Sonneville L, Licht R, Sergeant J, Van Spronsen F (2002a) Inhibition of prepotent responding and attentional flexibility in treated phenylketonuria. Dev Neuropsychol 22(2):481–499

Huijbregts SC, De Sonneville LM, Van Spronsen FJ, Licht R, Sergeant JA (2002b) The neuropsychological profile of early and continuously treated phenylketonuria: orienting, vigilance, and maintenance versus manipulation-functions of working memory. Neurosci Biobehav Rev 26(6):697–712

Jahja R, Van Spronsen FJ, De Sonneville LM, Van Der Meere JJ, Bosch AM, Hollak CE, Rubio-Gozalbo ME, Brouwers MC, Hofstede FC, De Vries MC, Janssen MC, Van Der Ploeg AT, Langendonk JG, Huijbregts SC (2016) Social-cognitive functioning and social skills in patients with early treated phenylketonuria: a PKU-COBESO study. J Inherit Metab Dis 39(3):355–362

Kuwabara S (2004) Guillain-Barre syndrome: epidemiology, pathophysiology and management. Drugs 64(6):597–610

Labelle Y, Puymirat J, Tanguay RM (1993) Localization of cells in the rat brain expressing fumarylacetoacetate hydrolase, the deficient enzyme in hereditary tyrosinemia type 1. Biochim Biophys Acta 1180(3):250–256

Larochelle J, Alvarez F, Bussieres JF, Chevalier I, Dallaire L, Dubois J, Faucher F, Fenyves D, Goodyer P, Grenier A, Holme E, Laframboise R, Lambert M, Lindstedt S, Maranda B, Melancon S, Merouani A, Mitchell J, Parizeault G, Pelletier L, Phan V, Rinaldo P, Scott CR, Scriver C, Mitchell GA (2012) Effect of nitisinone (NTBC) treatment on the clinical course of hepatorenal tyrosinemia in Quebec. Mol Genet Metab 107(1–2):49–54

Lindstedt S, Holme E, Lock EA, Hjalmarson O, Strandvik B (1992) Treatment of hereditary tyrosinaemia type I by inhibition of 4-hydroxyphenylpyruvate dioxygenase. Lancet 340(8823):813–817

Lock EA, Gaskin P, Ellis MK, Provan WM, Robinson M, Smith LL, Prisbylla MP, Mutter LC (1996) Tissue distribution of 2-(2-nitro-4-trifluoromethylbenzoyl) cyclohexane-1-3-dione (NTBC): effect on enzymes involved in tyrosine catabolism and relevance to ocular toxicity in the rat. Toxicol Appl Pharmacol 141(2):439–447

Macedo LG, Carvalho-Silva M, Ferreira GK, Vieira JS, Olegario N, Goncalves RC, Vuolo FS, Ferreira GC, Schuck PF, Dal-Pizzol F, Streck EL (2013) Effect of acute administration of L-tyrosine on oxidative stress parameters in brain of young rats. Neurochem Res 38(12):2625–2630

Masurel-Paulet A, Poggi-Bach J, Rolland MO, Bernard O, Guffon N, Dobbelaere D, Sarles J, De Baulny HO, Touati G (2008) NTBC treatment in tyrosinaemia type I: long-term outcome in French patients. J Inherit Metab Dis 31(1):81–87

McGough JJ, McCracken JT, Loo SK, Manganiello M, Leung MC, Tietjens JR, Trinh T, Baweja S, Suddath R, Smalley SL, Hellemann G, Sugar CA (2009) A candidate gene analysis of methylphenidate response in attention-deficit/hyperactivity disorder. J Am Acad Child Adolesc Psychiatry 48(12):1155–1164

Mitchell G, Larochelle J, Lambert M, Michaud J, Grenier A, Ogier H, Gauthier M, Lacroix J, Vanasse M, Larbrisseau A (1990) Neurologic crises in hereditary tyrosinemia. N Engl J Med 322(7):432–437

Moyle JJ, Fox AM, Arthur M, Bynevelt M, Burnett JR (2007) Meta-analysis of neuropsychological symptoms of adolescents and adults with PKU. Neuropsychol Rev 17(2):91–101

Nave KA (2010) Myelination and support of axonal integrity by glia. Nature 468(7321):244–252

Neve S, Aarenstrup L, Tornehave D, Rahbek-Nielsen H, Corydon TJ, Roepstorff P, Kristiansen K (2003) Tissue distribution, intracellular localization and proteolytic processing of rat 4-hydroxyphenylpyruvate dioxygenase. Cell Biol Int 27(8):611–624

Nieoullon A (2002) Dopamine and the regulation of cognition and attention. Prog Neurobiol 67(1):53–83

Pardridge WM (1998) Blood-brain barrier carrier-mediated transport and brain metabolism of amino acids. Neurochem Res 23(5):635–644

Pierach CA (1982) Hematin therapy for the porphyric attack. Semin Liver Dis 2(2):125–131

Pode-Shakked B, Shemer-Meiri L, Harmelin A, Stettner N, Brenner O, Abraham S, Schwartz G, Anikster Y (2013) Man made disease: clinical manifestations of low phenylalanine levels in an inadequately treated phenylketonuria patient and mouse study. Mol Genet Metab 110(Suppl):S66–S70

Pohorecka M, Biernacka M, Jakubowska-Winecka A, Biernacki M, KUSMIERSKA K, Kowalik A, Sykut-Cegielska J (2012) Behavioral and intellectual functioning in patients with tyrosinemia type I. Pediatr Endocrinol Diabetes Metab 18(3):96–100

Ramos AC, Ferreira GK, Carvalho-Silva M, Furlanetto CB, Goncalves CL, Ferreira GC, Schuck PF, Streck EL (2013) Acute administration of l-tyrosine alters energetic metabolism of hippocampus and striatum of infant rats. Int J Dev Neurosci Off J Int Soc Dev Neurosci 31(5):303–307

Rank JM, Pascual-Leone A, Payne W, Glock M, Freese D, Sharp H, Bloomer JR (1991) Hematin therapy for the neurologic crisis of tyrosinemia. J Pediatr 118(1):136–139

Russo PA, Mitchell GA, Tanguay RM (2001) Tyrosinemia: a review. Pediatr Dev Pathol Off J Soc Pediatr Pathol Paediatr Pathol Soc 4(3):212–221

Salgado-Pineda P, Delaveau P, Blin O, Nieoullon A (2005) Dopaminergic contribution to the regulation of emotional perception. Clin Neuropharmacol 28(5):228–237

Sassa S, Kappas A (1983) Hereditary tyrosinemia and the heme biosynthetic pathway. Profound inhibition of delta-aminolevulinic acid dehydratase activity by succinylacetone. J Clin Invest 71(3):625–634

Schlump JU, Perot C, Ketteler K, Schiff M, Mayatepek E, Wendel U, Spiekerkoetter U (2008) Severe neurological crisis in a patient with hereditary tyrosinaemia type I after interruption of NTBC treatment. J Inherit Metab Dis 31(Suppl 2):S223–S225

Scott CR (2006) The genetic tyrosinemias. Am J Med Genet Part C Semin Med Genet 142C(2):121–126

Smith QR (2000) Transport of glutamate and other amino acids at the blood-brain barrier. J Nutr 130(4S Suppl):1016S–1022S

Surtees R, Blau N (2000) The neurochemistry of phenylketonuria. Eur J Pediatr 159(Suppl 2):S109–S113

Thimm E, Herebian D, Assmann B, Klee D, Mayatepek E, Spiekerkoetter U (2011) Increase of CSF tyrosine and impaired serotonin turnover in tyrosinemia type I. Mol Genet Metab 102(2):122–125

Thimm E, Richter-Werkle R, Kamp G, Molke B, Herebian D, Klee D, Mayatepek E, Spiekerkoetter U (2012) Neurocognitive outcome in patients with hypertyrosinemia type I after long-term treatment with NTBC. J Inherit Metab Dis 35(2):263–268

Ucar HK, Tumgor G, Kor D, Kardas F, Mungan NO (2016) A case report of a very rare Association of Tyrosinemia Type I and pancreatitis mimicking neurologic crisis of Tyrosinemia Type I. Balkan Med J 33(3):370–372

Vaillancourt DE, Schonfeld D, Kwak Y, Bohnen NI, Seidler R (2013) Dopamine overdose hypothesis: evidence and clinical implications. Mov Disord Off J Mov Disord Soc 28(14):1920–1929

Van Ginkel WG, Jahja R, Huijbregts SC, Daly A, MacDonald A, De Laet C, Cassiman D, Eyskens F, Korver-Keularts IM, Goyens PJ, McKiernan PJ, Van Spronsen FJ (2016) Neurocognitive outcome in tyrosinemia type 1 patients compared to healthy controls. Orphanet J Rare Dis 11(1):87-016-0472-5

Van Spronsen FJ, Berger R, Smit GP, De Klerk JB, Duran M, Bijleveld CM, Van Faassen H, Slooff MJ, Heymans HS (1989) Tyrosinaemia type I: orthotopic liver transplantation as the only definitive answer to a metabolic as well as an oncological problem. J Inherit Metab Dis 12(Suppl 2):339–342

Van Spronsen FJ, Thomasse Y, Smit GP, Leonard JV, Clayton PT, Fidler V, Berger R, Heymans HS (1994) Hereditary tyrosinemia type I: a new clinical classification with difference in prognosis on dietary treatment. Hepatology (Baltimore, MD) 20(5):1187–1191

Van Spronsen FJ, Smit GP, Wijburg FA, Thomasse Y, Visser G, Heymans HS (1995) Tyrosinaemia type I: considerations of treatment strategy and experiences with risk assessment, diet and transplantation. J Inherit Metab Dis 18(2):111–114

Van Spronsen FJ, Hoeksma M, Reijngoud DJ (2009) Brain dysfunction in phenylketonuria: is phenylalanine toxicity the only possible cause? J Inherit Metab Dis 32(1):46–51

Van Vliet D, Van Dam E, Van Rijn M, Derks TG, Venema-Liefaard G., Hitzert MM, Lunsing RJ, Heiner-Fokkema MR and Van Spronsen FJ (2014) Infants with tyrosinemia type 1: should phenylalanine be supplemented? JIMD reports

Warren MJ, Cooper JB, Wood SP, Shoolingin-Jordan PM (1998) Lead poisoning, haem synthesis and 5-aminolaevulinic acid dehydratase. Trends Biochem Sci 23(6):217–221

Wilson CJ, Van Wyk KG, Leonard JV, Clayton PT (2000) Phenylalanine supplementation improves the phenylalanine profile in tyrosinaemia. J Inherit Metab Dis 23(7):677–683

Part IV

Screening

11 Diagnosing Hepatorenal Tyrosinaemia in Europe: Newborn Mass Screening Versus Selective Screening

Anibh M. Das, Sebene Mayorandan, and Nils Janzen

Abstract

Hepatorenal tyrosinaemia (HT1) is a serious condition that used to be fatal before the advent of nitisinone (NTBC, Orfadine®) as a therapeutic option. We have recently shown that selective screening is inadequate as initial symptoms are often uncharacteristic which leads to a considerable delay in diagnosis and treatment. This has a negative impact on morbidity and mortality as well as long-term outcome. For example, the odds ratio to develop hepatocellular carcinoma is 12.7 when treatment is initiated after the first birthday compared to start of treatment in the neonatal period. Timely diagnosis is only possible when neonatal mass screening is operational. HT1 meets all the criteria for neonatal mass screening at a clinical and analytical level. The natural course of the disease is well known, clinically there is a latent phase in most patients when presymptomatic treatment can be initiated. There are no mild phenotypes which do not require treatment. Using succinylacetone as the screening parameter a highly specific and sensitive test is available with acceptable financial burden. Neonatal mass screening for HT1 is acceptable to the target population as it can be performed simultaneously with the already existing screening

A.M. Das (✉)
Department of Paediatrics, Hannover Medical School, Carl Neuberg Strasse 1, D 30625 Hannover, Germany
e-mail: Das.Anibh@mh-hannover.de

S. Mayorandan
Department of Paediatrics, University Clinic Münster, Münster, Germany

N. Janzen
Screening Laboratory Hannover and Department of Clinical Chemistry, Hannover Medical School, Carl Neuberg Str. 1, D 30625 Hannover, Germany

© Springer International Publishing AG 2017

R.M. Tanguay (ed.), *Hereditary Tyrosinemia*, Advances in Experimental Medicine and Biology 959,
DOI 10.1007/978-3-319-55780-9_11

tests in dried blood, there are no false negative and false positive cases and the financial burden to the health system is moderate. An efficient treatment is available with nitisinone and protein-reduced diet supplemented with special amino acid mixtures. Despite compelling evidence in favour of a neonatal mass screening for HT1 only 57% of European centres taking part in our recent cross-sectional study have included HT1 in their newborn screening programme.

Keywords
Diet • Hepatocellular carcinoma • Hepatorenal tyrosinaemia • Liver transplantation • Newborn screening • Nitisinone • Selective screening • Succinylacetone • Monitoring • Tyrosine

Abbreviations

HT1 Hepatorenal tyrosinaemia
MS Mass spectrometry
PKU Phenylketonuria

11.1 Introduction

In central Europe, hepatorenal tyrosinaemia (HT1) is an ultra-rare disease with an incidence of 1:125,000. For many years, HT1 was clinically suspected and diagnosed by selective screening after a long odyssey. With the introduction of the tandem-mass spectrometry (MS) technique into newborn screening programmes in the 1990s, it was possible to apply screening to a wider range of metabolic diseases. HT1 is not a target disease of newborn mass screening in every country within Europe. At present, HT1 is part of the neonatal mass screening in (parts of) Austria, Belgium, Denmark, Hungary, Netherlands, Poland, Portugal and Spain (Bonham 2013; Loeber et al. 2012). Therefore, a considerable number of patients are still diagnosed via selective screening following a clinical and/or biochemical suspicion. This may have considerable implications for the age at diagnosis, start of treatment and long-term outcome.

It has previously been demonstrated that newborn mass screening using succinylacetone as the screening parameter is both specific and sensitive (Sander et al. 2006; Zytkovicz et al. 2013; Allard et al. 2004; Jakob et al. 1988; Magera et al. 2006). However, only few countries have included succinylacetone as a screening parameter in their newborn mass screening programme. In many European countries, there is much scepticism towards newborn mass screening.

In some countries, tyrosine is used as a screening parameter (phenylalanine/tyrosine ratio) for the diagnosis of phenylketonuria (PKU) and has been advocated as a screening parameter for HT1. In our hands, tyrosine is an unreliable screening parameter with many false positive and false negative results. This can be explained by the fact that tyrosine is part of many anabolic and catabolic pathways as shown in Fig. 11.1.

In a recent European multi-centre cross-sectional study, we were able to include data from 168 patients treated in 21 centres (Mayorandan et al. 2014). Participating centres came from Austria, Belgium, Czech Republic, France, Germany, Israel, Italy, Netherlands, Norway, United Kingdom, Spain, Switzerland and Turkey. We focussed on diagnostic, clinical, therapeutic and monitoring aspects as well as outcome in HT1. In the present chapter, we will deal with aspects of diagnosis based mainly on data from our recent study. Results of our study are compared with smaller regional studies (Masurel-Paulet et al. 2008; Vondrakova et al. 2010; Couce et al. 2010, 2011; Larochelle et al. 2012; Schiff et al. 2011; Coskun et al. 1991).

In order to include a disease in a newborn mass screening programme it has to be demonstrated that the criteria of Wilson and Jungner (1968) are met. We shall use slightly modified criteria adapted to inborn errors of metabolism (Andermann et al. 2011).

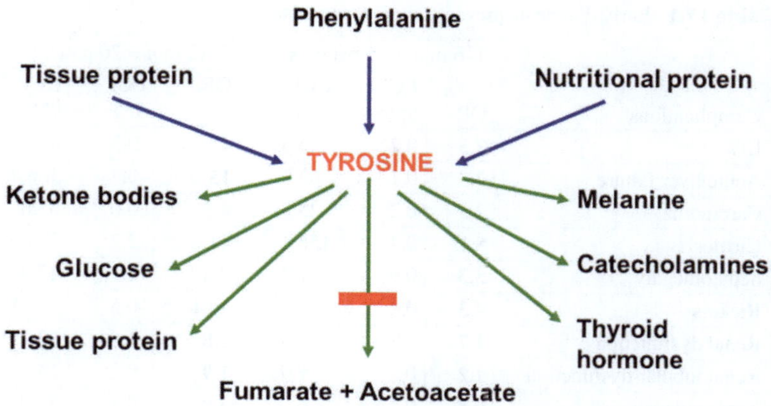

Fig. 11.1 Tyrosine metabolism: tyrosine as an unspecific parameter with multiple pathways

11.2 Screening

11.2.1 Neonatal Screening Versus Selective Screening

In our recent cross-sectional survey we addressed the question whether selective screening is sufficient to improve the outcome in HT1-patients (Mayorandan et al. 2014). We found that early diagnosis and treatment with nitisinone and a protein-reduced diet are essential to prevent hepatocellular carcinoma: The odds ratio to develop hepatocellular carcinoma was 12.7 when treatment started after the first year of life compared to patients already treated with NTBC in the neonatal period. The odds ratio to get liver cirrhosis was 40.5 in those patients treated beyond the first year of life compared to patients treated in the neonatal period. Similarly, the risk to develop acute liver disease, renal dysfunction or rickets increased when treatment started later in life (Table 11.1).

The majority of patients in our survey was diagnosed via selective screening (132/168). Presenting symptoms were not specific: Acute liver dysfunction, hepatomegaly, liver cirrhosis, nephromegaly, renal dysfunction, fever, diarrhoea, infections and feeding difficulties were observed (Fig. 11.2).

Laboratory parameters and imaging studies yielded unspecific findings as well. Many patients were asymptomatic in the first few months of life. Thus, a diagnosis by clinical suspicion is often delayed; the average age at diagnosis via selective screening was 15.5 months versus 0.58 months in patients diagnosed by newborn screening: much too late to prevent serious sequelae. These results clearly show that early diagnosis by clinical suspicion and subsequent selective screening is not feasible in most cases. Therefore, newborn mass screening is essential to initiate early treatment and improve the long-term outcome.

11.2.2 Criteria for Neonatal Mass Screening

The classical screening criteria of Wilson and Jungner (1968) were adapted for rare inborn errors of metabolism in the last years. At the levels of clinical services and laboratory testing, several criteria have to be met (Andermann et al. 2011). These criteria shall be applied to neonatal mass screening for HT1.

11.2.2.1 The Condition Sought Should Be a Common and/or Serious Health Problem and There Should Be a Defined Target Population

HT1 is definitely a serious condition, though with an incidence of 1:125 000 not common. The disease is more prevalent in consanguineous populations, however we do not think it is ethically justified to limit the screening programme for HT1 to certain ethnicities. We advocate screening

Table 11.1 Early diagnosis prevents clinical sequelae

	1–6 m $n = 45$ patients			7–12 m $n = 20$ patients			>12 m $n = 46$ patients		
Complications	OR	LCI 95%	UCI 95%	OR	LCI 95%	UCI 95%	OR	LCI 95%	UCI 95%
LTx	**2.5**	0.2	25.8	**4**	0.3	47.1	**12.7**	1.5	103
Acute liver failure	**2.5**	0.1	63.9	**15**	0.7	306.4	**2.4**	0.09	62.4
Carcinoma	**2.5**	0.2	25.8	**6.3**	0.6	65.6	**12.7**	1.5	103
Cirrhosis	**8.1**	0.4	156.1	**41.6**	2.2	779.9	**40.5**	2.3	704.1
hepatomegaly	**3.3**	0.9	11.3	**4.4**	1.1	17.7	**3.9**	1.1	13.3
Rickets	**4.3**	0.2	92.6	**10.1**	0.5	222.1	**19**	1.1	338.3
Renal dysfunction	**1.7**	0.2	9.8	**5.8**	1.0	33.4	**5.5**	1.1	26.6
Renal tubular dysfunction	**1.2**	0.1	7.9	**1.9**	0.2	14.9	**4.3**	0.8	21.6
Nephromegaly	**0.8**	0.1	6	**3.0**	0.4	20.2	**2.6**	0.4	13.8
Nephrocalcinosis	**4.3**	0.2	92.6	**5.7**	0.2	148.3	**2.5**	0.09	62.5
Impaired growth	**2,5**	0.1	63.9	**1.8**	0.03	95.6	**2.5**	0.09	62.5
Adipositas	**2.5**	0.1	63.9	**1.8**	0.03	95.6	**2.5**	0.09	62.5
Neurological concomitant disease	**1.4**	0.3	6.3	**0.5**	0.05	6.1	**0.2**	0.02	2.5
Neurological crises	**0.2**	0.02	2.5	**0.5**	0.05	6.1	**0.2**	0.02	2.5
Epilepsy/convulsion	**0.8**	0.1	3.4	**0.9**	0.1	5.5	**0.1**	0.02	1.7
ADS, behavioural disorders	**1.4**	0.3	6.3	**1.2**	0.1	8.2	**0.5**	0.08	3.2
Learning/language difficulties	**0.8**	0.05	13.5	**0.5**	0.02	15.2	**0.2**	0.01	6.6
Impaired psychomotor development	**0.7**	0.2	2.3	**1.5**	0.4	5.34	**0.7**	0.2	2.27
Death	**0.8**	0.01	42.5	**15**	0.7	306.4	**2.5**	0.09	62.5

in the general population simultaneously with the already existing screening procedure.

11.2.2.2 The Natural History of the Condition Should Be Adequately Understood and There Should Be a Latent Stage

If left untreated severe symptoms will result. Some of these symptoms were obvious in those patients included in our cross-sectional study who were diagnosed beyond the newborn period. Acute liver failure is the most devastating symptom, renal dysfunction is common, rickets as well as neurological symptoms will occur, a life-threatening complication is hepatocellular carcinoma (HCC) (Mayorandan et al. 2014).

In most patients, there is a latent phase of the disease where no symptoms occur (Mayorandan et al. 2014). Thus, if screening takes place within the first few days of life and treatment is instituted promptly in those newborns tested positive for HT1, symptoms can be prevented in most patients. Rare patients already display symptoms shortly after birth. One patient in our survey presented with nephromegaly and tubular dysfunction at age 12 days. These patients will benefit from initiation of treatment in the early symptomatic stage of disease. To our knowledge, there are no patients with a mild phenotype not requiring therapy.

11.2.2.3 There Should Be a Good Treatment Option

Nitisinone (NTBC, Orfadin®) has revolutionised the outcome in HT1. It has to be combined with a low-protein diet supplemented with a special amino acid mixture to prevent excessive elevation of tyrosine levels in plasma. The benefit for the first HT1 patients treated with nitisinone was so impressive that this compound was granted approval under exceptional circumstances.

In our survey, there was a clear benefit of early treatment (cf. Table 11.1). In line with previous

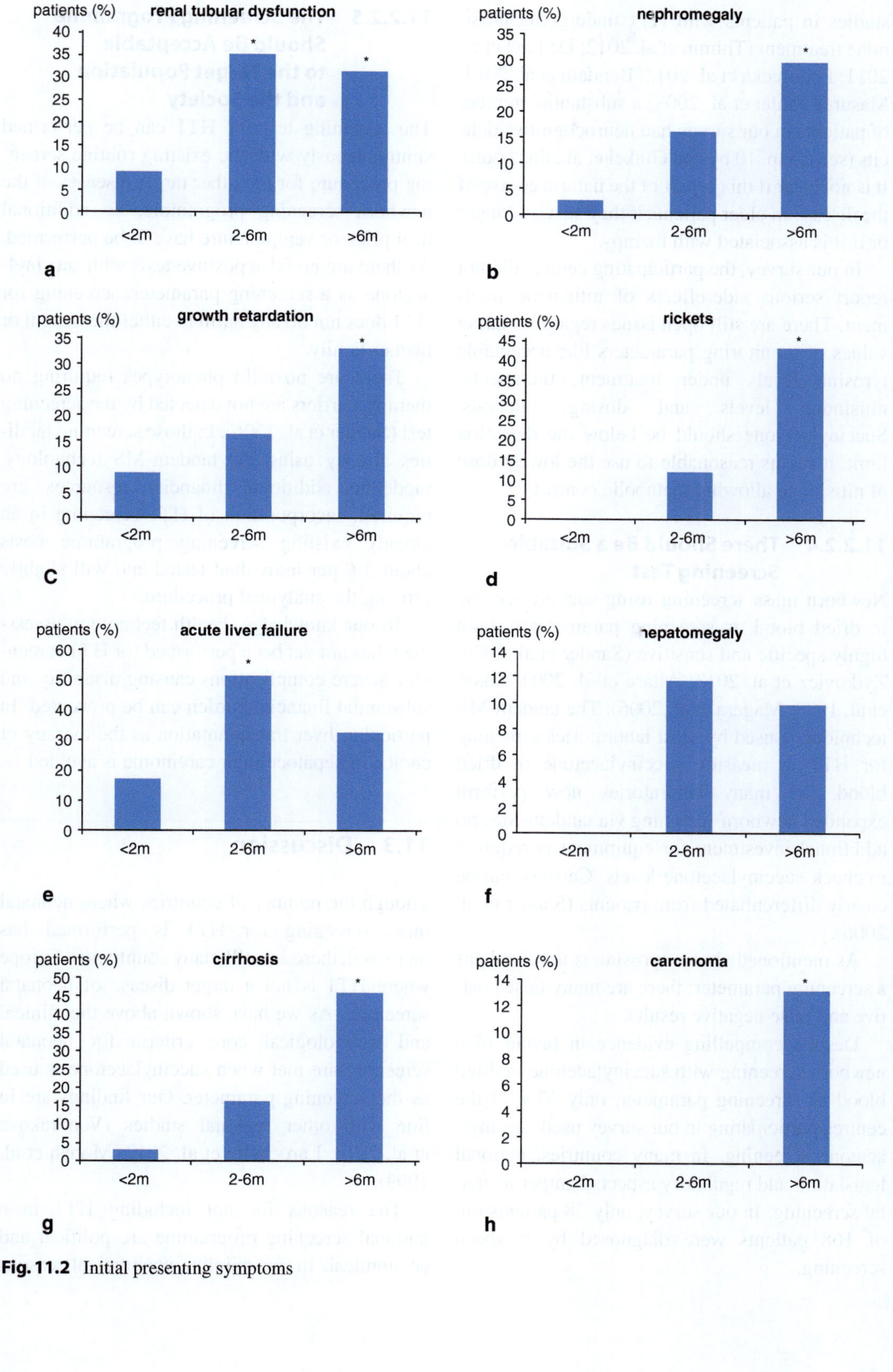

Fig. 11.2 Initial presenting symptoms

studies in patients with HT1 undergoing nitisinone treatment (Thimm et al. 2012; De Laet et al. 2011; Pohorecka et al. 2012; Bendadi et al. 2013; Masurel-Paulet et al. 2008) a substantial number of patients in our survey had neurocognitive deficits (see Chap. 10 by van Ginkel et al., this book). It is not clear if this is part of the natural course of the disease in older patients if they survive longer or if it is associated with therapy.

In our survey, the participating centres did not report serious side-effects of nitisinone treatment. There are still open issues regarding target values of monitoring parameters like acceptable tyrosine levels under treatment, therapeutic nitisinone levels and dosing intervals. Succinylacetone should be below the detection limit, it seems reasonable to use the lowest dose of nitisinone allowing metabolic control.

11.2.2.4 There Should Be a Suitable Screening Test

Newborn mass screening using succinylacetone in dried blood as screening parameter is both highly specific and sensitive (Sander et al. 2006; Zytkovicz et al. 2013; Allard et al. 2004; Jakob et al. 1988; Magera et al. 2006). The tandem-MS technique is used by most laboratories screening for HT1 to measure succinylacetone in dried blood. As many laboratories now perform expanded newborn screening via tandem-MS, no additional investment for equipment is required to check succinylacetone levels. Carriers can be clearly differentiated from patients (Sander et al. 2006).

As mentioned above, tyrosine is unreliable as a screening parameter; there are many false positive and false negative results.

Despite compelling evidence in favour of a newborn screening with succinylacetone in dried blood as screening parameter, only 57% of the centres participating in our survey used succinylacetone screening. In many countries, national legislation and regulatory aspects hamper neonatal screening. In our survey, only 28 patients out of 168 patients were diagnosed by newborn screening.

11.2.2.5 The Screening Programme Should Be Acceptable to the Target Population and the Society

The screening test for HT1 can be performed simultaneously with the existing routine screening procedure for the other target diseases of the newborn screening programme, no additional heel prick or venipuncture have to be performed. As there are no false positive tests with succinylacetone as a screening parameter, screening for HT1 does not do any harm to either the patient or his/her family.

There are no mild phenotypes requiring no therapy, carriers are not detected by the screening test (Sander et al. 2006). In those screening facilities already using the tandem-MS technology, moderate additional financial resources are required. Incorporation of HT1 screening in an already existing screening programme costs about 2 € per individual tested and will slightly prolong the analytical procedure.

To our knowledge, health technology assessment has not yet been performed for HT1 screening. Severe complications causing disability and substantial financial burden can be prevented. In particular, liver transplantation as the therapy of choice in hepatocellular carcinoma is avoided.

11.3 Discussion

Though the number of countries where neonatal mass screening for HT1 is performed has increased, there are still many countries in Europe where HT1 is not a target disease of neonatal screening. As we have shown above the clinical and technological core criteria for neonatal screening are met when succinylacetone is used as the screening parameter. Our findings are in line with other regional studies (Vondrakova et al. 2010; Larochelle et al. 2012; Mohan et al. 1999).

The reasons for not including HT1 in a national screening programme are political and economical. In our opinion, economical reasons for not performing neonatal screening for HT1 could be ruled out if health technology assessment of neonatal screening versus selective screening were performed due to high morbidity and mortality and expensive liver transplantation in patients developing hepatocellular carcinoma. In view of the inferiority of selective screening for HT1, further extension of neonatal mass screening to other European countries is essential to improve the overall outcome of patients and reduce both mortality and morbidity. However, there are still open issues like the development of neurocognitive deficiencies in patients with HT1. It is unclear if these problems are part of the natural course of the disease in patients growing older, are due to late diagnosis and treatment or are side effects of dietary and/or pharmacological treatment. Prospective studies are necessary to solve this question and address other unmet needs like target levels of tyrosine in plasma, nitisinone in blood and dosing intervals of nitisinone.

References

Allard P, Grenier A, Korson MS, Zytkovicz TH (2004) Newborn screening for hepatorenal tyrosinemia by tandem mass spectrometry: analysis of succinylacetone extracted from dried blood spots. Clin Biochem 37:1010–1015

Andermann A, Blancquaert I, Beauchamp S, Costea I (2011) Guiding policy decisions for genetic screening: developing a systematic and transparent approach. Publ Health Genomics 14:9–16

Bendadi F, de Koning TJ, Visser G, Prinsen HC, de Sain NG, Verhoeven-Duif N, Sinnema G, van Spronsen FJ, van Hasselt PM (2013) Impaired cognitive functioning in patiens with tyrosinemia type I receiving nitisinone. J Pediatr 164:398–401

Bonham JR (2013) Impact of new screening technologies: should we screen and does phenotype influence this decision? J Inherit Metab Dis 36:681–686

Coskun T, Ozalp I, Kocak N, Yüce A, Caglar M, Berger R (1991) Type I hereditary tyrosinaemia: presentation of 11 cases. J Inherit Metab Dis 14:765–770

Couce ML, Aldamiz-Echevarria L, Baldellou A, Blasco J, Bueno MA, Dalmau J, De La Vega A, Del Toro M, Diaz C, Lama R, Leao E, Marrero M, Navas VM, Pintos G (2010) Recomendaciones y manejo de la tirosinemia hereditaria Tipo I o Tirosinemia hepatorrenal. An Pediatr (Barc) 73(279):e1–e4

Couce ML, Dalmau J, del Toro M, Pintos-Morell G, Aldamiz-Echevarria (2011) Tyrosinemia type I in Spain: mutational analysis, treatment and long-term outcome. Pediatr Int 53:985–989

De Laet C, Munoz VT, Jaeken J, Francois B, Carton D, Sokal EM, Dan B, Goyens PJ (2011) Neuropsychological outcome of NTBC-treated patients with tyrosinaemia type 1. Dev Med Child Neurol 53:962–964

Jakob C, Dorland L, Wikkering B, Kok RM, de Jong APJM, Wadman SK (1988) Stable isotope dilution analysis of succinylacetone using eletrone capture negative ion mass fragmentography: an accurate approach to the pre-neonatal diagnosis of hereditary tyrosinemia type I. Clin Chim Acta 171:223–223

Larochelle J, Alvarez F, Bussières JF, Chevalier I, Dallaire L, Dubois J, Faucher F, Fenyves D, Goodyer P, Grenier A, Holme E, Laframboise R, Lambert M, Lindstedt S, Maranda B, Melancon S, Merouani A, Mitchell J, Parizeault G, Pelletier L, Phan V, Rinaldo P, Scott CR, Scriver C, Mitchell GA (2012) Effect of nitisinone (NTBC) treatment on the clinical course of hepatorenal tyrosinemia in Québec. Mol Genet Metab 107:49–54

Loeber JG, Burgard P, Cornel MC, Rigter T, Weinreich SS, Rupp K, Hoffmann GF, Vittozzi L (2012) Newborn screening programmes in Europe; arguments and efforts regarding harmonization. Part 1. From blood spot to screening result. J Inherit Metab Dis 35(4):603–611

Magera MJ, Gunawardena ND, Hahn SH, Tortorelli S, Mitchell G, Goodman SI, Rinaldo P, Matern D (2006) Quantative determination of succinylacetone in dried blood spots for newborn screening of tyrosinemia type I. Mol Genet Metab 88:16–21

Masurel-Paulet A, Poggi-Bach J, Rolland MO, Bernard O, Guffon N, Dobbelaere D, Sarles J, de Baulny HO, Touaty G (2008) NTBC-treatment in tyrosinaemia type I: long-term outcome in French patients. J Inherit Metab Dis 31:81–87

Mayorandan S, Meyer U, Gokcay G, Segarra NG, de Baulny HO, van Spronsen F, Zeman J, de Laet C, Spiekerkoetter U, Thimm E, Maiorana A, Dionisi-Vici C, Moeslinger D, Brunner-Krainz M, Lotz-Havla AS, Cocho de Juan JA, Couce Pico ML, Santer R, Scholl-Bürgi S, Mandel H, Bliksrud YT, Freisinger P, Aldamiz-Echevarria LJ, Hochuli M, Gautschi M, Endig J, Jordan J, McKiernan P, Ernst S, Morlot S, Vogel A, Sander J, Das AM (2014) Cross-sectional study of 168 patients with hepatorenal tyrosinaemia and implications for clinical practice. Orphanet J Rare Dis 9:107

Mohan N, McKiernan P, Preece MA, Green A, Buckels J, Mayer AD, Kelly DA (1999) Indications and outcome of liver transplantation in tyrosinaemia type 1. Eur J Pediatr 158(Suppl 2):S49–S54

Pohorecka M, Biernacka M, Jakubowska-Winecka A, Biernacki M, Kusmierska K, Kowalik A, Sykut-Cegielska J (2012) Behavioral and intellectual func-

tioning in patients with tyrosinemia type I. Pediatr Endocrinol Diabetes Metab 18:96–100
Sander J, Janzen N, Peter M, Sander S, Holtkamp U, Schwahn B, Mayatepek E, Trefz FK, Das AM (2006) Newborn screening or hepatorenal tyrosinemia: tandem mass spectrometric quantification of succinylacetone. Clin Chem 52:482–487
Schiff M, Broue P, Chabrol B, De Laet C, Habes D, Mention K, Sarles J, Spraul A, Valayannopoulos V, Baulny H (2011) Heterogneity of follow-up procedures in French and Belgian patients with treated hereditary tyrosinemia type I: results of a questionnaire and proposed guidelines. J Inherit Metab Dis 35:823–829
Thimm E, Richter-Werkle R, Kamp G, Molke B, Herebian D, Klee D, Mayatepek E, Spiekerkoetter U (2012) Neurocognitive outcome in patients with hypertyrosinemia type I after a long-term treatment with NTBC. J Inherit Metab Dis 35:263–268
Vondrakova A, Tesarova M, Magner M, Docekalova D, Chrastina P, Prochazkova D, Zeman J, Honzik T (2010) Clinical, biochemical and molecular characteristics in 11 Czech children with tyrosinemia type I. Cas Lek Cesk 149:411–416
Wilson JMG, Jungner G (1968) Principles and practice of screening for disease. Public Health Pap 34:26–35
Zytkovicz TH, Sahai I, Rush A, Odewale A, Johnson D, Fitzgerald E, Britton D, Eaton RB (2013) Newborn screening for hepatorenal tyrosinemia – I by tandem mass spectrometry using pooled samples: a four-year summery by the New England newborn screening program. Clin Biochem 46(7–8):681–684

12 Tyrosinemia Type I in Japan: A Report of Five Cases

Kimitoshi Nakamura, Michinori Ito, Yosuke Shigematsu, and Fumio Endo

Abstract

Tyrosinemia type I in Japan was reported for the first time in 1957 by Sakai et al. (Jikei Med J 2:1–10, 1957) and Kitagawa et al. (Proc Jpn Acad Ser B 88:192–200, 1957). Five cases of patients with tyrosinemia type I were reported to be definitively diagnosed in Japan. The first case was reported by Sakai et al. and Kitagawa et al. To the best of our knowledge, this was the first definite report in the world The second and third cases were those of a brother and a sister who underwent liver transplantation and who were the children of a Japanese-descent migrant worker; the fourth case was that of a girl who underwent liver transplantation after 2-(2-nitro-4-trifluoromethylbenzoyl)-1,3-cyclohexanedione (NTBC) treatment, which was reported by Hata et al.; and the fifth case was that of a patient who was administered NTBC, which was reported by Ito et al. These were of the subacute type, wherein residual activity was considerably present. When combined therapy with a low phenylalanine and tyrosine diet and NTBC administration is started after early diagnosis, patients can survive without liver transplantation. Development of liver cancer is not found in the cases in Japan, but performing liver transplantation without delay is necessary when liver cancer is found.

Keywords

HT1 Japan • NTBC

K. Nakamura (✉) • F. Endo
Department of Pediatrics, Graduate School of Medical Sciences, Kumamoto University, Kumamoto, Japan
e-mail: nakamura@kumamoto-u.ac.jp

M. Ito
Kagawa Children's National Hospital, Kagawa, Japan

Y. Shigematsu
Department of Health Science, Fukui University, Fukui, Japan

Abbreviations

AFP	alpha-fœtoprotein
ALP	alkaline phosphatase
APPT	activated partial thromboplastin time
FAH	fumarylacetoacetate hydrolase
HPT	heptoplastin T
γ-GTP	γ-glutamyl transferase

© Springer International Publishing AG 2017

R.M. Tanguay (ed.), *Hereditary Tyrosinemia*, Advances in Experimental Medicine and Biology 959,
DOI 10.1007/978-3-319-55780-9_12

Tyrosinemia type I in Japan was reported for the first time in 1957 by Sakai et al. (1957) and and Kitagawa et al (1957). Five cases of patients with tyrosinemia type I were reported to be definitively diagnosed in Japan. The first case was reported by Sakai et al. and Kitagawa et al. To the best of our knowledge, this was the first definite report in the world. The second and third cases were those of a brother and a sister who underwent liver transplantation and who were the children of a Japanese-descent migrant worker; the fourth case was that of a girl who underwent liver transplantation after 2-(2-nitro-4-trifluoromethylbenzoyl)-1,3-cyclohexanedione (NTBC) treatment, which was reported by Hata et al.; and the fifth case was that of a patient who was administered NTBC, which was reported by Ito et al. These were of the subacute type, wherein residual activity was considerably present. When combined therapy with a low phenylalanine and tyrosine diet and NTBC administration is started after early diagnosis, patients can survive without liver transplantation. Development of liver cancer is not found in the cases in Japan, but performing liver transplantation without delay is necessary when liver cancer is found.

12.1 Case 1: The First Tyrosinemia Type I Report

The initial case reported that occurred in Japan in 1955 was that of a 2-year-old boy. Based on the report by Sakai and Kitagawa (1957), he was born of a consanguineous marriage, and his birth weight was 3900 g. He had normal growth and development during infancy, but appetite loss and abdominal distension were noticed at the age of 2 years. Medical examination revealed that the spleen and liver were 5 cm and 4 cm upon palpation, respectively. A significant increase in urinary 4-hydroxyphenyllactate was shown after hospitalization. The child had diarrhea and weight loss and presented with hypophosphatemia, suggesting Fanconi rickets. Hepatosplenomegaly and appetite loss worsened, and he developed hepatic coma and subsequently died at 5 years of age. Pathological examination showed cirrhosis and liver cancer. From these courses, this case is thought to be the first report of tyrosinemia type I in the world.

12.2 Cases 2 and 3: Children of a Japanese-Descent Migrant Worker (Ueda et al. 2005)

The son of a Japanese-Brazilian father and a Brazilian mother was diagnosed with tyrosinemia type I, and underwent live donor liver transplant 5 months after birth. His younger sister was born weighing 4130 g and underwent blood amino acid analysis because her brother had tyrosinemia type I. The tyrosine and phenylalanine blood levels were 645.7 nmol/ml and 65.5 nmol/ml, respectively. She was treated with phenylalanine- and tyrosine-removed milk, but hypoproteinemia and abnormality of the coagulation system were found (Table 12.1) subsequently. The mother became a donor for the older brother, while the father had difficulty becoming a liver transplant donor because of his disease. The girl was offered the use of NTBC from abroad, which was not commercially available at that time. NTBC administration started 2 months after birth. Eruption, anemia, hypoproteinemia, and

Table 12.1 Clinical examination in administration

WBC 10.2x10^3 /μl	Alb 4.1 g/dl	PT 14.8 sec
RBC 4.67x10^6 /μl	UA 3.6 mg/dl	APTT 44.3 sec
Hb 13.4 g/ dl	BUN 5 mg/dl	HPT 24.8%
Ht 39.3%	Cr 0.3 mg/dl	Factor II 33.6%
Plt 15.1x10^4 /μl	Na 137 mEq/L	FactorV 105.0%
AST 42 IU/L	K 4.7 mEq/L	Factor VII 35.5%
Alt 26 IU/L	Cl 104 mEq/L	Factor VIII 120%
LDH 480 IU/L	Ca 9.8 mg/dl	Factor IX 39.9%
T-Bil 0.3 mg/dl	P 5.5 mg/dl	Factor X 43.2%
ALP 1154 IU/L g-	Mg 2.4 mg/dl	Factor XI 53.2%
GTP 101 IU/L	CRP 0.04 mg/dl	Factor XII 19.3%
TBA 153.1 μmol/L	IgG 818 mg/dl	Factor XIII >70%
T-Cho 161 mg/ dl	IgA 32 mg/dl	Serum tyrosine
TG 111 mg/ dl	IgM 99 mg/dl	350 μmol/L
TP 6.6 g/ dl	AFP 46,072 ng/ml	

liver function improved dramatically after starting NTBC. She was offered liver donation from an aunt on her mother's side at 1 year 9 months of age and underwent live donor liver transplant. Hypertyrosinemia was not observed after NTBC discontinuation after the liver transplantation.

12.3 Case 4: A Girl Reported by Hata and Shigematsu (2012)

A girl was born at 38 weeks gestational age at 2740 g without problem during the perinatal period or neonatal mass screening. Her sibling was normal. Poor weight gain was observed 1 month after birth, and epistaxis was noted 3 months after birth. Abdominal distension and edema occasionally developed 4 months after birth. Hypoproteinemia, coagulation system abnormality, and slight transaminase level increase were found in a blood examination. Alphafetoprotein (AFP) and tyrosine on amino acid analysis of the filter paper was 77,000 ng/ml and 540 nmol/ml, respectively. Urine organic acid analysis showed significant increase of succinylacetone level at 143 mmol/mol Cr. She was diagnosed with tyrosinemia type I because of elevation of AFP, tyrosine, and succinylacetone levels. She was administered phenylalanine- and tyrosine-removed milk and NTBC. She underwent partial liver transplantation from a live donor mother 5 months after birth.

12.4 Case 5: A Boy Reported by Ito et al. 2005

A boy was born without abnormality at 39 weeks and 4 days of gestation at 4074 g birth weight to parents without consanguineous marriage. Asphyxia or jaundice was not observed. No abnormality was found in his two siblings. The neonatal mass screening performed 5 days after birth was normal. He was administered vitamin K because the result of the hepaplastin test (HPT) 6 days after birth was 24%. The low level of HPT (14.5%) continued, and he was followed up as a case of congenital factor VII deficiency because the factor VII was found to be 8%. He was admitted to the hospital for examination because hepatosplenomegaly was found at 4 months of age. He was suspected to have tyrosinemia type I because of significant hepatosplenomegaly, mild liver dysfunction, high AFP level at 79,274 ng/ml, decreased coagulation factor, and increased urinary tyrosine excretion. He was transferred to a different hospital for close medical examination and treatment 6 months after birth. Bulbar conjunctiva did not show the jaundice, and heart and lung abnormality was not found upon admission. The abdomen was distended, and the liver and spleen were palpated to be 5.5 cm and 2.5 cm, respectively. No abnormal neurological finding was found.

12.5 Laboratory Findings

The laboratory findings at admission are shown in Table 12.1. A platelet count of $15.1 \times 10^4/\mu l$ decreased slightly. Biochemical examination of the blood revealed that alkaline phosphatase (ALP), γ-glutamyl transferase (γ-GTP), total bile acids, and AFP were 1154I U/L, 101I U/L, 153.1 μmol/l, and 46,072 ng/ml, respectively. Coagulation factors II, VII, IX, X, XI, and XII were low; prothrombin time (PT) and activated partial thromboplastin time (APTT) were prolonged; and HPT was low. Tyrosine level was high at 403 nmol/ml by blood amino acid analysis. Urinary organic acid analysis (Fig. 12.1) showed significantly increased excretion of tyrosine metabolites, such as 4-hydroxyphenylpyruvic acid, 4-hydroxyphenyl lactic acid, 4-hydroxyphenyl acetic acid, and significant succinylacetone level increase, and the patient was diagnosed with tyrosinemia type I. Fumarylacetoacetate hydrolase (FAH) activity of the liver biopsy specimen was 2.93 μmol/min/mg protein compared with FAH activity (5.14 μmol/min/mg protein) in the liver of patients with cirrhosis, and he was diagnosed with tyrosinemia type I. The pathological diagnosis of the liver biopsy specimen was cirrhosis.

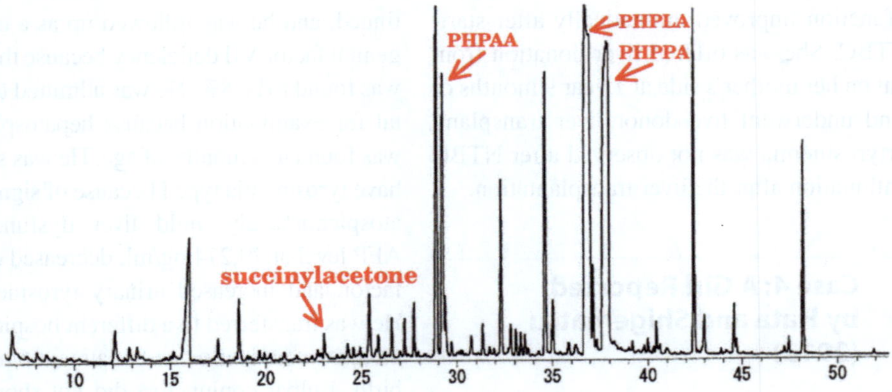

Fig. 12.1 Urinary organic acid analysis in case 5. Urinary organic acid analysis revealed elevation of succinylacetone, PHPAA, PHPLA and PHPPA

Table 12.2 Clinical examinations during NTBC administration

	Administration	Before	1 week	1 month	4 month	1 year	3 year	5 year	8 year
NTBC (mg/day)			9	9	10	15	15	20	30
Liver (cm)	5.5	5.0	5.0	4.5	3.0	2.5	2.0	2.0	0.5
Spleen (cm)	3.0	3.0	3.0	1.0	n.p.	n.p.	n.p.	n.p.	n.p.
Platelets (×10^4/μl)	15.1	25.0	27.3	46.6	33.3	23.6	24.1	20.1	23.4
HPT (%)	24.8	69.7	74.1	79.8	71.6	80.1	78.7	78.4	81.7
γ–GTP (IU/L)	101	128	113	97	32	18	17	15	15
TBA (μmol/l)	153.1	57.3	11.8	13.8	10.9	7.3	8.6	4.4	9.1
AFP (ng/ml)	46,072	2661	1475	463	91	8	5	4	4
Serum tyrosine (mg/dl)	7.30	0.82	1.06	0.32	4.70	5.60	8.91	8.11	6.33
EBC PBG synthase actibity (nkat/gHb)	–	0.030	0.900	1.070	0.850	0.830	0.720	–	–
Serum succinylacetone (mmol/l)	–	2.30	1.20	0.26	<0.10	0.12	<0.10	–	–
Urine 5–Aminolevulinic acid (mmol/mol Cr)	–	64.0	8.5	3.8	5.6	5.0	5.0	–	–
Urine Succinylacetone (mmol/mol Cr)	–	20.0	<1	<1	<1	<1	<1	–	–

n.p.: not palpable
(–): not evaluated

12.6 Clinical Course (Table 12.2)

NTBC therapy was administered after the chemical diagnosis by using organic acid analysis because the family consented. Professor Lindstedt from the Gotenberg University and Swedish Orphan Company provided NTBC. Before NTBC the administration, he was treated with a low tyrosine phenylalanine diet. PT, APTT, hepaplastin test, blood coagulation factors, and platelet count of 25.0 × 10^4/μl improved after the diet therapy. Biochemical test showed that the total bile acids and AFP at 57.3 μmol/l and 2331 ng/dl, respectively, decreased; however, γ-GTP level of 128 IU/l remained high. Porphobilinogen (PBG) synthase activity in red blood cells was decreased at 0.03 nkat/g Hb (normal range, 0.58–1.25 nkat/g Hb), and plasma succinylacetone of 2.30

μmol/l (normal range, <0.1 μmol/l) increased. Urinary succinylacetone of 20.0 mmol/mol Cr (normal range, 1 or less) and urine aminolevulinic acid level of 64.0 mmol/mol Cr (normal range, 0–3) were high. One week after NTBC administration, PBG synthase activity in the blood was within the normal range, and urinary succinylacetone was lower than detection limit. Two months after NTBC administration, blood coagulation system examination, γ-GTP, plasma succinylacetone, and total bile acids were within the normal range. However, AFP levels became normal after 1 year (less than 10 ng/dl), and urine aminolevulinic acid decreased by 8.5 mmol/mol Cr after 1 week, but did not decrease to normal range (3.0 mmol/mol Cr). Hepatosplenomegaly gradually improved after NTBC administration, and the spleen was not palpated 1 month after NTBC administration, and the liver surface became smooth 8 months after, and hepatomegaly improved to approximately 2 cm beyond the rib bottom 1 year later. Abdominal MR imaging did not exhibit any neoplastic lesion in the liver. a special formula was used in addition to NTBC therapy to keep the patient's serum tyrosine concentration to 3–5 mg/dl. The NTBC dose is 0.6–1 mg/kg/day and does not show the adverse effects.

12.7 Newborn Screening of Tyrosinemia Type I

A pilot study of the newborn screening of tyrosinemia type I was conducted by Shigematsu et al. (2007). He measured tyrosine level from a filter paper sample. If the tyrosine level was 200 nmol/ml or more, succinylacetone of the filter paper was measured as the second testing. The cut-off value of succinylacetone was 5.0 nmol/ml. Of the 27,905 neonate samples, 499 (1.79%) were assayed for the second test because of high concentrations with tyrosine. No positive sample with succinylacetone concentrations level above the cut-off value was found. Further screening was continued, but no sample with high succinylacetone concentrations above the cut-off value was found after screening approximately 500,000 specimens. Therefore, the tyrosinemia type I incidence in Japan seemed to be extremely low, and newborn screening of tyrosinemia type I by measuring succinylacetone does not seem to be realistic.

12.8 Discussion

A total of 5 cases of tyrosinemia type I were found in Japan and were of the subacute type, wherein residual activity was considerably present. When combined therapy with a low phenylalanine and tyrosine diet and NTBC administration is started after early diagnosis, patients can survive without liver transplantation. We should start NTBC administration initially while considering the indication of liver transplantation(Lindstedt et al. 1992; Kelsey et al. 1993; Grompe et al. 1995). However, development of liver cancer cannot be prevented in some cases even after NTBC administration (Michell et al. 2001), and performing liver transplantation is necessary in severe cases without improvement after NTBC administration. Development of liver cancer is not found in the cases in Japan, but performing liver transplantation without delay is necessary when liver cancer is found (Nakamura et al. 2015).

References

Grompe M et al (1995) Pharmacological correction of neonatal lethal hepatic dysfunction in a murine model of hereditary tyrosinemia type 1. Nat Genet 10:453–460

Hata I, Shigematsu Y (2012) A case of hypertyrosinemia I. Annual report of the Grant-in-aid for the research on rare and intractable diseases, Health and Labour Sciences Research. (Japanese)

Ito M, Matsuda J, Shinahara H (2005) A case of hypertyrosinemia I with favorable prognosis after administration of long-term NTBC. Bull Spec Formula 41:27–30. (Japanese)

Kelsey G et al (1993) Rescue of mice homozygous for lethal albino deletions: implications for an animal model for the human liver disease tyrosinemia type 1. Genes Dev 7:2285–2297

Lindstedt S et al (1992) Treatment of hereditary tyrosinemia type 1 by inhibition of 4-hydroxyphenylpyruvate dioxygenase. Lancet 340:813–817

Michell GA et al (2001) Hypertyrosinemia. In: Scriver CR et al (eds) The metabolic and molecular bases of inherited diseases, 8th edn. McGraw Hill, New York, pp 1777–1805

Nakamura K, Matsumoto S, Mitsubuchi H et al (2015) Diagnosis and treatment of hereditary tyrosinemia in Japan. Pediatr Int 57:37–40

Sakai K, Kitagawa T (1957) An atypical case of tyrosinosis. Part 1. Jikei Med J 2:1–10

Shigematsu K, Hata I (2007) Annual report for the research on rare and intractable diseases. Health Labour Sci Res Grants:86–88

Ueda A, Ito T, Okubo Y et al (2005) A case of hypertyrosinemia I. Bull Spec Formula 41:23–26

13

Newborn Screening for Hereditary Tyrosinemia Type I in Québec: Update

Yves Giguère and Marie-Thérèse Berthier

Abstract

Hereditary tyrosinemia type I (HTI) is a rare autosomal recessive disorder caused by a fumarylacetoacetate hydrolase (FAH) deficiency. If untreated, its acute form is characterized by hepatic failure, renal dysfunction and neurological crisis, and may lead to death. Due to a genetic founder effect in the French-Canadian population, the prevalence of HTI is increased in the province of Quebec (1/19 819), with the IVS12 + 5G>A (1062 + 5G>A) splice site mutation responsible for more than 90% of mutated alleles. Universal newborn screening for (HT1) was thus established in 1970, and close to four million infants have been tested so far, allowing to identify 185 of the 190 affected newborns. During the 1970–1997 period, 2,249,000 newborns were screened at 3–7 days of life on dried filter paper blood spots by tyrosine (Tyr) concentration followed by indirect colorimetric semi-quantitative and quantitative (Q) succinylacetone (SA) testing (red blood cells δ-aminolevulinate dehydratase inhibition), with immunoreactive FAH as the confirmatory test. This approach allowed to identify 118 of 123 affected newborns. In 1998, owing to earlier hospital discharge and increased rate of breastfeeding, four cases were missed within the same year as the discriminating power of blood Tyr became inadequate. Thus, the screening algorithm was modified: indirect semi-quantitative SA

Y. Giguère (✉)
Programme de dépistage néonatal sanguin, CHU de Québec – Université Laval, 10, rue de l'Espinay, Québec, QC, G1L 3L5, Canada

Department of Molecular Biology, Medical Biochemistry and Pathology, Faculty of Medicine, Université Laval, Québec City, QC, Canada
e-mail: yves.giguere@crchudequebec.ulaval.ca

M.-T. Berthier
Programme de dépistage néonatal sanguin, CHU de Québec – Université Laval, 10, rue de l'Espinay, Québec, QC, G1L 3L5, Canada

© Springer International Publishing AG 2017

R.M. Tanguay (ed.), *Hereditary Tyrosinemia*, Advances in Experimental Medicine and Biology 959,
DOI 10.1007/978-3-319-55780-9_13

measurement became the first-tier test between 1998 and 2014, and direct SA measurement by tandem mass spectrometry (MS/MS) was implemented in 2014, followed by indirect quantitative SA measurement as second tier test. Confirmation is performed by plasmatic amino acid profile and molecular testing. During the 1998–2016 period, more than 1,5 million neonates have been tested (90% sampled between 24 and 48 h of life): 67 of the 67 HTI cases were identified. Both indirect and direct SA measurement as the initial HTI screening test proved to be highly sensitive and specific, with positive and negative predicting value of 79% and 100% respectively.

Keywords
Hereditary tyrosinemia type I • Genetic founder effect • Newborn screening • Tyrosine • Succinylacetone

13.1 Screening for HT1 in Quebec: Background

Hereditary tyrosinemia type I (HT1, McKusick 27,670), also called hepatorenal tyrosinemia, is a rare autosomal recessive inborn error of metabolism caused by a deficiency of fumarylacetoacetate hydrolase (FAH), in the last step of tyrosine catabolic pathway (Fig. 13.1). FAH deficiency leads to an accumulation of fumarylacetoacetate (FAA), maleylacetoacetate (MAA), and succinylacetone (SA), which are responsible for the clinical symptoms of the disorder. The worldwide incidence of HTI is estimated to be 1:100,000–125,000 (Mitchell et al. 2001; Halvorsen 1980). Close to 100 different mutations in the *FAH* gene have been identified so far, many of which have specific ethno-geographic distributions (Angileri et al. 2015).

The French-Canadian population of Quebec, representing about 6,5 of the 8,1 million inhabitants, descents from about 8500 French settlers from a few villages in northern France who arrived in Nouvelle-France between 1608 and 1759. The migration of those settlers and their descendants led to genetic founder effects for various disorders, including HT1, which are reflected in the geographical distribution of genetic diseases in the province (Laberge et al. 2005). Indeed, due to a complex founder effect (De Braekeleer and Larochelle 1990; Laberge 1969), the frequency of HT1 is much higher in the province of Quebec where it has been previously estimated at 1:16,786, with the highest frequency of 1:1846 in the Saguenay Lac-Saint-Jean region (SLSJ) in the northeastern part of Quebec, and an estimated carrier rate between 1:20 and 1:31 (De Braekeleer and Larochelle 1990; Laberge and Dallaire 1967). In the SLSJ region of Quebec, owing to the founder effect, about 95% of mutated alleles are due to a splice site mutation (IVS12 + 5G–>A; 1062 + 5G>A) in the *FAH* gene (Grompe et al. 1994; Poudrier et al. 1996).

HTI may present either as an acute or a chronic form. In the acute form, the prenatal and perinatal periods are unremarkable, but symptoms appear in the neonatal period. It is characterized by acute liver failure, peripheral porphyria-like neurological crises and coagulation disorders. In the absence of treatment, neonates affected by the acute form usually present with hepatic decompensation and porphyria-like neurological crises, caused by the accumulation of 5-aminolevulinic acid (ALA) due to ALA dehydratase inhibition by SA (Scott 2006), which is often precipitated by a catabolic state. In the chronic form, the child presents with failure to thrive and develops chronic

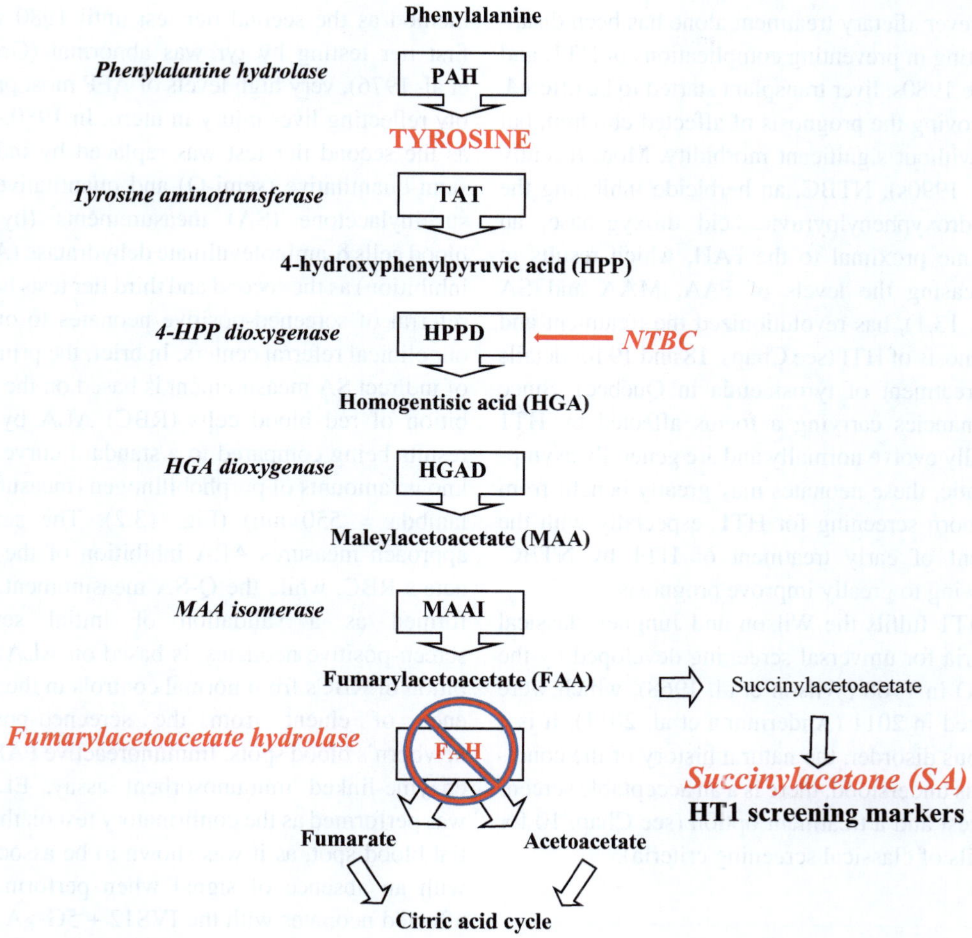

Fig. 13.1 Catabolic pathway of tyrosine
NTBC: 2-[2-nitro-4-(trifluoromethyl) benzoyl] cyclohexane-1,3-dione

liver disease, renal tubular dysfunction and rickets due to renal Fanconi syndrome. Untreated patients are at high risk of developing liver cirrhosis and hepatocellular carcinoma (Scott 2006). The acute form is referred to as the "French Canadian" type, while the chronic form is referred to as the "Scandinavian type", but this classification may be misleading as both forms may coexist.

As affected neonates are usually asymptomatic in the first month of life and usually present with unspecific symptoms, in the absence of newborn screening, diagnosis may go unrecognized and will often be delayed, leading to serious complications and poor prognosis. Indeed, in severely affected neonates that were not screened and identified asymptomatically, the age of onset of symptoms is an important prognostic factor: before inception of treatment by 2-(2-nitro-4-trifluoromethylbenzoyl)-1,3 cyclohexanedione (NTBC) in the mid-90's, 1-year mortality was 60% in neonates who became symptomatic before 2 months of age, compared to 4% in children presenting signs and symptoms after 6 months of age (van Spronsen et al. 1994).

Treatment of HT1 has evolved greatly over the last decades. Before the 1980s, the only treatment proposed was a diet restricted in both phenylalanine and tyrosine, with the hope of obtaining successful therapy, in line with the success of early dietary treatment against phenylketonuria.

However, dietary treatment alone has been disappointing in preventing complications of HT1, and in the 1980s, liver transplant started to be offered, improving the prognosis of affected children, but not without significant morbidity. More recently (mid 1990s), NTBC, an herbicide inhibiting the 4-hydroxyphenylpyruvic acid dioxygenase, an enzyme proximal to the FAH, which results in decreasing the levels of FAA, MAA and SA (Fig. 13.1), has revolutionized the treatment and prognosis of HTI (see Chaps. 18 and 19 for details on treatment of tyrosinemia in Québec). Since pregnancies carrying a foetus affected by HT1 usually evolve normally and are generally asymptomatic, these neonates may greatly benefit from newborn screening for HT1, especially with the advent of early treatment of HT1 by NTBC, allowing to greatly improve prognosis.

HT1 fulfils the Wilson and Jungner classical criteria for universal screening developed by the WHO in 1968 (Wilson et al. 1968), which were revised in 2011 (Andermann et al. 2011): it is a serious disorder, the natural history of the condition is understood, there is a an acceptable screening test and a treatment option (see Chap. 10 for details of classical screening criteria).

13.2 Screening for HT1 in Quebec: Historical Perspective 1970–1997

With the above considerations in mind, in the context of an increased frequency of HTI in the province of Quebec, and following the public health success of newborn screening for PKU (which began in October 1969 in Quebec), universal newborn screening for HT1 was implemented as a pilot project in the fall of 1970 and offered without interruption in the province since then. This represents close to four million newborns screened for HT1 over the last 45 years.

During the 1970–1997 period, 2,249,000 newborns were screened for HT1 at 3–7 days of life by tyrosine (tyr) concentration from dried filter paper blood spots as the first tier test, followed by α-fetoprotein (AFP) performed by a fluorimetric method as the second tier test until 1980 when first tier testing by tyr was abnormal (Grenier et al. 1976), very high levels of AFP most probably reflecting liver injury in utero. In 1980, AFP as the second tier test was replaced by indirect semi-quantitative (semi-Q) and quantitative (Q) succinylacetone (SA) measurements (by red blood cells δ-aminolevulinate dehydratase (ALA) inhibition) as the second and third tier tests before referral of screened-positive neonates to one of our clinical referral centers. In brief, the principle of indirect SA measurement is based on the inhibition of red blood cells (RBC) ALA by SA, results being compared to a standard curve with known amounts of porphobilinogen (measured at lambda = 550 nm) (Fig. 13.2). The semi-Q approach measures ALA inhibition of the neonate's RBC, while the Q-SA measurement, performed as a validation of initial semi-Q screen-positive neonates, is based on ALA inhibition of RBCs from normal controls in the presence of eluent from the screened-positive newborn's blood spots. Immunoreactive FAH (by enzyme-linked immunosorbent assay, ELISA) was performed as the confirmatory test on the initial blood spot, as it was shown to be associated with an absence of signal when performed in affected neonates with the IVS12 + 5G->A FAH mutation (Laberge et al. 1990).

The high specificity of SA measurements as the second and third tier screening tests, although indirect through ALA dehydratase inhibition, led to very rare false-positive results, while allowing to identify neonates affected with HT (Table 13.1). However, owing to earlier hospital discharge of postpartum mothers and their babies and increasing rates of breastfeeding, even if the tyr cut-off levels to identify screen-positive neonates were gradually lowered from 414 μmol/L in 1970 to 200 μmol/L in 1997, the sensitivity, specificity and discriminating power of first tier tyr testing became insufficient. In 1997, four cases of HTI with tyr < 200 μmol/L (tyr levels of 198, 193, 168, and 166 mol/L) were missed within a year, as they did not proceed to second tier indirect semi-Q and Q-SA measurement (but showed later to have both increased semi-Q and Q-SA

Fig. 13.2 Principle of blood spot indirect succinylacetone measurement by the inhibition of red blood cells δ-aminolevulinate dehydratase from succinlyacetone

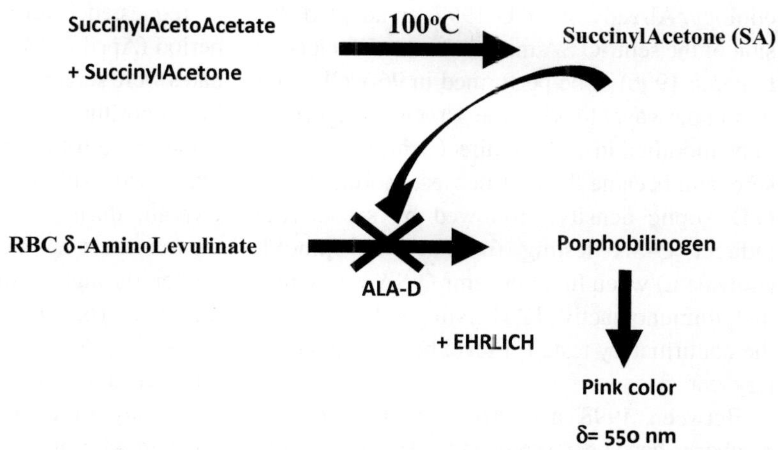

levels). Global figures for the 1970–1997 period are shown in Table 13.1.

Thus, if the presence of increased levels of tyr from blood spots may suggest HTI, tyr measurement in the first 24–48 h of life is neither a sensitive, nor a specific screening marker of HT1, and normal screening tyr level did rule out a diagnosis of HT1 in the new context of early discharge of mothers and infants in the mid-1990s.

13.3 Screening for HT1 in Quebec: Historical Perspective 1998-...

Following early discharge of neonates after birth, tyr did not qualify anymore as an adequate first tier screening test for HT1. In addition, elevated tyr is also observed in tyrosinemia type II and III, transient tyrosinemia of the newborn, prematurity and hepatocellular dysfunction of almost any

Table 13.1 Global performance of newborn screening for hereditary tyrosinemia type I in Quebec between 1970 and January 2016

	1970–1997 (Tyrosine as the first tier)	1998–2016[a] (Succinylacetone as the first tier)	1970–2016[a]
Newborn tested	2,249,000	1,516,517	3,765,517
Time of sampling	3–7 days of life	>90% before 72 h of life	>90% between 24 h and 7 days of life
Total number of cases	123	67	190
Cases detected	118	67	185
Cases missed (acute form)	5	0	5
Sensitivity	95.9%	100%	97,4%
Positive predictive value	88.1% (after SA measurement as 2nd tier)	79.1%	85.0%
False positive rate	0.01% (after SA measurement as second tier)	0.0001%	0.001%
Incidence	1:18,285	1:22,634	1:19,819

[a]Data as of January 1st, 2016. Between 1998 and 2012, first tier testing was performed by indirect measurement of SA through ALA dehydratase inhibition, while direct (MS/MS) SA measurement using hydrazine monohydrate as per the PerkinElmer non-derivatized NeoBase kit® and assay solution is performed by tandem mass spectrometry as the first tier screening test since 2014 (after a transition between 2012 and 2014 where both indirect and direct SA measurements were performed in parallel as first tier screening test)
SA succinylacetone

etiology. Already in 1996–1997, an adapted version of the semi-Q SA measurement (Grenier and Laberge 1996) to be performed in 96-well plates was underway, allowing the screening algorithm to be modified in 1998: indirect semi-Q SA measurement became the first-tier test (normal >0.20 O.D.; optic density), followed by second tier indirect Q-SA testing (normal <2.5 μmol/L equivalent) when first tier semi-Q SA was abnormal, immunoreactive FAH testing still serving as the confirmatory test. Tyr level became an auxiliary test.

Between 1998 and April 2014, 1,366,294 newborns have been screened for HTI using indirect semi-Q SA testing as the first tier; over 90% of blood spots were sampled between 24 and 48 h of age. During this period, only 152 (0.0001%) neonates showed abnormal results after indirect semi-Q SA, 85 of them normalized after indirect Q-SA and FAH measurement, while 53 of the remaining 67 with abnormal by indirect Q-SA and FAH referred to a specialized clinic were confirmed to be affected by HTI (positive predictive value = 79,1%; incidence of HT1 = 1:25,779). No cases of acute form of HT1 were missed. Of note, between 1998 and September 2011, while tyr was still measured by fluorimetry (before we implemented tandem mass spectrometery (MS/MS) measurement), the average blood spot tyr level of confirmed HT1 cases (n = 44) was 189 ± 67 μmol/L compared with 151 ± 51 μmol/L in normal newborns (n = 1,150,888), showing significant overlap, confirming poor discriminating power of tyr.

In 2011, the newborn screening program was enhanced by the acquisition of tandem mass spectrometers. Between 2012 and 2014, direct SA measurement using MS/MS was performed using hydrazine monohydrate as per the PerkinElmer non-derivatized Neobase kit® in parallel with the first tier indirect semi Q-SA methods. In April 2014, both the indirect semi-Q and FAH measurement were removed from the screening algorithm. Since then, HT1 is screened by direct SA measurement by MS/MS as the first tier test (normal <0.7 μmol/L), followed by indirect Q-SA measurement when SA is positive by MS/MS. In case of discordance, a repeat sample is requested. During the following 20-month period (April 2014–January 2016), 150,223 neonates were screened for HT1 with the new screening algorithm, and 14/19 (positive predictive value = 73,6%) screen-positive neonates were diagnosed with HT1 (incidence of 1:10,730). Overall, during the 17-year interval since SA measurement has been used as the first tier test (either through indirect or direct SA measurement), 1,516,517 newborns have been screened for HT1: 86 screen-positive newborns were referred to a specialized referral center at about 10–15 days of life, while 67 were confirmed to be affected with the disorder, resulting to a positive predictive value = 79,1% (incidence of TYR1: 1:22,634). Since 1998, SA measurement as first tier screening test, either indirect through ALA dehydratase inhibition or by direct MS/MS measurement, proved to be highly sensitive and specific, and to our knowledge no cases were missed (negative predictive value = 100%). Data for the April 1998–January 1st 2016 period and global figures since 1970 are shown in Table 13.1.

13.4 Conclusion

HT1 is a rare disease presenting with non-specific signs and symptoms and may be overlooked in the investigation of an affected child. Owing to a founder effect, there is an increased incidence of HTI in the province of Quebec. Since inception of universal newborn screening for HT1 in the province in the fall of 1970, the program screened close to four million infants and identified 185 of the 190 cases of HT1 over the last 45 years, with a global frequency of HTI of 1:19,819. Interestingly, this represents more affected neonates and a higher frequency compared to classical phenylketonuria (Fig. 13.3).

Over the years, the newborn screening program had to adapt to changes in the management of the postpartum period of mothers and their babies in the health care system: when earlier hospital discharge of mothers and their newborns was instituted, tyrosine did not have sufficient sensitivity and specificity as a first tier screening test and was replaced by SA measurement.

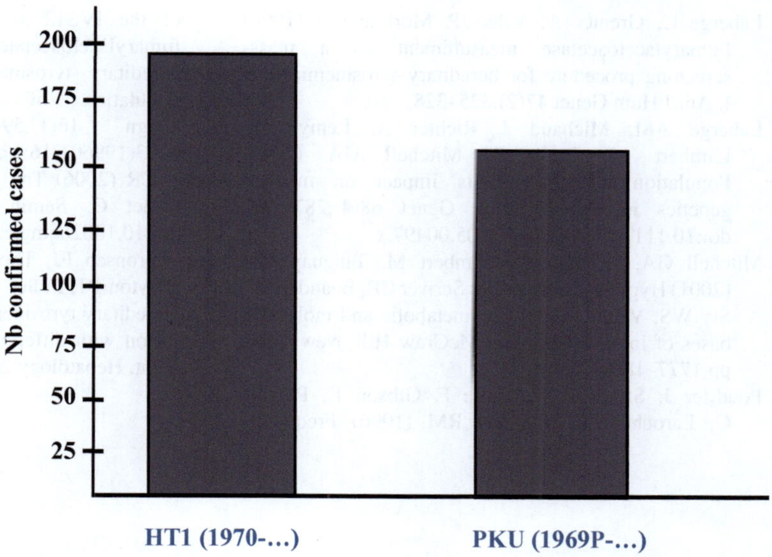

	HT1 (1970-...)	PKU (1969P-...)
Nb tested	3 765 517	3 810 497
Nb cases	190	158
Frequency	1/19 810	1/24 117
N/100 000	5,05	4,15

Fig. 13.3 Comparison of the number of newborns tested, cases confirmed and frequency of disorder between tyrosinemia type 1 (HT1) and classical phenylketonuria (PKU) by the Quebec Newborn screening program from October 1969 to January 1st 2016

For decades, the province of Quebec was the only jurisdiction offering HT1 universal newborn screening. More recently, the introduction of NTBC as an effective treatment, combined with the implementation of MS/MS technology in newborn screening, led to the addition of HT1, using SA as a sensitive and specific primary screening marker of the disease, to the screening panel of many jurisdictions worldwide.

References

Andermann A, Blancquaert I, Beauchamp S, Costea I (2011) Guiding policy decisions for genetic screening: developing a systematic and transparent approach. Public Health Genomics 14(1):9–16. doi:10.1159/000272898

Angileri F, Bergeron A, Morrow G, Lettre F, Gray G, Hutchin T, Ball S, Tanguay RM (2015) Geographical and ethnic distribution of mutations of the fumarylacetoacetate hydrolase gene in hereditary tyrosinemia type 1. JIMD Rep 19:43–58. doi:10.1007/8904_2014_363

De Braekeleer M, Larochelle J (1990) Genetic epidemiology of hereditary tyrosinemia in Quebec and in Saguenay-Lac-St-Jean. Am J Hum Genet 47(2):302–307

Grenier A, Laberge C (1996) Neonatal screening for tyrosinemia type I and early sampling. In: Proceedings of the Third International Society for Neonatal Screening, Boston, Oct 20–23

Grenier A, Belanger L, Laberge C (1976) alpha1-Fetoprotein measurement in blood spotted on paper: discriminating test for hereditary tyrosinemia in neonatal mass screening. Clin Chem 22(7):1001–1004

Grompe M, St-Louis M, Demers SI, al-Dhalimy M, Leclerc B, Tanguay RM (1994) A single mutation of the fumarylacetoacetate hydrolase gene in French Canadians with hereditary tyrosinemia type I. N Engl J Med 331(6):353–357. doi:10.1056/NEJM199408113310603

Halvorsen S (1980) Screening for disorders of tyrosine metabolism. In: Bickel H, Guthrie R, Hammerson G (eds) Neonatal screenings for inborn errors of metabolism. Springer, New York

JMG W, Jungner G, World Health Organization (1968) Principles and practice of screening for disease, Public Health Papers, vol 34. World Health Organization, Geneva

Laberge C (1969) Hereditary tyrosinemia in a French Canadian isolate. Am J Hum Genet 21(1):36–45

Laberge C, Dallaire L (1967) Genetic aspects of tyrosinemia in the Chicoutimi region. Can Med Assoc J 97(18):1099–1101

Laberge C, Grenier A, Valet JP, Morissette J (1990) Fumarylacetoacetase measurement as a mass-screening procedure for hereditary tyrosinemia type I. Am J Hum Genet 47(2):325–328

Laberge AM, Michaud J, Richter A, Lemyre E, Lambert M, Brais B, Mitchell GA (2005) Population history and its impact on medical genetics in Quebec. Clin Genet 68(4):287–301. doi:10.1111/j.1399-0004.2005.00497.x

Mitchell GA, Grompe M, Lambert M, Tanguay RM (2001) Hypertyrosinemia. In: Scriver CR, Beaudet AL, Sly WS, Valle D (eds) The metabolic and molecular bases of inherited disease. McGraw Hill, New York, pp 1777–1806

Poudrier J, St-Louis M, Lettre F, Gibson K, Prevost C, Larochelle J, Tanguay RM (1996) Frequency of the IVS12 + 5G–>A splice mutation of the fumarylacetoacetate hydrolase gene in carriers of hereditary tyrosinaemia in the French Canadian population of Saguenay-Lac-St-Jean. Prenat Diagn 16(1):59–64. doi:10.1002/(SICI)1097-0223(199601)16:1<59::AID-PD810>3.0.CO;2-D

Scott CR (2006) The genetic tyrosinemias. Am J Med Genet C: Semin Med Genet 142C(2):121–126. doi:10.1002/ajmg.c.30092

van Spronsen FJ, Thomasse Y, Smit GP, Leonard JV, Clayton PT, Fidler V, Berger R, Heymans HS (1994) Hereditary tyrosinemia type I: a new clinical classification with difference in prognosis on dietary treatment. Hepatology 20(5):1187–1191

14 Hepatorenal Tyrosinemia in Mexico: A Call to Action

Isabel Ibarra-González, Cecilia Ridaura-Sanz, Cynthia Fernández-Lainez, Sara Guillén-López, Leticia Belmont-Martínez, and Marcela Vela-Amieva

Abstract

Hepatorenal tyrosinemia is a treatable metabolic disease characterized by progressive liver failure, renal damage and pronounced coagulopathy. Its clinical diagnosis is difficult because of its low prevalence and heterogeneous symptoms. In developed countries, expanded newborn screening, based on succinylacetone quantification by tandem mass spectrometry, has been very valuable in the early detection of hepatorenal tyrosinemia, providing the opportunity for rapid treatment of affected patients. In developing countries without systematic expanded newborn screening, however, diagnosis and treatment of this disease remain major challenges, as genetic diseases in these countries are not a health priority and there are few referral centers for infants with inherited errors of metabolism. This chapter describes the diagnosis, follow-up and outcome of 20 Mexican patients with hepatorenal tyrosinemia. This chapter also constitutes a call to action to pediatricians, gastroenterologists, geneticists and other health professionals, and to academic organizations, health authorities and patient advocacy groups, to promote early patient detection and treatment, reducing the unacceptably high mortality rate (75%) in Mexican infants with this potentially deadly but eminently treatable condition.

Keywords

Hepatorenal tyrosinemia • Tyrosinemia • Developing countries

I. Ibarra-González
Instituto de Investigaciones Biomédicas, UNAM, México 03100, México

C. Ridaura-Sanz
Departamento de Patología, Instituto Nacional de Pediatría, Ciudad de México, México

C. Fernández-Lainez • S. Guillén-López
L. Belmont-Martínez • M. Vela-Amieva (✉)
Laboratorio de Errores Innatos del Metabolismo y Tamiz, Instituto Nacional de Pediatría, Av. IMAN #1, piso 9, Insurgentes-Cuicuilco, CP 04530 Ciudad de México, México
e-mail: dravelaamieva@yahoo.com

© Springer International Publishing AG 2017
R.M. Tanguay (ed.), *Hereditary Tyrosinemia*, Advances in Experimental Medicine and Biology 959, DOI 10.1007/978-3-319-55780-9_14

Abbreviations

GC/MS	Gas chromatography
HCC	Hepatocellular carcinoma
HPLC	High performance liquid chromatography
IEM	Inherited error of metabolism
MS/MS	Tandem mass spectrometry
SA	Succinylacetone

14.1 Hepatorenal Tyrosinemia in a Tertiary Care Center in Mexico

Hepatorenal tyrosinemia is a complex inherited error of metabolism (IEM). It is generally recommended that patients with this condition be managed in a tertiary referral hospital by multidisciplinary teams, comprising metabolic specialists, biochemists, gastroenterologists, dietitians, geneticists, pathologists and nurses, among others, to provide optimum care and improve the quality of life of affected children and their families (http://www.eurordis.org/sites/default/files/publications/Declaration_Centres%20of%20Expertise-nov08.pdf). In developing countries such as Mexico, however, there are obstacles to comprehensive medical care of patients with hepatorenal tyrosinemia and other IEMs. These include disparities in the availability of health services and access to treatment; the low priority of genetic metabolic diseases; and economic and political instability among others (Guigliani 2010). Despite these difficulties and limitations, the experience of institutions, such as the National Institute of Pediatrics in Mexico, may be useful for other developing countries. This chapter describes the diagnosis, follow-up and outcome of a cohort of Mexican patients with hepatorenal tyrosinemia.

The National Institute of Pediatrics is a 45-year-old non-profit institution with federal government support dedicated to the specialized medical attention of patients with complex diseases. Most of the patients referred are from the Mexico City metropolitan area and surrounding states. Over 12,000 new patients are referred annually, and there are about 212,000 medical appointments per year (pediatria.gob.mx/transfocal_estadis.html). Our laboratory (*Laboratorio de Errores Innatos del Metabolismo y Tamiz*) is exclusively devoted to the diagnosis and management of patients with inherited errors of intermediary metabolism, particularly disorders in amino acid, organic acid and carbohydrate metabolism. Over the last 20 years, our laboratory has diagnosed and treated 20 patients with hepatorenal tyrosinemia (Velázquez et al. 2000; Ibarra-González et al. 2014a, b). Although patients were originally tested using classical qualitative chemical tests, biochemical markers of IEMs in body fluid are quantitated by specialized tests. These include high performance liquid chromatography (HPLC) for amino acid analyses, gas chromatography coupled to mass spectrometry (GC/MS) for organic acid analyses, and tandem mass spectrometry (MS/MS) to analyze amino acids and acylcarnitines in dried blood spots (Velázquez et al. 2000; Ibarra-González et al. 2014a). The professional staff includes three biochemists, one geneticist, one metabolic dietitian, two pediatricians and two technicians, all of whom have received specialized training in the management of patients with IEM. Furthermore, the laboratory staff closely communicates with all medical departments of the institution, primarily with the neurology, gastroenterology, intensive care unit and emergency room departments.

14.2 Clinical Characteristics of Mexican Patients with Hepatorenal Tyrosinemia

Patients with a suspected IEM involving the liver were referred to our laboratory to test for hepatorenal tyrosinemia. Blood and urine samples were collected and selective screening tests were performed (HPLC, GC/MS and MS/MS). The first 15 patients were diagnosed with hepatorenal tyrosinemia based on the presence of succinylacetone (SA) in urine, whereas the last five patients were diagnosed by elevated SA in blood, as well

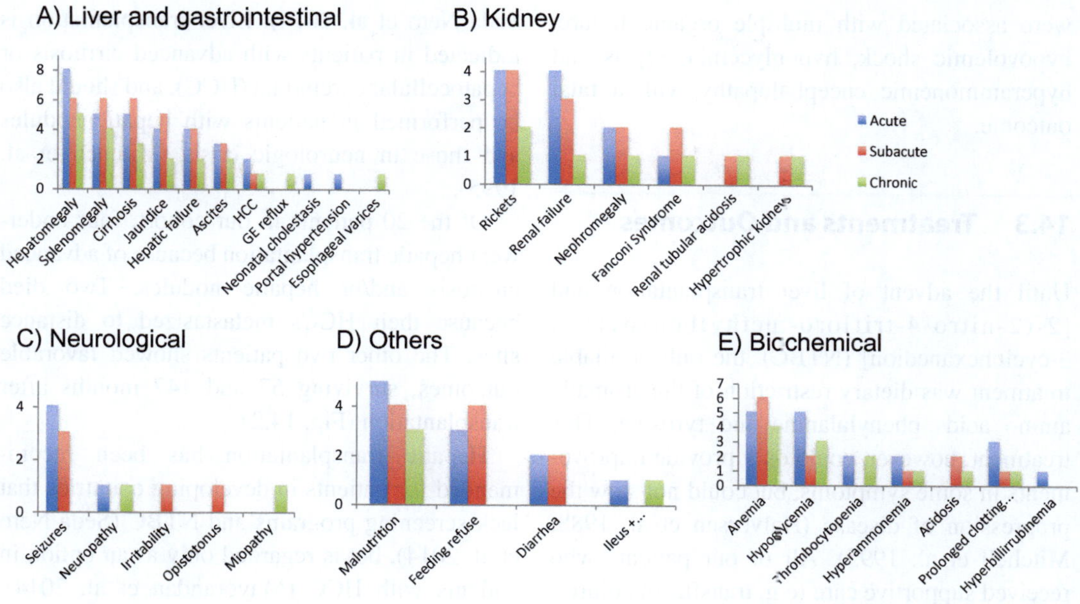

Fig. 14.1 Biochemical and clinical data observed at diagnosis of Mexican patients with hepatorenal tyrosinemia (n = 20)

as SA in urine. Of the 20 patients, 8 (40%) presented with the acute form of hepatorenal tyrosinemia, characterized by clinical onset during the first 6 months of life with acute liver failure; 6 (30%) with subacute hepatorenal tyrosinemia, characterized by less severe presentation at ages ranging between 6 and 12 months; and 6 (30%) with the chronic form, occurring after the first year of life (Angileri et al. 2015). The signs and symptoms of these different types of hepatorenal tyrosinemia are summarized in Fig. 14.1. Similar to other case series, the acute form of hepatorenal tyrosinemia was the most common one (Couse et al. 2011; Zeybek et al. 2015).

All 20 patients presented with the classical features of hepatorenal tyrosinemia, including progressive hepatic impairment (Mitchell et al. 2001). The mean age at onset of clinical symptoms was 10.18 months (range 0.3–60 months), and the mean age at diagnosis was 19.48 months (1.3–60.9 months). The youngest patient, a boy diagnosed at age 1.3 months, showed symptoms since birth with irritability, including poor feeding and vomiting; later, he developed hepatomegaly, edema and ascites with severe abnormalities of hepatic synthetic function, renal tubular disorder and fulminant sepsis, dying at the age of 1 month 14 days. The oldest patient at diagnosis was a 5 year old boy whose symptoms, including hepatomegaly, growth retardation and renal failure, began at age 3 years. All patients had a delayed diagnosis, with a mean gap between the commencement of symptoms and diagnosis of 9.3 months (10.8, 9.2, and 7.4 months in patients with the acute, subacute and chronic forms of hepatorenal tyrosinemia, respectively). These data are similar to those reported in other countries without newborn screening programs for this disease (Zeybek et al. 2015; Rainman et al. 2012).

The frequency of neurological crises, characterized by painful paresthesia and autonomic signs (e.g., hypertension, tachycardia, ileus paralysis) (Mitchell et al. 2001), varies among populations; it has been documented in 65% of Turkish patients (Zeybek et al. 2015), 42% of French-Canadian patients (Mitchell et al. 1990), and 6% of Spanish patients (Couce et al. 2011). Similarly, 10% of our patients experienced neurological crises. Seizures (Fig. 14.1), including status epilepticus, were documented in 35% of our patients, and were observed in patients with advanced stage disease. Most of these seizures were associated with multiple organic failure, hypovolemic shock, hypoglycemia, sepsis and hyperammonemic encephalopathy with a fatal outcome.

14.3 Treatments and Outcomes

Until the advent of liver transplantation and [2-(2-nitro-4-trifloro-methylbenzoyl)-1, 3-cyclohexanedion] (NTBC), the only available treatment was dietary restriction of the aromatic amino acids phenylalanine and tyrosine. This treatment, however, could only provide improvements in some symptoms, but could not slow the progression of disease (Halvorsen et al. 1988; Mitchell et al. 1995). All of our patients who received supportive care (e.g. transfusion, diuretics, pain management) or nutritional treatment, died (Fig. 14.2).

Since the first liver transplant for hepatorenal tyrosinemia was performed in 1976 (Fisch et al. 1978), several studies have been performed with different survival rates (Mayorandan et al. 2014; Seda Neto et al. 2014). Liver transplantation is indicated in patients with advanced cirrhosis or hepatocellular carcinoma (HCC), and should also be performed in patients with hepatic nodules and those in neurologic crisis (Mitchell et al. 1995).

Of the 20 patients in our cohort, four underwent hepatic transplantation because of advanced cirrhosis and/or hepatic nodules. Two died because their HCCs metastasized to distance sites. The other two patients showed favorable outcomes, surviving 57 and 147 months after transplantation (Fig. 14.2).

Hepatic transplantation has been recommended for patients in developing countries that lack screening programs and NTBC (Seda Neto et al. 2014), but is regarded only as an option in patients with HCC (Mayorandan et al. 2014). Hepatic transplants, however, are expensive and limited to some centers. Pediatric hepatic transplantation was introduced only recently in Mexico (Varela-Fascinetto et al. 2011), and is not accessible to all patients because of the high demand and few experienced hospitals.

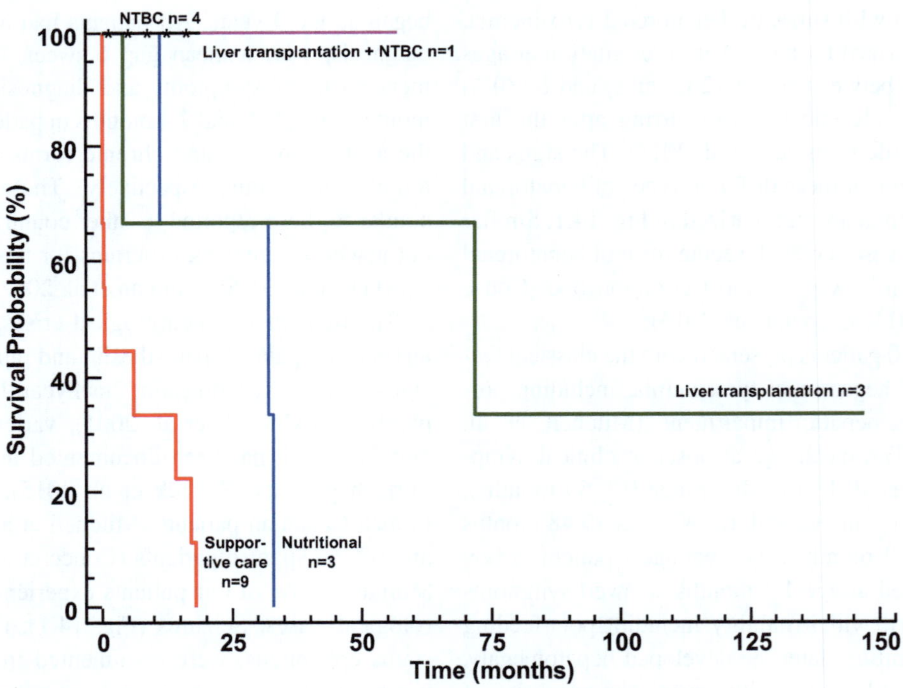

Fig. 14.2 Kaplan-Meier survival curve of Mexican patients with hepatorenal tyrosinemia

Treatment with NTBC was introduced for the first time in 1992 (Linstedt et al. 1992), and its combination with a diet restricting tyrosine and phenylalanine has changed the natural history of this disease (Linstedt et al. 1992; Bartlett et al. 2014; Larochelle et al. 2012). NTBC was first introduced into Mexico in 2010 and has been used to treat four patients with hepatorenal tyrosinemia. Early introduction of pharmacological therapy, during the first 30 days of life, has been associated with a better outcome (Larochelle et al. 2012; Mayorandan et al. 2014). Unfortunately, the five Mexican patients were first treated at ages 4, 5, 10, 28 and 37 months, putting them at risk of developing HCC and other renal and liver complications. To date, one of these five NTBC treated patients has undergone orthotopic hepatic transplantation because of the presence of hepatic nodules. Despite these patients starting pharmacological therapy relatively late, they showed remarkable clinical and biochemical improvements, with serum and urinary SA reduced to normal values within 1 week (Fig. 14.3a). Alpha-fetoprotein levels diminished slower than SA, becoming normal 2.5 months to 1 year after the beginning of NTBC therapy (Fig. 14.3b), these findings are similar to previous results (Raimann et al. 2012).

Nutritional treatment in our center, which is based on treatment protocols, is supervised by a specialized metabolic dietitian (Acosta and Matalon 2012). Therapy consists of a diet restricted in tyrosine and phenylalanine, supplemented with free tyrosine and phenylalanine medical formula that provides the protein that patients need. In order to facilitate adherence and variety, a serving list in which foods can be exchanged based on the phenylalanine and tyrosine content, is given to patients. In Mexico, only one brand of the latter is currently available, and patients at the National Institute of Pediatrics without health insurance are able to receive this formula for free or at very low cost. Intake of energy, protein, minerals and vitamins recommended for these patients are higher than those recommended for the general Mexican population. Nutritional and biochemical assessments are performed weekly during the first year of treatment and every 1–3 months thereafter. Tests include clinical status, dietary intake, anthropometric measurements, and plasma levels of amino acids, with the latter determining future diet. Patients receive phenylalanine supplements when levels of this amino acid are below reference concentrations for age. The main goal of nutritional treatment is to maintain tyrosine plasma concentrations below 500 μmol/L.

14.4 Histopathological Findings of Mexican Patients with Hepatorenal Tyrosinemia on Necropsy and Biopsy

Six patients in our institution who died of hepatorenal tyrosinemia were autopsied. All showed major hepatic and renal damage (Figs. 14.4–14.6), similar to histopathological findings in other studies (Perry 1967; Russo and O'Regan 1990; Mitchell et al. 1995, 2001). Pancreatic alterations were also documented, with the presence of pancreatic islet hypertrophy being the most common, observed in all autopsied patients (Fig. 14.7); acute pancreatitis was observed in one patient, with HCC. Details are shown in Table 14.1.

We previously reported a diagnosis concordance between hepatic biopsy and biochemical results in 8 of 11 Mexican patients diagnosed with hepatorenal tyrosinemia. Biopsies of very young patients require more careful interpretation, as initial histopathological findings of hepatorenal tyrosinemia may be non-specific, including cholestasis, giant cell transformation, fibrosis, leukocyte infiltrates and hemosiderin deposits. These findings may result in a misdiagnosis of neonatal hemochromatosis, neonatal hepatitis or Wilson's disease (Fernandez-Lainez et al. 2014).

14.5 Challenges, Opportunities and Perspectives

Hepatorenal tyrosinemia type 1 still shows a devastating natural history in Mexican patients. There is little knowledge of this disease, and even

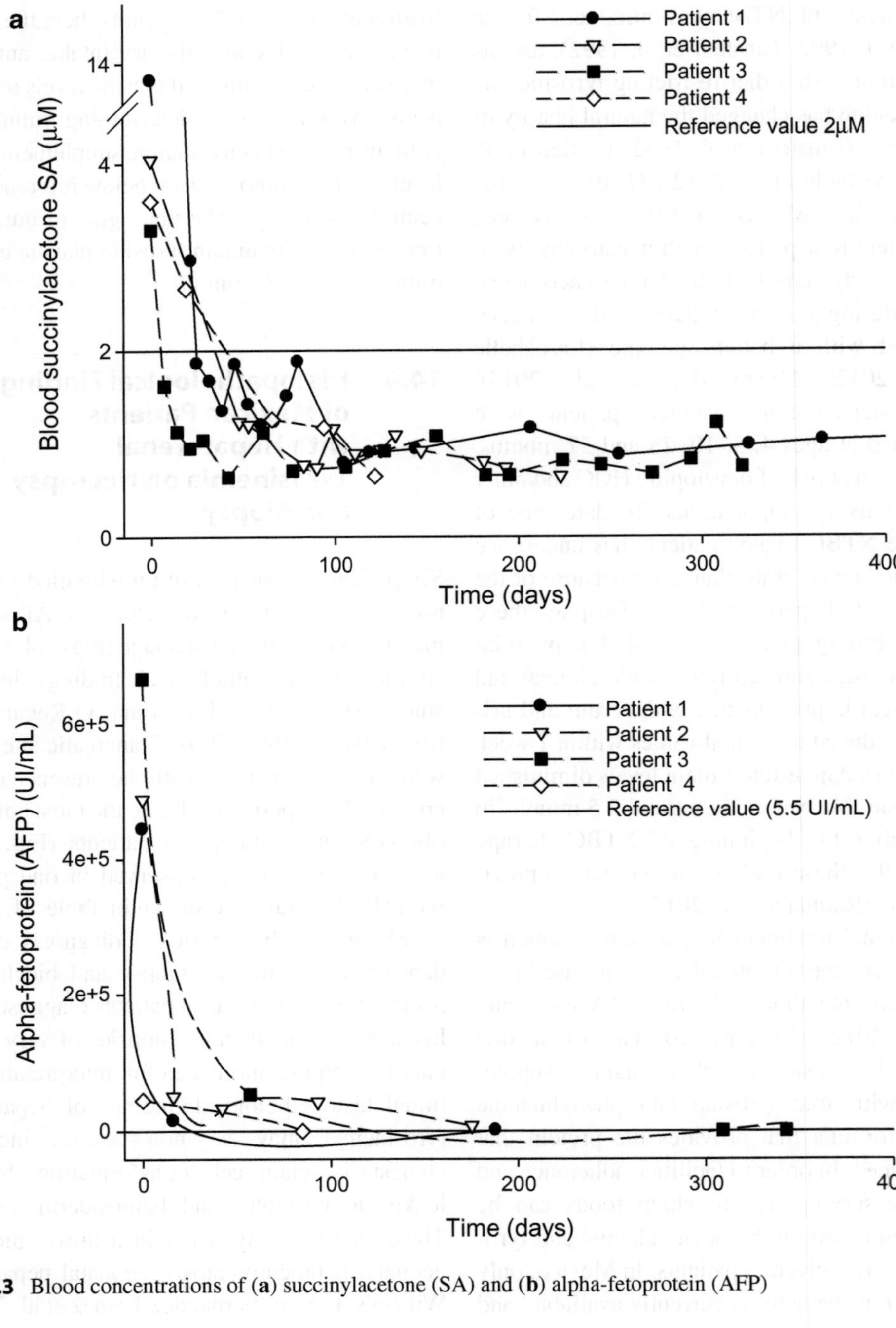

Fig. 14.3 Blood concentrations of (**a**) succinylacetone (SA) and (**b**) alpha-fetoprotein (AFP)

its birth prevalence in Mexico is unknown. It has been estimated that this disease affects 1:100,000 newborns (Angileri et al. 2015). Thus, as 2.46 million babies are born annually (http://www.inegi.org.mx/lib/olap/consulta/general_ver4/MDXQueryDatos.asp) in Mexico, at least 24 are born with hepatorenal tyrosinemia each year. Most of these infants are misdiagnosed, treated inappropriately and die. Newborn screening programs, combined with greater availability of NTBC and infant formulas containing low phenylalanine and tyrosine concentrations, as well as

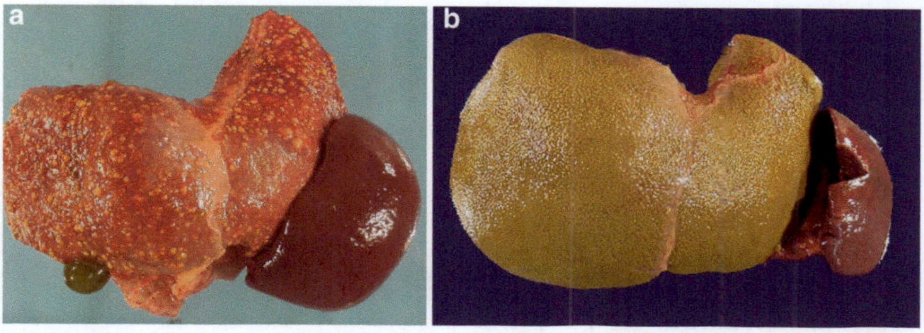

Fig. 14.4 Macroscopical liver damage in hepatorenal tyrosinemia autopsies

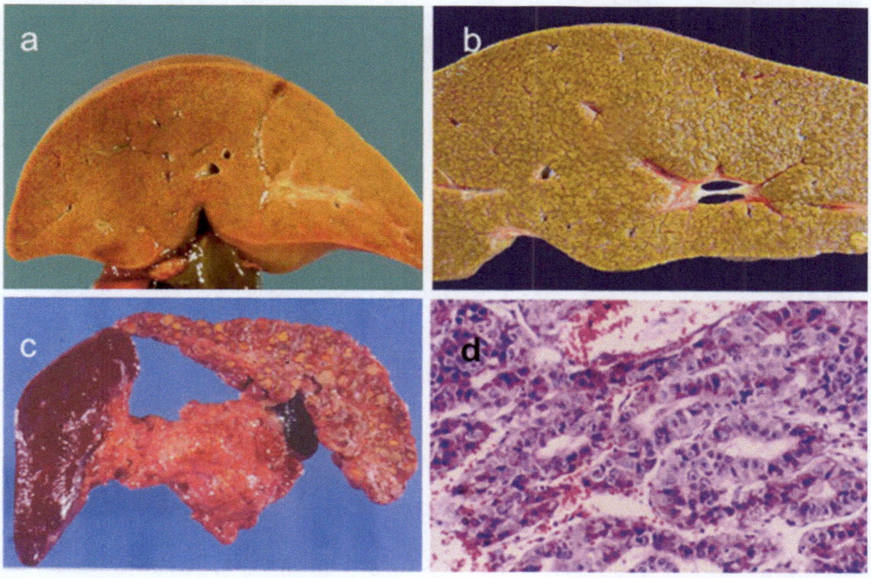

Fig. 14.5 The pathological liver spectrum of hepatorenal tyrosinemia in autopsy specimens. Panel (**a**). Hepatitis and steatosis. Panel (**b**). Micronodular cirrhosis. Panel (**c**). Macronodular cirrhosis with typical hepatocellular carcinoma lesions. Panel (**d**). Histologic section of hepatocarcinoma, hematoxylin-eosin staining

proper monitoring, genetic counseling, and adherence to international clinical practice guidelines can result in earlier detection, delay progression, and improve the quality of life of patients. Sadly, none of our patients was screened as a newborn for this disease. The current health system in Mexico, as in other developing countries, is complex and fragmented, with wide variations in practice (Vela-Amieva et al. 2009). Although newborns have been screened in Mexico since 1985, only screening for congenital hypothyroidism was mandatory (Borrajo 2007). Recently the Mexican neonatal screening panel has been expanded to include four other diseases: phenylketonuria, congenital hyperplasia, galactosemia and biotinidase deficiency. However, except for some isolated states, like Tabasco, Yucatan and Nuevo Leon, in which expanded screening by MSMS is routinely performed for all newborns, other metabolic diseases, such as hepatorenal tyrosinemia, are screened only in private laboratories by request (Therrell et al. 2015).

In addition to expanded screening programs and treatment availability, the outcomes of patients with

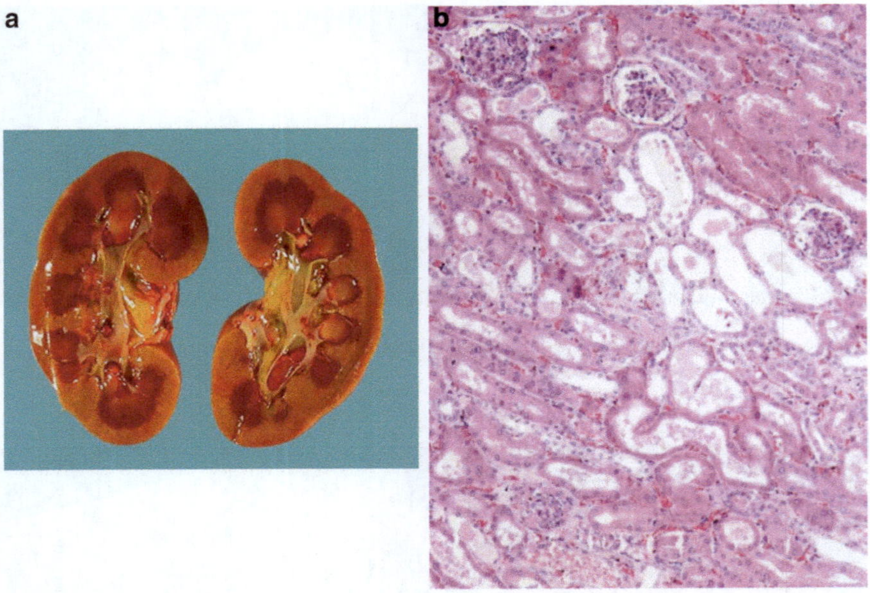

Fig. 14.6 Autopsy photograph showing renal pathology of hepatorenal tyrosinemia

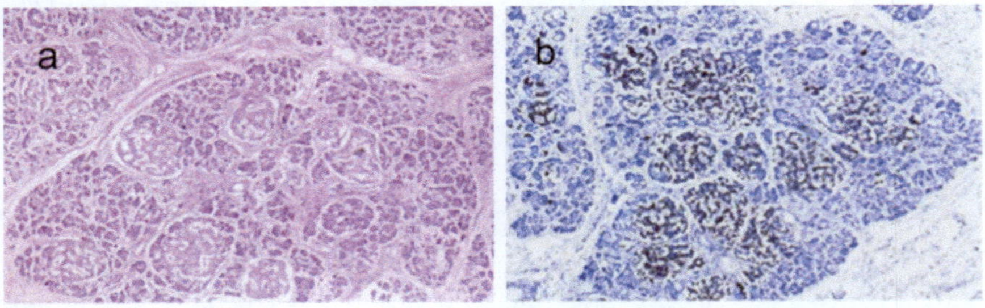

Fig. 14.7 Characteristic pancreatic islet hypertrophy in hepatorrenal tirosinemia. Panel (**a**). Pancreas specimen with marked hyperplasia and hypertrophy of the islets of Langerhans, hematoxylin-eosin staining. Panel (**b**). Same specimen with chromogranin technique that enhances abnormal islets

hepatorenal tyrosinemia in Mexico can be improved by multidisciplinary teams. The information expressed in this chapter is a call to action to pediatricians, gastroenterologists, geneticists and other health professionals, as well as to academic organizations, public health programs, policy makers and patient advocacy groups. More attention must be paid to this potentially deadly but eminently treatable condition, to reduce the gap in outcomes between developing and developed countries.

Table 14.1 Main histopathological findings on necroscopy of Mexican patients with hepatorenal tyrosinemia

Age	Sex	Liver	Renal	Pancreas	Other organs
9 months	M	Micro and macronodular cirrhosis	Tubulopathy with micro vacuolar epithelium degeneration; intersticial nephritis; focal glomerulosclerosis; cholemic nephrosis; bilateral nephromegaly	Islet cell hyperplasia	
15 months	F	Submasive necrosis with centrelobulillar fibrosis; extensive ductular proliferation and hepatocellular focal regeneration	Tubulopathy calcinosis	Islet cell hyperplasia	Cerebellum granular necrosis[a]
		Minimum cholestasis, focal steatosis; hemorrhagic perivenular necrosis with mixed inflammation[a]			
9 months	M	Post-hepatic cirrhosis with giant cells	Tubulopathy; distal tube dilatation	Islet cell hyperplasia	Cardiomegaly
					Cerebral atrophy
6 months	F	Submasive necrosis	Tubulopathy; cholemic nephrosis	Islet cell hyperplasia	
9 years	F	Multinodular cirrhosis; dysplasia	Tubular necrosis; hepatocarcinoma metastasis	Acute focal pancreatitis	Pulmonar hepatocarcinoma metastasis
		Hepatocellular carcinoma; cholestasis			
4 months	M	Micronodular cirrhosis; steatosis	Tubulopathy	Islet cell hyperplasia	
			Cholemic nephrosis		

[a]These last findings were also compatible with toxic hepatitis secondary to anticonvulsants drugs

References

Acosta PB, Matalon KM (2012) Nutrition management of patients with inherited disorders of aromatic amino acid metabolism. In: Acosta PB (ed) Nutrition management of patients with inherited metabolic disorders. Jones and Bartlett Publishers, Sudbury, pp 153–174

Angileri F, Bergeron A, Morrow G, Lettre F, Gray G, Hutchin T, Ball S, Tanguay RM (2015) Geographical and ethnic distribution of mutations of the fumarylacetoacetate hydrolase gene in hereditary tyrosinemia type 1. JIMD Rep 19:43–58

Bartlett DC, Lloyd C, McKiernan PJ, Newsome PN (2014) Early nitisinone treatment reduces the need for liver transplantation in children with tyrosinaemia type 1 and improves post-transplant renal function. J Inherit Metab Dis 37(5):745–752

Borrajo GJ (2007) Newborn screening in Latin America at the beginning of the 21st century. J Inherit Metab Dis 30(4):466–481

Couce ML, Dalmau J, del Toro M, Pintos-Morell G, Aldámiz-Echevarría L, Spanish Working Group on Tyrosinemia type 1 (2011) Tyrosinemia type 1 in Spain: mutational analysis, treatment and long-term outcome. Pediatr Int 53(6):985–989

Declaration of common principles on centres of expertise and European reference networks for rare diseases. http://www.eurordis.org/sites/default/files/publications/Declaration_Centres%20of%20Expertise-nov08.pdf

Fernández-Lainez C, Ibarra-González I, Belmont-Martínez L, Monroy-Santoyo S, Guillén-López S, Vela-Amieva M (2014) Tyrosinemia type I: clinical and biochemical analysis of patients in Mexico. Ann Hepatol 13(2):265–272

Fisch RO, McCabe ER, Doeden D, Koep LJ, Kohlhoff JG, Silverman A, Starzl TE (1978) Homotransplantation of the liver in a patient with hepatoma and hereditary tyrosinemia. J Pediatr 93(4):592–596

Giugliani R (2010) Inborn errors of metabolism in Latin America: challenges and opportunities. J Inherit Metab Dis 33(Suppl 2):S315–S320

Halvorsen S, Kvittingen EA, Flatmark A (1988) Outcome of therapy of hereditary tyrosinemia. Acta Paediatr Jpn 30(4):425–428

Ibarra-González I, Fernández-Lainez C, Reyes-González DI, Belmont-Martínez L, Guillén-López S, Monroy-Santoyo S, Vela-Amieva M (2014a) Inborn errors of intermediary metabolism in critically ill mexican newborns. JIEMS 2:1–7. doi:10.1177/2326409814529649

Ibarra-González I, Fernández-Lainez C, Belmont-Martínez L, Guillén-López S, Monroy-Santoyo S, Vela-Amieva M (2014b) Characterization of inborn errors of intermediary metabolism in Mexican patients. An Pediatr (Barc) 80(5):310–316. [article in Spanish]

Larochelle J, Alvarez F, Bussieres JF, Chevalier I, Dallaire L, Dubois J, Faucher F, Fenyves D, Goodyer P, Grenier A, Holme E, Laframboise R, Lambert M, Lindstedt S, Maranda B, Melancon S, Merouani A, Mitchell J, Parizeault G, Pelletier L, Phan V, Rinaldo P, Scott CR, Scriver C, Mitchell GA (2012) Effect of nitisinone (NTBC) treatment on the clinical course of hepatorenal tyrosinemia in Quebec. Mol Genet Metab 107(1–2):49–54

Lindstedt S, Holme E, Lock EA, Hjalmarson O, Strandvik B (1992) Treatment of hereditary tyrosinaemia type I by inhibition of 4-hydroxyphenylpyruvate dioxygenase. Lancet 340(8823):813–817

Mayorandan S, Meyer U, Gokcay G, Segarra NG, de Baulny HO, van Spronsen F, Zeman J, de Laet C, Spiekerkoetter U, Thimm E, Maiorana A, Dionisi-Vici C, Moeslinger D, Brunner-Krainz M, Lotz-Havla AS, Cocho de Juan J, Couce Pico ML, Santer R, Scholl-Bürgi S, Mandel H, Bliksrud YT, Freisinger P, Aldamiz-Echevarria LJ, Hochuli M, Gautschi M, Endig J, Jordan J, McKiernan P, Ernst S, Morlot S, Vogel A, Sander J, Das AM (2014) Cross-sectional study of 168 patients with hepatorenal tyrosinaemia and implications for clinical practice. Orphanet J Rare Dis 9:107

Mitchell GA, Larochelle J, Lambert M, Michaud J, Grenier A, Ogier H, Gauthier M, Lacroix J, Vanasse M, Larbrisseau A, Paradis K, Weber A, Lefevre Y, Melançon S, Dallaire L (1990) Neurologic crisis in hereditary tyrosinemia. N Engl J Med 322(7):432–437

Mitchell GA, Lambert M, Tanguay RM (1995) Hypertyrosinemia. In: Scriver CR, Beaudet AL, Sly WS, Valle D (eds) The metabolic and molecular bases of inherited disease, 7th edn. McGraw-Hill, New York, pp 1077–1106

Mitchell GA, Grompe M, Lambert M, Tanguay RM (2001) Hypertyrosinemia. In: Scriver CR, Beaudet AL, Sly WS, Valle D (eds) The metabolic and molecular bases of inherited disease, 8th edn. McGraw-Hill, New York, pp 1177–1805

Perry TL (1967) Tyrosinemia associated with hypermethioninemia and islet cell hyperplasia. Can Med Assoc J 97(18):1067–1075

Raimann E, Cornejo V, Arias C, Cabello JF, Castro G, Fernández E, de la Parra A (2012) Clinical follow up of Chilean patients with tyrosinemia type 1 treated with 2-(2-nitro-4-trifluoromethylbenzoyl)-1,3-ciclohexanedione (NTBC). Rev Med Chil 140(2):169–175. [article in Spanish]

Russo P, O'Regan S (1990) Visceral pathology of hereditary tyrosinemia type I. Am J Hum Genet 47(2):317–324

Seda Neto J, Leite KM, Porta A, Fonseca EA, Feier FH, Pugliese R, Miura IK, Chapchap P, Porta G (2014) HCC prevalence and histopathological findings in liver explants of patients with hereditary tyrosinemia type 1. Pediatr Blood Cancer 61(6):1584–1589

Therrell BL, Padilla CD, Loeber JG, Kneisser I, Saadallah A, Borrajo GJ, Adams J (2015) Current status of newborn screening worldwide: 2015. Semin Perinatol 39(3):171–187

Varela-Fascinetto G, Hernández-Plata JA, Nieto-Zermeño J, Alcántar-Fierros JM, Fuentes-García V, Castañeda-Martínez P, Valencia-Mayoral P, Salgado-Ramírez JM (2011) Pediatric liver transplant program at Hospital Infantil de Mexico Federico Gomez. Rev Investig Clin 63(Suppl. 1):57–61. [article in Spanish]

Vela-Amieva M, Belmont-Martínez L, Ibarra-González I, Fernández-Lainez C (2009) Variabilidad interinstitucional del tamiz neonatal en México. Bol Med Hosp Infant Mex 66(5):431–439. [article in Spanish]

Velázquez A, Vela-Amieva M, Cicerón-Arellano I, Ibarra-González I, Pérez-Andrade ME, Olivares-Sandoval Z, Jiménez-Sánchez G (2000) Diagnosis of inborn errors of metabolism. Arch Med Res 31(2):145–150

Zeybek AC, Kiykim E, Soyucen E, Cansever S, Altay S, Zubarioglu T, Erkan T, Aydin A (2015) Hereditary tyrosinemia type 1 in Turkey: twenty year single-center experience. Pediatr Int 57(2):281–289

Hereditary Tyrosinemia Type 1 in Turkey

15

Ayse Cigdem Aktuglu-Zeybek, Ertugrul Kiykim, and M. Serif Cansever

Abstract

Hereditary tyrosinemia type 1 (HT1, OMIM 276700) is a rare autosomal recessively inherited inborn error of metabolism in the tyrosine catabolic pathway due to deficiency of the enzyme fumarylacetoacetate hydrolase. The clinical features of HT1 are widely heterogenous even within the same family members. Clinical features includes acute or chronic liver disease with increased risk of hepatocellular carcinoma, hypophosphatemic rickets due to renal tubular dysfunction, glomerulosclerosis, failure to thrive, neurological porphyria-like crisis, hypertrophic cardiomyopathy and hypoglycemia due to hyperinsulinism. Currently, the treatment in HT1 consists of two principles: inhibition of the formation of toxic metabolites by nitisinone [2-(2-nitro-4-trifluoromethylbenzoyl)-1,3-cyclohexanedione; NTBC] and reduction of tyrosine levels by dietary treatment. In this chapter besides presenting the data for 42 patients that had been followed up by Pediatric Metabolic Diseases and Nutrition Unit, Cerrahpasa Medical Faculty, Istanbul University, we also evaluated the data abstracted from the previously published case studies in order to better understand the disease course and gain further insight in the current diagnosis and treatment for HT1 in Turkey.

Keywords

Tyrosinemia type 1 • Nitisinone • Turkey

Abbreviations

ALAD	δ-aminolevulinic acid dehydratase
ALT	Alanine aminotransferase
AST	Aspartate aminotransferase
d-ALA	δ-aminolevulinic acid
FAA	Fumaryl acetoacetate
FAH	Fumarylacetoacetate hydrolase
HCC	Hepatocellular carcinoma
HT1	Hereditary tyrosinemia
HTIV	Hereditary tyrosinemia type 1

A.C. Aktuglu-Zeybek (✉) • E. Kiykim
Cerrahpasa Medical Faculty, Department of Pediatrics, Division of Metabolic Diseases and Nutrition, University of Istanbul,
34098, Kocamustafapasa Fatih, Istanbul, Turkey
e-mail: dracaz@istanbul.edu.tr

M. Cansever
Cerrahpasa Medical Faculty, Central Laboratories, Metabolic Diseases Unit, University of Istanbul,
34098, Kocamustafapasa Fatih, Istanbul, Turkey

© Springer International Publishing AG 2017

R.M. Tanguay (ed.), *Hereditary Tyrosinemia*, Advances in Experimental Medicine and Biology 959,
DOI 10.1007/978-3-319-55780-9_15

LDLT	Living donor liver transplantation
MAA	Maleyl acetoacetate
MRI	Magnetic resonance imaging
NBS	Newborn screening
OLT	Orthotopic liver transplantation
PBGS	Porphobilinogen synthase
SA	Succinyacetone
SAA	Succinylacetate
TAT	Tyrosine aminotransferase
WISC-R	Wechsler Intelligence Scale for Children-Revised

15.1 Introduction

Hereditary tyrosinemia type 1 (HT1, OMIM 276700) is a rare autosomal recessively inherited inborn error of metabolism in the tyrosine catabolic pathway due to deficiency of the enzyme fumarylacetoacetate hydrolase (FAH) (Chakrapani et al. 2012). FAH catalyzes the final step in the tyrosine degradation and blockage leads to accumulation of the upstream toxic metabolites and their derivatives fumarylacetoacetate (FAA), maleylacetoacetate (MAA), succinylacetoacetate (SAA) and succinylacetone (SA) that are responsible for the tissue damage with progressive hepatic, renal and neurological findings (McKiernan et al. 2015; Jorquera and Tanguay 2001; Tanguay et al. 1990; Sassa and Kappas 1983). The human *fah* gene is mapped to chromosome 15q. Overall 95 mutations have been reported within the *fah* gene (Angileri et al. 2015) (and see Morrow et al., Chap. 3).

The clinical features of HT1 are widely heterogenous even within the same family members. Three main clinical forms have been described based on the age at symptom onset: the *acute* form, which presents in the first 6 months of life; the *subacute* form, presenting within 6 months and 1 year of age; the *chronic* form, which appears after the first year of age (van Spronsen et al. 1994; Chakrapani et al. 2012). Clinical features includes acute or chronic liver disease with increased risk of hepatocellular carcinoma, hypophosphatemic rickets due to renal tubular dysfunction, glomerulosclerosis, failure to thrive, neurological porphyria-like crisis, hypertrophic cardiomyopathy and hypoglycemia due to hyperinsulinism (Chakrapani et al. 2012; Mitchell et al. 1990; Mohamed et al. 2013; Arora et al. 2006; Baumann et al. 2005). Patients who survive beyond infancy can develop chronic renal failure (Kvittingen et al. 1991).

Currently, the treatment in HT1 consists of two principles: inhibition of the formation of toxic metabolites by nitisinone [2-(2-nitro-4-trifluoromethylbenzoyl)-1, 3-cyclohexanedione; NTBC] and reduction of tyrosine levels by dietary treatment (Holme and Lindstedt 1992).

Before the introduction of nitisinone, the natural history of the disease usually resulted in death. The treatment consisted of symptomatic treatment with tyrosine restricted diet until liver transplantation. Low tyrosine (tyr) and phenylalanine (phe) diet was not effective in the acute form and had limited success in the chronic form, neither did it prevent hepatocellular carcinoma (HCC) (van Spronsen et al. 1994). The 1 year survival rate was 38% when the smptoms started <2 months of age, 74% between 2 and 6 months of age and 96% >6 months of age. The main causes of death were liver failure, HCC and porphyria-like neurological crises (van Spronsen et al. 1994).

Nitisinone is a potent inhibitor of 4-hydroxyphenylpyruvate dioxygenase (Lindstedt et al. 1992) leading to the suppression of MAA, FAA and SA accumulation and the accumulation of tyrosine (Mckiernan 2006). Nitisinone had significantly improved survival in HT1, with a clinical response of 90% and a decrease in the risk of early development of HCC in those who began treatment at an early age. The maximal benefit of the drug is achieved when the treatment is started in the early stages of life (Holme and Lindstedt 1998; Lock et al. 1998; Santra and Baumann 2008; Larochelle et al. 2012; Nakamura et al. 2015; Couce et al. 2011; El-Karaksy et al. 2011; McKiernan et al. 2015). Universal newborn screening (NBS) with SA extracted from newborn dried blood spots, as a biochemical marker, is feasible and has been established in some countries. Early detection with NBS can allow the treatment to be started before the development of clinical symptoms (Larochelle et al. 2012; McKiernan et al. 2015).

In Turkey, where nationwide NBS for HT1 is not mandatory, diagnosis still depends on clinical suspicion and laboratory investigations. Pediatric Metabolic Diseases and Nutrition Unit, Cerrahpasa Medical Faculty, Istanbul University is one of the first Metabolic Diseases Center in Turkey initiating nitisinone therapy in HT1 patients, since 1993.

In this chapter, besides presenting the data for 42 patients that had been followed up by Pediatric Metabolic Diseases and Nutrition Unit, Cerrahpasa Medical Faculty, Istanbul University we also evaluated the data abstracted from the previously published case studies in order to better understand the disease course and gain further insight in the current diagnosis and treatment for HT1 in Turkey.

15.2 Methods

Information from the medical records of HT1 patients diagnosed at the Pediatric Metabolic Diseases Unit of Cerrahpasa Medical Faculty, Istanbul University between December 1993 and May 2016 was collected and included detailed history, family history including consanguinity and siblings with HT1 or with suggestive symptoms, demographic information, gender, age at first clinical symptom, age at diagnosis, clinical-biochemical and radiological data. HT1 diagnosis based on either by presence of elevated SA in urine/blood samples or mutation analysis. None of the patients were screened at birth for HT1. All patients were treated with a tyrosine and phenylalanine restricted diet under the supervision of a clinical dietician to maintain a plasma level below 400 μmol/L as recommended. Forty patients were treated with nitisinone according to a standard protocol starting nitisinone at 1–2 mg/kg/day, q.d. or b.i.d. at diagnosis. The nitisinone dose was adjusted afterwards according to blood nitisinone levels determined either by plasma or dried blood samples. Plasma levels were considered adequate between 30 and 60 μmol/L. Two patients were only treated with diet due to unavailability of nitisinone.

For these 42 patients, clinical and biochemical data at diagnosis and follow-up including liver and renal functions, blood alpha-fetoprotein (AFP), urine or blood SA, urine delta-aminoalevulinic acid (δ-ALA) (when available), plasma Tyr levels, liver and renal imaging (by ultrasound, CT or MRI) findings, growth parameters, compliance with the treatment, and adverse effects were evaluated. If available, the histopathological findings from the liver biopsies were recorded.

15.2.1 Published Cases

The following terms were used to search for HT1 cases the PubMed database from 1966 to May 2016: Tyrosinemia type 1 or hereditary tyrosinemia type 1 or fumarylacetoacetate hydrolase deficiency or hepatorenal tyrosinemia and Turkish or Turkey with filters for case reports, clinical trials, and review articles. The search yielded 22 publications.

15.2.2 Statistical Analysis

Data are displayed as mean and standard deviation and/or median and range for continuous variables and as frequency and percentage for categorical variables. Non-parametric Mann–Whitney U-test was used for abnormally distributed data. P-values <0.05 were considered statistically significant.

15.2.3 Description of Clinical Characteristics of Turkish Patients with HT1

Information was collected on 112 patients with HT1. Of these, 42 cases were followed by Pediatric Metabolic Diseases Unit of Cerrahpasa Medical Faculty, while information on 69 cases was extracted from the literature.

Forty-two HT1 patients diagnosed at Pediatric Metabolic Diseases Unit of Cerrahpasa Medical Faculty, aged between 15 days and 100 months at the time of diagnosis. Twenty-six were male and 16 were female. None of the patients was detected by expanded newborn screening for HT1 but two patients were diagnosed by selective screening in the newborn period due to an affected sibling history. All patients were symptomatic at the age of diagnosis, even diagnosed with selective screening. Mean age at onset of clinical symptoms was 8.8 months (range 0–54 months) and the mean age at diagnosis was 18.1 months (range 0.43–100 months). Fifteen of the 42 cases presented as acute HT1, 13 with subacute HT1 and 14 with chronic HT1. Very early onset of symptoms (<2 months of age) was detected in five patients. The majority of the cases were male (62%). Clinical diagnosis was based on presense of SA in urine/blood samples for all cases. All but one patient excreted SA in their urine at the time of diagnosis. One patient did not excrete SA in his urine despite a mild increase in plasma SA level and the diagnosis was made via mutation analysis (c.1A > G/c. (1096–1098) del TCG, compound heterozygous mutation). No enzymatic studies were performed.

On family history, consanguinity was noted in 25 patients (59.5%). Prematurity was noted in six patients without any other risk factor for premature labor. Three patients were noted to be large for gestational age due to gestational diabetes and one to be small for gestational age.

Table 15.1 Chief clinical symptoms at diagnosis

Complaints	n (%)
Pallor	32 (76.2)
Anorexia	31 (73.8)
Abdominal distention	31 (73.8)
Irritability/abdominal pain	27 (64.3)
Abnormal urine odor	22 (52.4)
Growth retardation	17 (40.7)
Fever	13 (31)
Jaundice	11 (26.2)
Nasal bleeding	11 (26.2)
Hematemesis/melena	9 (21.4)
Vomiting	7 (16.7)
Diarrhea	4 (9.5)

Table 15.2 Main clinical manifestations at diagnosis

Clinical manifestations	n (%)
Hepatomegaly	41 (97.6)
Splenomegaly	36 (87.5)
Abdominal distention	31 (73.8)
Anemia	32 (76.2)
Rickets	27 (64.3)
Ascites	17 (40.5)
Growth retardation	14 (33.3)
Jaundice	11 (26.2)
Epistaxis	11 (26.2)
Gastrointestinal bleeding	9 (21.4)
Petechia/ecchymosis	5 (12)
Lethargy	3 (7.1)

The onset of clinical symptoms was at 2. ± 1.2, 4.2 ± 2.7 and 20.4 ± 16.2 months in acute, subacute and chronic HT1 patients, respectively. The median interval between the first complaints and diagnosis was 1.3 months (range, 0–3.1) in acute, 5.3 months (range 0.5–11) in subacute and 21.5 months (range, 1–67) in chronic HT1. The main clinical manifestations at diagnosis were hepatomegaly (97.6%), splenomegaly (87.5%), hepatic dysfunction (82%) and renal tubular dysfunction (59.5%) (Tables 15.1 and 15.2). On initial anthropometric evaluation six patients were found to be ≤2 SD for weight and five were ≤2 SD for length. Biochemical parameters at diagnosis revealed a marked increase in alpha-fetoprotein (AFP) level for all ages, with a high variability (between 35 and 624,000 ng/ml). AFP level of chronic HT1 patients was significantly lower than the acute and subacute HT1 patients ($P < 0.0001$ and $P < 0.0001$, respectively). Plasma tyrosine, phenylalanine and methionine levels were all increased (Table 15.3), while erythrocyte porphobilinogen synthase (ePBGS) was markedly decreased. Altered coagulation parameters due to hepatic dysfunction were observed all but four patients of 42 patients (90%).

Tubulopathy was detected 36/42 (85%) of patients with one or more components such as generalized aminoaciduria, metabolic acidosis, proteinuria, glucosuria and hypophosphatemic rickets of varying severity. Urinary phospahte results were available for 13 patients and 12 had

Table 15.3 Biochemical parameters at diagnosis

Parameter	Normal	n	Mean	Range	SD
Plasma					
AFP (ng/ml)	<13	42	99,222	35–624,000	137,647
ALT (IU/L)	0–40	41	34.2	11–102	21.2
AST (IU/L)	0–40	41	70.7	25–186	31.7
GGT(IU/L)	3–25	40	119.8	19–341	80.6
Total bilirubin (mg/dl)		36	1.97	0.41–6.10	1.27
Direct bilirubin (mg/dl)		36	0.99	0.06–3.3	0.8
Ca (mg/dl)	8.4–10.8	40	9.08	7.51–11.1	0.87
P (mg/dl)	2.7–5.5	40	2.99	1.2–6.1	1.29
ALP (IU/L)	60–525	41	1407	170–4430	1062.9
Total protein (g/dL)	5.6–8	39	5.64	3.4–8.3	1.17
Albumin (g/dL)	3.2–5.4	39	3.31	1.9–4.7	0.76
Hemoglobin (g/dL)		40	9.75	6.4–19	2.27
Thrombocyte (/mm^3)	150,000–400,000	40	129,025	33,000–577,000	82349
Coagulation					
PT (s)	10.4–14	40	31.2	13.3–70	15.5
PT activity (%)	70–130	40	36.2	11.34–90	21.4
INR	0.85–1.15	40	2.74	1.05–6.15	1.38
aPTT (s)	26–40.8	40	66.88	29.9–144	28.8
Plasma					
Tyr (μmol/L)	50–130	40	410.7	23–1095	286.1
Phe (μmol/L)	40–120	40	133.2	15–415	99.9
Met (μmol/L)	20–50	40	368.9	13.3–1590	381.9
ePBGS (nkat/g Hb)	0.58–1.25	17	0.071	0.00–0.32	0.10
SAp	<0.1	20	35.5	0.73–136	33.23
Urine					
SA (mmol/mol creatinin)	<1	19	503.3	1.40–1800	560.3
DALA (mmol/mol creatinin)	0–3	16	123.6	11–350	11.7
SA (qualitative)	Undetectable	42	Increased		

AFP α-fetoprotein, *ALP* alkaline phosphatase, *ALT* alanine transaminase, *aPTT* activated partial thromboplastin time, *AST* Aspartate transaminase, *Ca* calcium, *DALA* δ-aminolevulinic acid, *ePBGS* erythrocyte porphobilinogen synthase, *GGT* γ-glutamine transaminase, *INR* international normalized ratio, *Met* methionine, *P* phosphorus, *Phe* phenylalanine, *PT* prothrombin time, *SA* succinylacetone, *SAp* plasma succinylacetone, *Tyr* tyrosine

abnormal tubulary phosphate excretion. Clinical, radiological and biochemical rickets was detected in 27 patients (64.3%). Hypoglycemia was detected in 11 patients, two were hyperinsulinemic and required temporary slow rate i.v. dextrose infusion. Hepatic ultrasonography showed hepatomegaly in 40 children and liver nodules in 28 (66.7%). Multiple hypoechoic nodules were detected in 20 and hyperechoic nodules were detected in 21 patients. Five patients had macronodular appearance detected either on ultrasonography or MRI. Increase in renal echogenity was detected on renal ultrasonography in 24 children and nephrocalcinosis was detected in five children on presentation. Liver biopsy was performed in 13 children and all had active cirrhosis and one also had hepatic macrovesicular steatosis. None of the 15 patients with initial echocardiography, had restrictive cardiomyopathy. One had mild interventricular septal hypertrophy and mild mitral insufficiency, one had small ventricular defect, one had secundum atrial septal defect and one had patent foramen ovale and mild pulmonary stenosis.

Information on 69 cases extracted from the literature regarding the age of diagnosis (43/69) yielded a mean age at the diagnosis of 15.3 months (range 0,06–108 months (Dursun et al. 2011; Coskun et al. 1991; Bay et al. 2012; Yagci et al. 2015; Onenli Mungan et al. 2016). Diagnosis was based on the presence of SA in urine. Five patients had additional confirmatory enzyme analysis in fibroblasts and 44 had mutation analysis. One patient did not excrete SA in his urine and the diagnosis was made via mutation analysis (Dursun et al. 2011). Two cases were detected by selective newborn screening and five cases were detected by selective screening, because of an affected sibling. The average age at diagnosis was 10.5 months (range 1–45 months) in those diagnosed via selective screening (Dursun et al. 2011; Onenli Mungan et al. 2016; Rootwelt et al. 1994).

Data was available for 48 patients regarding the distribution of HT1 subtype; 24 of the cases were acute HT1 (34.8%), 8 were subacute (11.6%) and 14 (20.3%) were chronic. Very early onset of symptoms (<2 months) was noted in two cases. Information regarding the sex of the patients was available in 29/69 cases and most majority was female (17/29; 58%) (Coskun et al. 1991; Sener 2005a, b; Yagci et al. 2015; Onenli Mungan et al. 2016).

Abdominal distention (30/44), hepatomegaly (30/44), splenomegaly (13/44), failure to thrive (15/43), rachitis (15/43), irritability (5/34), abdominal distention (30/44), diarrhea (9/44), polyuria-polydipsia (6/44), vomitting (4/44), melena (5/44), ascites (9/43), intermittent hypertension (3/44), and jaundice (2/44) were the most common symptoms and disease manifestations described in Turkish HT1 patients in the literature (Coskun et al. 1991; Dursun et al. 2011; Bay et al. 2012; Onenli Mungan et al. 2016).

In the group of patients (35/69) with plasma tyrosine level reported, the level ranged between 143 and 1385 μmol/L. Data was available for 26 patients regarding AFP level at diagnosis that revealed a marked increase with a wide variability (37–855,000 ng/ml), for 42 patients regarding ALT and AST (10–144 and 13–312 IU/L, respectively). Altered coagulation parameters due to hepatic dysfunction were observed all but four patients of 31 patients reported (87%) (Coskun et al. 1991; Dursun et al. 2011; Bay et al. 2012).

In the group of patients with mutational analysis (40/69), we found 12 different mutations. In accordance with the literature the most common mutation was IVS6-1G>T (26%) followed up by D233V (22%), and IVS3-3C>G (10%) mutations. Eight patients were homozygous for IVS6-1G>T mutation. IVS12+5G>A (three patients), V166G (two patients), R237X (two patients), IVS9+2T>C (two patients) c.191delA (one patient), c.(440–441) del8nt (one patient), R174X (one patient), N232K(one patient), N334Tfsx (one patient), V259D (one patient), A134D (one patient) mutations were also described in Turkish patients (Rootwelt et al. 1994, 1996; Dursun et al. 2011; Onenli Mungan et al. 2016).

15.2.3.1 Adherence to Therapy

Among 42 patients followed in our clinic only two patients were treated only with low tyrosine and phenylalanine diet. Although the dietary compliance was good, patients died due to severe hepatic dysfunction and massive gastrointestinal bleeding. The average time for the rest of the 40 patients in treatment was 70.4 months (range, 2–255 months) with a median of 128 months. During this period, dietary compliance was good in 13 patients (Tyr <400 μmol/L), moderate in 13 patients (Tyr 400–600 μmol/L) and poor in 16 patients (Tyr >400 μmol/L). No information about the dietary compliance was available in the published cases, except one with corneal pseudodendritic lesions with a plasma Tyr level of 840 μmol/L (Gulmez Sevim et al. 2015).

15.2.3.2 Response to Nitisinone Treatment

In our patient's group, nitisinone treatment was started at <6 months of age in 16 patients (38.1%), 7–12 months in nine patients (21.4%), 13–24 months in four patients (9.5%) and after 24 months in 11 patients (26.2%). The mean interval between diagnosis and nitisinone treatment was 18 days (range, 0–120 days). Long-term follow up of the patients was carried out for a mean period of 70.5 months (range 2–255 months).

The mean dose of nitisinone was 1.2 mg/kg/day (range 0.6–2 mg/kg/day) with an average plasma level of 41 μmol/L. The dose was adjusted according to the plasma nitisinone level that was considered to be adequate, between 30 and 60 μmol/L. No urinary SA excretion was detected under nitisinone treatment, but interruption of the treatment led to re-excretion.

Adherence to nitisinone therapy was very good in all 42 patients except two chronic HT1 patients. Good metabolic control was achieved in 35/40 patients with normalization of the hypoprothrombinemia and decrease in AFP. Despite good adherence to therapy, four patients did not respond to nitisinone treatment; one subacute HT1 patient died due to severe hepatic dysfunction (patient S1, Table 15.4), one chronic HT1 patient died due to hepatic dysfunction and variceal bleeding as a result of portal hypertension (patient C5), and one subacute patient had a sucessful liver transplant at 16 months of age (patient S7). One chronic HT1 patient was under evaluation for LT due to partial response and HCC suspicion at the time of writing.

Among 36 patients who had been followed for >48 months, AFP normalized (<13 μg/L) in 7/24 (29.1%) within the first year of therapy, in 11/24 (45.9%) at 12–24 months, and in 6/24 (25%) at 25–48 months of therapy.

Table 15.4 Cause of death in hereditary tyrosimenia type 1 patients

Patient no	HT1 subtype	Treatment	Length of follow-up (months)	Cause of death
A1	Acute	Nitisinone+low tyr+Phe diet	5.65	Septic shock
A2	Acute	Nitisinone+low tyr+Phe diet	10	Aspiration pneumonia
S1	Subacute	Nitisinone+low tyr+Phe diet	2.7	Non-responder; hepatic insufficiency with increasing jaundice and intractable coagulopathy despite 5 months of therapy
S2	Subacute	Nitisinone+low tyr+Phe diet	12	Hepatocellular carcinoma
S3	Subacute	Low tyr+Phe diet	2	Massive bleeding due to hepatic insufficiency
S4	Subacute	Nitisinone+low tyr+Phe diet	12	Porphyria like attack during interruption of treatment
S5	Subacute	Nitisinone+low tyr+Phe diet	46	Acute liver rejection after LDLT due to hepatocellular carcinoma
C1	Chronic	Low tyr+Phe diet	2	Hepatic insufficiency and massive esophageal variceal bleeding
C2	Chronic	Nitisinone+low tyr+Phe diet	12	Hepatic insufficiency and massive esophageal variceal bleeding after permanent cessation of NTBC treatment (parents' decision)
C4	Chronic	Nitisinone+low tyr+Phe diet	55	Hepatocellular carcinoma after permanent cessation of NTBC treatment (parents' decision)
C5	Chronic	Nitisinone+low tyr+Phe diet	2.2	Hepatic insufficiency and massive esophageal variceal bleeding (partial responder)
C6	Chronic	Nitisinone+low tyr+Phe diet	4	Metastatic hepatocellular carcinoma

HT1 hereditary tyrosimenia type 1

Six patients (three subacute and three chronic HT1 patients) had persistent high AFP with normal liver function tests despite nitisinone treatment. All were diagnosed with hepatocellular carcinoma or hepatic adenoma. One subacute HT1 patient died during follow up (patient S2; Table 15.4), five had liver transplantation (patient S5, S7, C3, C7 and C10).

Secondary increase in AFP after normalisation was also detected in five patients (one acute, two subacute, two chronic). Two patients responded to increase of nitisinone dose (The acute and subacute HT1 patient) while two patients did not (patient S6 and C3, Table 15.5) and both were liver transplanted (both patients were proved to have HCC) After termination of nitisinone treatment patient C4 (Table 15.4) died due to hepatocellular carcinoma, despite normalization of AFP level while under nitisinone treatment. None of the patients with HCC had normal plasma AFP level.

Eight patients underwent liver transplantation, all living donor liver transplantations (LDLT) (Table 15.5). The median age at transplantation was 81.8 months (range, 16–172 months) and the median age of treatment at the time of transplantation was 52.1 months (range, 4–159 months). Suspected HCC with normal liver function tests was the reason for LDLT in two subacute (patients S6 and S5) and three chronic HT1 patients (patients C3, C7 and C10). Non-compliance with both nitisinone and dietary treatment along with HCC suspicion was the reason for LDLT in patient C3. He had neither maintained adequate nitisinone nor Tyr (4.7 μmol/L and 567 μmol/L, respectively). Serious difficulty in adherence to dietary treatment was the reason LDLT for two chronic HT1 patients (patient C8 and C9); both maintained adequate nitisinone level (54.7 and 79.1 μmol/L, respectively) but had high plasma Tyr concentration (430 and 693 μmol/L, respectively). Hepatocellular carcinoma was detected incidentally in one subacute HT1 patient who underwent liver transplantation due to partial respons to nitisinone therapy with high alpha fetoprotein level and slight increase in PT and aPTT time. All patients but one are alive after liver transplantation (Table 15.5).

Hepacellular carcinoma was reported in 7/32 of patients reported by Dursun et al. (2011). All of these patients were reported to be treated irregularly and inadequetely with nitisinone and there was no data available concerning nitisinone plasma levels or urinary SA excretion. Among these seven patients five who carried N232K, D233V, V259D and IVS3-3C>G mutations developed liver cancer between 10 and 12 years of age. On patient with IVS6-1G>T mutation had normal AFP level although histological test results of liver tissue was compatible with neoplasm.

None of the 36 patients in our cohort with tubulopathy had glomerular involvement, nor developed renal insufficiency. Two patients required temporary supplementation of bicarbonates. Rickets was cured in all 27 patients. No data was available about the renal tubular or glomerular functions in the other published Turkish cases treated with nitisinone.

No porphyria-like neurologic crisis was detected under regular nitisinone treatment in our patients' group. Dursun et al. also reported disappearance of neurologic crisis after nitisinone treatment (Dursun et al. 2011). But interrupted nitisinone treatment for approximately 8 months resulted in death due to severe polyneuropathy with phrenic paresis and respiratory insufficiency in one subacute HT1 patient (S4), and severe abdominal pain due to porphyria-like attack for 2 and 3 months of interruption, respectively, in two of our subacute HT1 patients. Onenli Mungan et al. also reported a 9 months old patient with HT1 detected by selective newborn screening represented with neurologic crisis after 1 month discontinuation of nitisinone (Onenli Mungan et al. 2016).

The only patient with left ventricular septal hypertrophy in our patients' group responded well to nitisinone therapy with disappearance of the hypertrophy. Cardiomyopathy was not reported during the treatment. Dursun et al. also reported reversal of restrictive cardiomyopathy under nitisinone therapy (Dursun et al. 2011).

Height and weight development normalized in all but two patients who remained at <2 SDS for

Table 15.5 Hepatic transplantation in hereditary tyrosimenia type 1 patients

Patient no	HT1 subtype	Age at diagnosis	Time between diagnosis and treatment (months)	Reason for LT	Age at LT (months)	Type of LT	Result
S5	Subacute	9	0	Suspicion of HCC	55	LDLT	Died (acute liver rejection) (HCC was detected at transplantation)
S6	Subacute	12	26	Suspicion of HCC	172	LDLT	Alive (HCC was detected at transplantation)
S7	Subacute	9	0	Partial responder	16	LDLT	Alive (HCC was detected at transplantation)
C3	Chronic	27	30	Non-compliance with the treatment and suspicion of HCC	90	LDLT	Alive (HCC was detected at transplantation)
C7	Chronic	50	57	Suspicion of HCC	132	LDLT	Alive (Hepatic dysplasia detected at transplantation)
C8	Chronic	30	15	Non-compliance with dietary treatment	52	LDLT	Alive
C9	Chronic	39	90	Non-compliance with the dietary treatment	91	LDLT	Alive
C10	Chronic	43	1	Suspicion of HCC	47	LDLT	Alive (HCC was detected at transplantation)

LT liver transplantation, *LDLT* living donor liver transplantation

height and five remained at <2 SDS for weight in our patients' group.

Overall, the primary causes of death in patients with HT1 patients under nitisinone therapy are hepatocellular carcinoma (three patients; patient S2, C4 and C6), hepatic insufficiency (patient S1, patient C1), cirrhosis and variceal bleeding (patient C5), porphyria like attack (patient S4), acute rejection after liver transplantation (patient S5), sepsis (patient A1), aspiration pneumonia (patient A2) (Table 15.4). Both patients who were only treated with low tyrosine and phenylalanine diet, died due to hepatic insufficiency.

Among 34 Turkish patients under treatment with nitisinone therapy, one died after liver transplantation, one due to hepatocellular carcinoma, one due to porphyria like neurologic crisis due to discontinuation of therapy (Dursun et al. 2011; Onenli Mungan et al. 2016). No etiology was reported in five patients that died on the follow-up (Dursun et al. 2011).

15.2.3.3 Adverse Effects of Nitisinone

No adverse effect required interruption of nitisinone treatment. Two patients in our patients' group complained of foreign body sensation in the eyes at plasma tyrosine concentration 865 and 1290 μmol/L with subepithelial corneal opacities. Eye symptoms resolved with strict adherence to diet and decrease of plasma tyrosine <400 μmol/L. No cutaneous lesions were detected. Two patients had transient leukopenia and two had transient thrombopenia without clinical consequences.

No adverse effect of nitisinone was reported on 69 cases extracted from the literature except one patient who presented with corneal pseudodentritic lesions due to increase in plasma tyrosine level and responded to strict compliance to the dietary treatment at a 4 week follow-up (Gulmez Sevim et al. 2015).

Eleven patients in our patients' group had cognitive evaluation: two undertook the age-appropriate Wechsler Scale IQ test (WISC-R; Wechsler Intelligence Scale for Children for age 7–17 years), eight were assessed with the Stanford-Binet Intelligence Scale, and one was assessed with the Cattell Culture Fair Intelligence Test. The mean total IQ was 85 (range 50–115). Low IQ score was associated with special education attendance. WISC-R was repeated at a 1 year and 7 months interval in one patient and Stanford-Binet IQ test was repeated at a 3 year interval in one patient. Both revealed a decline from 74 to 73 and 91 to 88, respectively. Two patients were assessed with both Stanford-Binet Intelligence Scale and WISC-R at a 3 year interval, which revelead a decline from 95 to 84 and 104 to 100, respectively. Eleven patients were evaluated with the Denver II test. The mean total score was 92 (range, 70–100).

15.3 Discussion

This case series represents the largest analysis of data and longterm outcome of HT1 patients in Turkey. Turkey, with a high rate of consanguineous marriages, has a high estimated prevelence of inborn errors of metabolism (Ozalp et al. 1990; Tuncbilek and Ozguc 2007). HT1 has a birth incidence of approximately 1 in 100,000 in most countries but the exact incidence of HT1 in Turkey is still unknown. Consanguinity was noted in 25/42 of our patients (59.5%).

Diagnosis still depends on the clinical suspicion and laboratory investigations as HT1 is not a part of the nationwide screening program, in Turkey. None of the patients was detected by expanded newborn screening for HT1 but two patients were diagnosed by selective newborn screening in the first week of life, due to an affected sibling history (Aktuglu Zeybek et al. 2015; Dursun et al. 2011; Onenli Mungan et al. 2016). Anorexia, pallor, abdominal distension, irritability with abdominal pain were the most common complaints detected in our patients' group. Interestingly babies were most commonly diagnosed with infantile colic at the onset of symptoms. Due to severity of the disease the interval between the onset of clinical symptoms is shortest in acute HT1 patients an longest in the chronic HT1 patients. In all patients reported the initial clinical manifestations ranged from asymptomatic hepatomegaly to severe hepatic insufficiency. Rickets, splenomegaly, diarrhea and anemia were the other most common clinical findings. Neurologic crisis with porphyria like symptoms and restrictive cardiomyopathy was rarely reported but both responded to nitisinone therapy (Coskun et al. 1991; Dursun et al. 2011; Onenli Mungan et al. 2016).

Laboratory diagnosis of HT1 depends on urinary excretion of SA and mutation analysis. As plasma tyrosine that may not always be elevated is not recommended for diagnosis and newborn screening (Dhillon et al. 2011; Morrissey et al. 2011; Zytkovicz et al. 2013). Mild increase in ALT, AST and ALP with hepatic synthesis dysfunctions and elevated AFP are usually the most common laboratory findings in Turkish HT1 patients as reported in the literature (Coskun et al. 1991; Dursun et al. 2011; Bay et al. 2012). The laboratory findings were most striking in acute and subacute HT1 patients.

All Turkish patients reported, excreted SA in their urine except one in our group and one that was reported by Dursun et al. (2011). Detection of increased plasma SA and mutation analysis confirmed the diagnosis in these patients. This finding has previously reported in the literature

(Haagen and Duran 1987; Rinaldo et al. 2006; Cassiman et al. 2009; Blackburn et al. 2016). Haagen and Duran (1987) proved that this finding was due to the low sensitivity of the test method used, but Cassiman et al. (2009) considered that this finding might be due to the residual activity of the FAH enzyme and Blackburn et al. (2016) hypothesized that the p.R142G variant can bind and catabolize SAA efficiently, resulting in undetectable levels of SA. Collecting a 24 h urine sample (all urine portions freezed at −20 °C, separetely) might also increase the chance for SA detection in urine samples where plasma SA and mutation analysis is not available (Aktuglu Zeybek et al. 2015). No significant difference in urinary SA excretion was noted between acute, subacute and chronic HT1 patients.

The tyrosine and phenylalanine restricted diet alone was not effective in HT1 treatment (Coskun et al. 1991). Most common complications in untreated patients are hepatic failure, cirrhosis and hepatocellular carcinoma (Coskun et al. 1991; van Spronsen et al. 1994). Nitisinone has significantly improved survival and quality of life of Turkish HT1 patients, as reported in the literature (Lindstedt et al. 1992; Dursun et al. 2011; Masurel-Paulet et al. 2008; Aktuglu Zeybek et al. 2015; Simoncelli et al. 2015; McKiernan 2006, McKiernan et al. 2015). The recommended initial nitisinone dose ranges from 1 to 2 mg/kg/day, in one or two doses (Jenkins 2002; Roth 2007). Afterwards, the dose must be adjusted according to the plasma nitisinone level that is estimated to be sufficient to eleminate SA excretion (Counce 2011; El-Karaksy et al. 2010; Masurel-Paulet et al. 2008). Initial nitisinone dose ranged from 0.6 to 2 mg/kg/day given in two divided doses in our patients group. On the follow up, the mean nitisinone dose was 1.2 mg/kg/day, which was slightly higher than the recommended dose, 1 mg/kg/day (El-Karaksy 2011; Couce 2011). The mean nitisinone blood level on the follow-up was 39 μmol/L and was sufficient to eliminate SA excretion (Masurel-Paulet et al. 2008; El-Karaksy et al. 2011; Couce 2011).

Despite its effectiveness, still some patients do not (nonresponder) or partially respond to treatment (partial responder). No predictive factors have been identified for this lack of response (van Spronsen et al. 1994). Holme and Lindstedt (2000) described improvement in liver functions in 90% of cases treated with nitisinone before 6 months of age, but eight patients did not respond to treatment. One subacute HT1 patient in our group was found to be nonresponder to nitisinone treatment. The subacute HT1 patient (patient S1) was diagnosed at 8 months of age: although he was treated with nitisinone (2 mg/kg/day), he died due to progressive hepatic insufficiency in the second month of treatment (Table 15.4). Two patients (one subacute and one chronic HT1 patient) had partial response to nitisinone treatment with partial recovery in the coagulation profile, and were under evaluation for LT. The chronic HT1 patient (Patient C5) was diagnosed at 29 months of age and despite nitisinone treatment, the patient died at 10 months of treatment due to progressive cirrhosis and massive esophageal variceal bleeding. The subacute HT1 patient underwent succefull living donor liver transplantation (Table 15.5, patient S7). Unresponsiveness to nitisinone has already been reported but still the reasons for lack of response have not been clearly identified (van Spronsen et al. 1994) Methionine level can be an indicator for non/partial response. There was no significant difference in plasma methionine level between responders (mean, 356.3 μmol/L; range, 13.3–1590 μmol/L) and non/partial-responders in our patients group at presentation (mean, 465 μmol/L; range, 70–820 μmol/L) but the increased methionine did not decrease to normal (20–50 μmol/L) despite maintenance of adequate nitisinone and tyrosine. There was a significant difference between both groups during follow up: mean plasma methionine was 15.9 μmol/L (range, 12–39 μmol/L) in responders, while it was 265 μmol/L (range, 79–470 μmol/L) in non/partial-responders ($P < 0.001$). No further data was available regarding responsiveness and non/partial responsiveness in Turkish HT1 patients.

Although nitisinone had significantly improved survival in both our cohort and in Turkish patients described in the literature, the overall survival rate of the 40 patients treated with nitisinone was 75% (30/40) and 88% (22/24), respectively. This rate is higher than in pre-nitisinone reports (van Spronsen et al. 1994). Still, the survival rate in the present cohort was lower than in the French, Spanish and Quebec series, with rates of 97.8%, 100%, 96%, respectively (Masurel-Paulet et al. 2008; Couce et al. 2011; Larochelle et al. 2012). The most striking difference between the present study and these studies was that nitisinone treatment was initiated after 6 months of age in 69% of the present patients and 60.6% of Turkish patients described (Dursun et al. 2011; Bay et al. 2012; Onenli Mungan et al. 2016). Temporary interruption of the treatment due to various reasons (e.g. health insurance problems) was common, leading to irregular and inadequate treatment with nitisinone. Permanent interruption of nitisinone treatment resulted in death in four patients. One patient died due to HCC, one patient due to hepatic insufficiency and two due to porphyria-like neurologic crisis. Although latest data suggest that early nitisinone treatment is the key factor in the good outcome of the neonatally treated patients, and that a combination of neonatal screening and early nitisinone treatment is recommended, even in early diagnosis patients, drug interruption can lead to increased mortality rate, especially due to HCC (Larochelle et al. 2012; Morrissey et al. 2011; Zytkovicz et al. 2013).

Liver cancer has been reported as an important risk for patients with HT1 treated with nitisinone (Holme and Lindstedt 1998; Seda Neto et al. 2014; Bahador et al. 2015). Development of HCC is the main risk for patients with the chronic form or who have been treated with nitisinone after 2 years of age (Holme and Lindstedt 2000; van Spronsen et al. 2005). The patients with persistent high AFP and/or has a slow AFP decline without reaching to normal are under particular risk (Koelink et al. 2006; Larochelle et al. 2012; Mayorandan et al. 2014). All six patients in our study group with persistent high AFP along with normal liver function tests are proved to have hepatocellular carcinoma. Secondary AFP increase was detected in four patients: two responded to dose increase while two patients did not and HCC was detected at liver transplantation. The present data support the findings of previous reports (Koelink et al. 2006; Larochelle et al. 2012). Close follow up of serum AFP, even of minor changes, is important for early detection of HCC and hepatocellular dysplasia especially in late-diagnosed patients. Also, disruption of the hepatic architecture was a common finding along with nodularity even at the time of presentation, especially in late diagnosed patients. Nitisinone treatment was not effective in complete reversion of the hepatic lesions and normalization of the architecture, except in two patients. De novo hyperechoic and hypoechoic nodule formation were also detected despite nitisinone treatment without increase in AFP. As all of our patients with HCC had hypoechoic nodules detected on ultrasound, although malign transformation was not definitely proven on abdominal MRI hypoechoic de novo nodule formation should always be an alert for HCC as HCC has been reported in HT1 patients without clear increase of AFP (van Ginkel et al. 2015).

Renal tubular dysfunction with increased echogenicity was common in the present patients, as previously described (Couce et al. 2011; Mayorandan et al. 2014).

The findings resolved with nitisinone treatment although asymptomatic nephrocalcinosis persisted in 5/5 patients. None of the patients developed renal insufficiency during nitisinone treatment.

Episodes of acute neuropathy that clinically resemble porphyric crises occur in up to 50% of untreated children (Mitchell et al. 1990; van Spronsen et al. 1994). Succinylacetone acts as a competitive inhibitor of the enzyme delta aminolevulinic acid dehydratase in the haem biosynthetic pathway. This enzymatic block leads to increase in delta aminolevulinic acid and PBG. Neurological crises in HT1 are identical to those occurring in porphyria and lead poisoning, in which delta aminolevulinic acid is also increased (Russo et al. 2001). The mortality is up to 65% if left untreated (Chakrapani et al. 2012). Normalization of SA in blood, correction of the complete inhibition of PBGS in erythrocytes during nitisinone treatment protects against porphyric crisis (Lindstedt et al. 1992). Schlump et al. reported that interruption of nitisinone treatment can cause severe neurological crisis in patients with HT1 (Schlump et al. 2008). In the present cohort, interrupted nitisinone treatment resulted in porphyria-like attacks in three patients, and death in one patient due to respiratory insufficiency. Kalkanoglu and Ucar reported two Turkish HT1 cases of acute pancreatitis mimicking neurologic crisis due to discontinuation of nitisinone treatment (Kalkanoglu and Coskun 1999; Ucar et al. 2016), resolved after continuation of nitisinone. Onenli Mungan also reported a case with severe neurologic crisis after discontinuation of nitisinone only for a month (Onenli Mungan et al. 2016).

Cardiomyopathy has been reported as a frequent finding in HT1 (André et al. 2005; Arora et al. 2006; Mohamed et al. 2013; Seda Neto et al. 2014). Neither the mechanism nor the natural history of this complication is understood. It is suggested that the cardiomyopathy may be due to direct cardiotoxicity of circulating Tyr metabolites during a critical period of vulnerability, and nitisinone reduces the level of these metabolites (Arora et al. 2006). We had only one patient with septal hypertrophy and Dursun reported one patient, both responded to nitisinone therapy.

Compliance with low Tyr low Phe diet is challenging in the follow-up of patients with HT1. The recommended level of Tyr, 200–400 μmol/L, is difficult to achieve. Noncompliance increases as the child grew older. Even patients with good and moderate compliance had periods of bad control. The serious dietary non-compliance problems led two patients to prefer LDLT in our group.

Variation of plasma tyrosine seems to be an important pathogenic factor in abnormal intellectual development and attention disorder in HT1 patients under long-term treatment with nitisinone (De Laet et al. 2011; Bendadi et al. 2014; Pohorecka et al. 2012; van Ginkel et al. 2015). Mouse model studies revealed that lower learning and cognitive differences caused by tyrosinemia type 1 and not by the treatment with NTBC (Hillgartner et al. 2016).

Eleven patients had cognitive evaluation and the mean total IQ was 85 (range, 50–115). Low IQ score was associated with special education. Repeated IQ evaluation revealed a decline in average IQ score in all four patients. These findings were consistent with recent studies (De Laet et al. 2011; Bendadi et al. 2014; Pohorecka et al. 2012; van Ginkel et al. 2015). Attention deficit and learning difficulties were also common, but behavior was not evaluated in the present cohort. Although exact mechanism for intellectual impairment still remains unknown it is hypothesized that elevated plasma tyrosine concentrations under NTBC treatment may be associated with neurocognitive functioning (van Ginkel et al. 2015). Mouse models revealed that tyrosinemia type I and Not treatment with NTBC causes slower learning and altered behaviour in mice (Hillgartner et al. 2016).

Genetic data documents obtained from the literature revealed the most prevelant mutation in Turkish patients were homozygosity for IVS6-1G>T (26%) followed up by D233V (22%). IVS6-1G>T has previously been reported as the most common mutation in Mediterranean populations (Bergman et al. 1998; Couce et al. 2011; Dursun et al. 2011; Mayorandan et al. 2014; Angileri et al. 2015). D233V which was the second most common seen mutation is specific to Turkish population (Dursun et al. 2011; Rootwelt et al. 1994, 1996). IVS12+5G>A was detected in three patients (Rootwelt et al. 1994; Dursun et al. 2011). V166G and V259D mutations, probably causing dysfunction via misfolding, was reported in three Turkish patients (Dursun et al. 2011). No clear genotype/phenotype correlation was found for the most common mutations (IVS6-1G>T, D233V, IVS3-3C>G) (Bergman et al. 1998; Dursun et al. 2011), although Counce reported that hepatic presentation was more common and the renal involvement less frequent in HT1 patients carrying IVS6-1G>T mutation than in patients with other mutations (Counce et al. 2011).

15.4 Conclusion

Nitisinone treatment is effective and improves both the short-term and the long-term prognosis of HT1. Still, early diagnosis on newborn screening is needed because late diagnosis along with delay in treatment carries the risk of persistence of hepatic disease and HCC. In countries where

HT1 is not part of newborn screening, it is important to be able to recognize the clinical and laboratory findings in order to prevent delay in diagnosis. More importantly, adding HT1 screening to the nationwide screening program in Turkey is necessary for early diagnosis and facilitation of early treatment, to decrease the mortality and morbidity of the disease.

References

Aktuglu Zeybek AC, Kiykim E, Soyucen E, Cansever S, Altay S, Zubarioglu T, Erkan T, Aydin A (2015) Hereditary tyrosinemia type 1 in Turkey: twenty-year single-center experience. Pediatrics International 57(2):281–289

André N, Roquelaure B, Jubin V et al (2005) Successful treatment of severe cardiomyopathy with NTBC in a child withtyrosinaemia type I. J Inherit Metab Dis 28(1):103–106

Angileri F, Bergeron A, Morrow G et al (2015) Geographical and ethnic distribution of mutations of the fumarylacetoacetate hydrolase gene in hereditary tyrosinemia type 1. JIMD Rep 19:43–58

Arora N, Stumper O, Wright J et al (2006) Cardiomyopathy in tyrosinaemia type I is common but usually benign. J Inherit Metab Dis 29(1):54–57

Bahador A, Dehghani SM, Geramizadeh B et al (2015) Liver transplant for children with hepatocellular carcinoma and hereditary tyrosinemia type 1. Exp Clin Transplant 13(4):329–332

Baumann U, Preece MA, Green A et al (2005) J Inherit Metab Dis 28(2):131–135

Bay A, Karaoglu O, Sivasli E et al (2012) An infant with prolonged circumcision bleeding and unexplained coagulopathy. Indian J Hematol Blood Transf 28(3):181–183

Bendadi F, de Koning TJ, Visser G et al (2014) Impaired cognitive functioning in patients with tyrosinemia type 1 receiving nitisinone. J Pediatr 164:398–401

Bergman AJ, van den Berg IE, Brink W et al (1998) Spectrum of mutations in the fumarylacetoacetate hydrolase gene of tyrosinemia type 1 patients in northwestern Europe and Mediterranean countries. Hum Mutat 12(1):19–26

Blackburn PR, Hickey RD, Nace RA et al (2016) Silent tyrosinemia type i without elevated tyrosine or succinylacetone associated with liver cirrhosis and hepatocellular carcinoma. Hum Mutat 37:1097–1105. [Epub ahead of print]

Cassiman D, Zeevaert R, Holme E et al (2009) A novel mutation causing mild, atypical fumarylacetoacetase deficiency (Tyrosinemia type I): a case report. Orphanet J Rare Dis 4:28

Chakrapani A, Gissen P, McKierrnan P (2012) Disorders of tyrosine metabolism. In: Saudubray JM, van den Berghe G, Walter JH (eds) Inborn metabolic diseases, diagnosis and treatment, 5th edn. Springer, Heidelgerg, pp 265–276

Coskun T, Ozalp I, Koçak N et al (1991) Type I hereditary tyrosinaemia: presentation of 11 cases. J Inherit Metab Dis 14(5):765–770

Counce ML, Dalmau J, del Toro M et al (2011) Tyrosinemia type 1 in Spain: mutational analysis, treatment and long-term outcome. Pediatr Int 53(6):985–989

De Laet C, Munoz VT, Jaeken J et al (2011) Neuropsychological outcome of NTBC-treated patients with tyrosinaemia type 1. Dev Med Child Neurol 53:962–964

Dhillon KS, Bhandal AS, Aznar CP et al (2011) Improved tandem mass spectrometry (MS/MS) derivatized method for the detection of tyrosinemia type I, amino acids and acylcarnitine disorders using a single extraction process. Clin Chim Acta 412:873–879

Dursun A, Ozgül RK, Sivri S et al (2011) Mutation spectrum of fumarylacetoacetase gene and clinical aspects of tyrosinemia type I disease. JIMD Rep 1:17–21

El-Karaksy H, Rashed M, El-Sayed R et al (2010) Clinical practice. NTBC therapy for tyrosinemia type 1: how much is enough? Eur J Pediatr 169:689–693

El-Karaksy H, Fahmy M, El-Raziky M et al (2011) Hereditary tyrosinemia type 1 from a single center in Egypt: clinical study of 22 cases. World J Pediatr 7(3):224–231

Gulmez Sevim D, Gumus K, Cavanagh HD (2015) Corneal pseudodendritic lesions masquerading as herpetic keratitis in a patient with tyrosinemia type I. Eye Contact Lens 43:e7–e9. [Epub ahead of print]

Haagen AAM, Duran M, (1987) Absence of increased succinylacetone in the urine of a child with hereditary tyrosinaemia type I. J Inherit Metab Dis 10 (S2):323–325

Hillgartner MA, Coker SB, Koenig AE et al (2016) Tyrosinemia type I and not treatment with NTBC causes slower learning and altered behavior in mice. J Inherit Metab Dis 39:673–682. [Epub ahead of print]

Holme E, Lindstedt S (1992) Neonatal screen for hereditary tyrosinemia type I. Lancet 340:850

Holme E, Lindstedt S (1998) Tyrosinaemia type 1 and NTBC (2-(2-nitro- 4-trifluoromethylbenzoyl)-1,3-cyclohexanedione). J Inherit Metab Dis 21:507–517

Holme E, Lindstedt S (2000) Nontransplant treatment of tyrosinemia. Clin Liver Dis 4:805–814. 34

Jenkins J (2002) Orphadin. Cited 30 June 2016. Available from URL: http://www.accessdata.fda.gov/drugsatfda_docs/label/2002/21232lbl.pdf

Jorquera R, Tanguay RM (2001) Fumarylacetoacetate, the metabolite accumulating in hereditary tyrosinemia, activates the ERK pathway and induces mitotic abnormalities and genomic instability. Hum Mol Genet 10:1741–1752

Kalkanoglu HS, Coskun T (1999) Neurological crisis mimicking acute pancreatitis in tyrosinemia type I. Turk J Pediatr 41:501–504. 38

Koelink CJL, van Hasselt P, van der Ploeg A, et al (2006) Tyrosinemia type I treated by NTBC: how does AFP predict liver cancer? Mol Genet Metab 89(4):310–315

Kvittingen EA, Talseth T, Halvorsen S et al (1991) Renal failure in adult patients with hereditary tyrosinaemia type I. J Inherit Metab Dis 14(1):53–62

Larochelle J, Alvarez F, Bussières JF et al (2012) Effect of nitisinone (NTBC) treatment on the clinical course of hepatorenal tyrosinemia in Québec. Mol Genet Metab 107:49–54

Lindstedt S, Holme E, Lock EA et al (1992) Treatment of hereditary tyrosinaemia type I by inhibition of 4-hydroxyphenylpyruvate dioxygenase. Lancet 340:813–817

Lock EA, Ellis MK, Gaskin P et al (1998) From toxicological problem to therapeutic use: the discovery of the mode of action of 2-(2-nitro-4-trifluoromethylbenzoyl)-1,3-cyclohexanedione (NTBC), its toxicology and development as a drug. J Inherit Metab Dis 21:498–506

Masurel-Paulet A, Poggi-Bach J, Rolland MO et al (2008) NTBC treatment in tyrosinaemia type I: long-term outcome in French patients. J Inherit Metab Dis 31(1):81–87

Mayorandan S, Meyer U, Gokcay G et al (2014) Cross-sectional study of 168 patients with hepatorenal tyrosinaemia and implications for clinical practice. Orphanet J Rare Dis 1:9. 107

Mckiernan PJ (2006) Nitisinone in the treatment of hereditary tyrosinemia type 1. Drugs 66:743–750

McKiernan PJ, Preece MA, Chakrapani A (2015) Outcome of children with hereditary tyrosinaemia following newborn screening. Arch Dis Child 100(8):738–741

Mitchell G, Larochelle J, Lambert M et al (1990) Neurologic crisis in hereditary tyrosinemia. N Engl J Med 322:432–437

Mohamed S, Kambal MA, Al Jurayyan NA et al (2013) Tyrosinemia type 1: a rare and forgotten cause of reversible hypertrophic cardiomyopathy in infancy. BMC Res Notes 6:362

Morrissey MA, Sunny S, Fahim A et al (2011) Newborn screening for Tyr-I: two years' experience of the New York State program. Mol Genet Metab 103:191–2.22

Nakamura K, Matsumoto S, Mitsubuchi H et al (2015) Diagnosis and treatment of hereditary tyrosinemia in Japan. Pediatr Int 57(1):37–40

Onenli Mungan N, Yıldızdas D, Kor D et al (2016) Tyrosinemia type 1 and irreversible neurologic crisis after one month discontinuation of nitisone. Metab Brain Dis 31:1181–1183. [Epub ahead of print]

Ozalp I, Coskun T, Tokol S et al (1990) Inherited metabolic disorders in Turkey. J Inherit Metab Dis 13:732–738

Pohorecka M, Biernacka M, Jakubowska-Winecka A et al (2012) Behavioral and intellectual functioning in patients with tyrosinemia type 1. Pediatr Endocrinol Diabetes Metab 18:96–100

Rinaldo P, Hahn SH, Matern D (2006) Inborn errors of amino acid, organic acid, and fatty acid metabolism. In: Burtis CA, Ashwood ER, Bruns DE (eds) Tietz textbook of clinical chemistry and molecular diagnostics. Elsevier, St. Louis, pp 2218–2219

Rootwelt H, Berger R, Gray G et al (1994) Novel splice, missense, and nonsense mutations in the fumarylacetoacetase gene causing tyrosinemia type 1. Am J Hum Genet 55(4):653–658

Rootwelt H, Hoie K, Berger R et al (1996) Fumarylacetoacetase mutations in tyrosinaemia type I. Hum Mutat 7(3):239–243

Roth KS (2007) Tyrosinemia (Emedicine Website). Cited 30 June 2016. Available from URL: http://www.emedicine.com/ped/TOPIC2339.HTM

Russo P, Mitchell G, Tanguay R (2001) Tyrosinemia: a review. Pediatr Dev Pathol 4:212–221

Santra S, Baumann U (2008) Experience of nitisinone for the pharmacological treatment of hereditary tyrosinaemia type 1. Expert Opin Pharmacother 9:1229–1236

Sassa S, Kappas A (1983) Hereditary tyrosinemia and the heme biosynthetic pathway. Profound inhibition of delta-aminolevulinic acid dehydratase activity by succinylacetone. J Clin Invest 71:625–634

Schlump JU, Perot C, Ketteler K et al (2008) Severe neurological crisis in a patient with hereditary tyrosinaemia type I after interruption of NTBC treatment. J Inherit Metab Dis 31(Suppl. 2):S223–S225

Seda Neto J, Leite KM, Porta A et al (2014) HCC prevalence and histopathological findings in liver explants of patients with hereditary tyrosinemia type 1. Pediatr Blood Cancer 61(9):1584–1589

Sener RN (2005a) Tyrosinemia: computed tomography, magnetic resonance imaging, diffusion magnetic resonance imaging, and proton spectroscopy findings in the brain. J Comput Assist Tomogr 29(3):323–325

Sener RN (2005b) Brain magnetic resonance imaging in tyrosinemia. Acta Radiol 46(6):618–620

Simoncelli M, Samson J, Bussières JF et al (2015) Cost-consequence analysis of nitisinone for treatment of tyrosinemia type I. Can J Hosp Pharm 68(3):210–217

Tanguay RM, Valet JP, Lescault et al (1990) Different molecular basis for fumarylacetoacetate hydrolase deficiency in the two clinical forms of hereditary tyrosinemia (type I). Am J Hum Genet 47(2):308–316

Thimm E, Richter-Werkle R, Kamp G et al (2012) Neurocognitive outcome in patients with hypertyrosinemia type 1 after long-term treatment with NTBC. J Inherit Metab Dis 35:263–268. 43

Tuncbilek E, Ozguc M (2007) Application of medical genetics in Turkey. Turk J Pediatr 49:353–359. 25

Ucar HK, Tumgor G, Kor D et al (2016) A case report of very rare association of Tyrosinemia type I and pan-

creatitis mimicking neurologic crisis of Tyrosinemia type I. Balkan Med J 33(3):370–372
van Ginkel WG, Gouw AS, van der Jagt EJ et al (2015) Hepatocellular carcinoma in tyrosinemia type 1 without clear increase of AFP. Pediatrics 135(3):e749–e752
van Spronsen FJ, Thomasse Y, Smit GP et al (1994) Hereditary tyrosinemia type I: a new clinical classification with difference in prognosis on dietary treatment. Hepatology 20(5):1187–1191
Van Spronsen FJ, Bijleveld CM, Van Maldegem BT et al (2005) Hepatocellular carcinoma in hereditary tyrosinemia type I despite 2-(2nitro-4-3-trifluoromethylbenzoyl)-1,3-cyclohexanedione treatment. J Pediatr Gastroenterol Nutr 40:90–93
Yagci MA, Tardu A, Karagul S et al (2015) Living donor liver transplantation with vena cava replacement. Transplant Proc 47(5):1453–1457
Zytkovicz TH, Sahai I, Rush A et al (2013) Newborn screening for hepatorenal tyrosinemia-I by tandem mass spectrometry using pooled samples: a four-year summary by the New England newborn screening program. Clin Biochem 46:681–684

Part V

Management and Future

From Weed Killer to Wonder Drug

16

Edward A. Lock

Abstract

The discovery that a natural product leptospermone had herbicidal activity formed the starting point for chemical synthesis to find more activity and selectivity. A series of molecules called triketones were found to possess good activity and 2-(2-nitro-4-trifluoromethylbenzoyl)-cyclohexane-1,3-dione (NTBC) was selected for toxicology testing. NTBC fed at low doses to rats and dogs caused keratopathy, which on cessation of the diet recovered. Mice, rabbits and monkeys fed NTBC did not show this response. Research discovered that NTBC caused tyrosinaemia which was due to inhibition of the enzyme 4-hydroxyphenylpyruvate dioxygenase in both mammals and plants thereby finding a novel target for killing plants. NTBC was also used sucessfully as a drug to treat a rare inborn error of metabolism, tyrosinaemia type I, in collaboration with Professor's Sven Lindstedt and Elisabeth Holme. Understanding the mechanism of toxicity of NTBC led to novel herbicide discovery and saved the lives of children with acute tyrosinaemia type I.

Keywords

Tyrosinaemia type I • 4-hydroxyphenylpyruvate dioxygenase • Nitisinone • Alcaptonuria

Abbreviations

AKU	Alkaptonuria
HGA	Homogentisic acid
HGD	Homogentisic acid dioxygenase
HPPD	4-hydroxyphenylpyruvate dioxygenase

E.A. Lock (✉)
School of Pharmacy & Biomolecular Sciences, Liverpool John Moores University, Byrom Street, Liverpool L3 3AF, UK
e-mail: e.lock@ljmu.ac.uk

© Springer International Publishing AG 2017

R.M. Tanguay (ed.), *Hereditary Tyrosinemia*, Advances in Experimental Medicine and Biology 959, DOI 10.1007/978-3-319-55780-9_16

16.1 Introduction

I wish to dedicate this chapter to two former collaborators Professor Emeritus Sven Lindstedt (Fig. 16.1) and Professor Emeritus Elisabeth Holme (Fig. 16.1). Without their enthusiasm, commitment and knowledge, Nitisinone would not have been developed as an orphan drug for the treatment of tyrosinaemia type 1.

16.2 Obituaries

Sven Lindstedt M.D., Ph.D. studied medicine at Lund and obtained his PhD in 1957. He then spent time in the USA and returned to the Karolinska Institute and Karolinska hospital's clinical chemistry laboratory where Sven establish his own research on carnitine, vitamin B6 and the metabolic disease tyrosinaemia. Sven was appointed Professor of Medicinal Chemistry in Lund in 1966. Two years later, he became Professor and Consultant in Clinical Chemistry at the University of Gothenburg and Sahlgrenska University Hospital, where he worked for the rest of his career. The ground-breaking research on tyrosinaemia type 1 was his team's discovery of the primary cause of the disease which lead in 1991 to the use of an herbicide as an effective treatment against this deadly disease. The drug Nitisinone is now accepted worldwide for the treatment of tyrosinaemia type 1. In addition to his research as the Director of Clinical Chemistry at Sahlgrenska Hospital, he introduced modern automated analytical instruments and advanced computerisation to enable a large number of samples to be analysed. Sven also acquired a mass spectrometer to help analyse samples from children with congenital metabolic diseases. Sven was a leader in his field who will be sadly missed. He died aged 88 in 2015 leaving a wife and five children. Kindly translated from Swedish by Professor Anders Lindahl.

Elisabeth Holme M.D., Ph.D. also studied medicine at Lund and obtained her M.D. in 1975 and started full timework at Sahlgrenska University Hospital where she specialised in Clinical Chemistry. In 1982, she finished her doctoral thesis and appointed an associate senior consultant in 1985, then a senior consultant in 1992 at Sahlgrenska University Hospital. In 1984, she became associate Professor in Clinical Chemistry and in 2009 appointed Professor in Clinical Chemistry at the Department of Clinical Chemistry and Transfusion Medicine at the Institute of Biomedicine at the Sahlgrenska Academy at Gothenburg University. However, she had earned her Professorship a lot earlier!

Fig. 16.1 Photographs of Sven Lindstedt and Elisabeth Holme, from the internet

Elisabeth became Professor Emeritus in July 2014, but continued working as a senior consultant at the hospital. Elisabeth made major contributions in the field of mitochondrial disease and fatty acid oxidation defects as well as other fields of inborn errors of metabolism. However, her work that had the largest impact was that which she did with Sven Lindstedt demonstrating the dramatic effects of Nitisinone in the treatment of tyrosinaemia. Elisabeth was also an enthusiastic member of the Society for the Study of Inborn Errors of Metabolism. She organised a meeting in Gothenburg in 1997 and did a great deal to encourage and help younger members of the Society. Elisabeth died aged 68 in 2015 having lost her husband in 2009 and leaving two children. Adapted from Kollberg and Clayton (2015).

It was a great pleasure to work with Sven and Elisabeth to enable Nitisinone to be used for the treatment of tyrosinaemia type 1; we had some most enjoyable discussions regarding the best approach to enable the chemical to be given to children. I found my visits to Gothenburg scientifically stimulating and rewarding.

16.3 The Discovery of 4-Hydroxyphenyl Pyruvate Dioxygenase Inhibitors

Naturally occurring diketone and triketone alkaloids are produced by a number of myrtaceous plants (Helleyer 1968) and lichens (Romagni et al. 2000). These compounds act as allelopathic agents, synthesised to prevent the growth of surrounding plants and or microbes. The discovery of the triketone herbicides began in 1977 when a scientist by the name of Reed Gray working for Stauffer Chemicals at the Western Research Center, California noticed that few weeds grew under the bottlebrush plant *Callistemon citrinus* than in surrounding areas. Because of this observation, he separated compounds from bottlebrush plant extracts and showed that grass became stunted and white in colour when exposed to these extracts. This enabled him to isolate and identify the structure of the herbicidally active chemical that was an acyl syncarpic acid, leptospermone (Fig. 16.2a). This compound is a natural

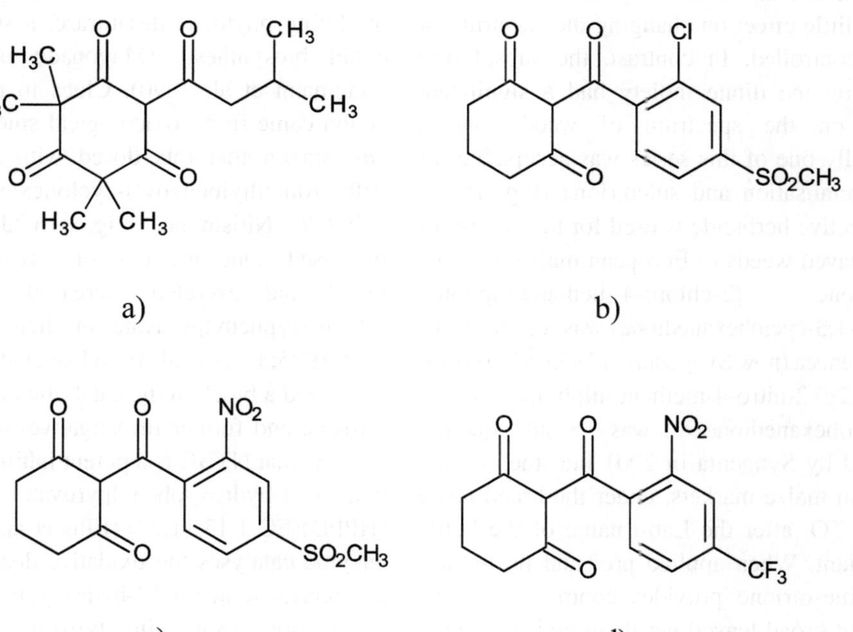

Fig. 16.2 The chemical structures of 4-hydroxyphenylpyruvate dioxygenase inhibitors (**a**) Leptospermone; (**b**) Sulcotrione; (**c**) Mesotrione and (**d**) NTBC/Nitisinone

product but its biological activity was not known. Further testing of leptospermone in the greenhouse showed it to be a moderate herbicide against a narrow range of weed species. For grass weeds, it is active at an application rate of 1000 g/ha or more (see Beaudegnies et al. 2009). Combining the syncarpic acid unit of leptospermone with a benzoyl moiety led to a significant boost in both the spectrum of weeds controlled and overall herbicidal potency e.g. 2-chlorobenzoyl syncarpic acid showed herbicidal activity against a wide range of broad-leaved and grass weeds at 500 g/ha and the compound showed the same bleaching symptoms as leptospermone. Optimisation of the substituents on the benzoyl moiety showed that substitution on the 2 position was critical for herbicidal activity and 2,4-substitution was preferred with 2-nitro-4-trifluromethylbenzoyl syncarpic acid showing very good herbicidal activity in the greenhouse at 62.5 g/ha or less (see Beaudegnies et al. 2009). Subsequent work using 2-substituted and 2, 4-substituted benzoylcyclohexanedione-1, 3-diones showed good herbicidal activity. The effects of substitution on the benzoyl group modulated the overall herbicidal potency, but tended to have little effect on changing the spectrum of weeds controlled. In contrast, the substitution pattern on the dione moiety had a significant impact on the spectrum of weed control. Eventually one of this series was acceptable for commercialisation and sulcotrione (Fig. 16.2b) as a selective herbicide is used for the control of broad-leaved weeds in European maize markets. Sulcotrione (2-chloro-4-methanesulphonyl benzoyl-1,3-cyclohexanedione) was registered for use by Zeneca (now Syngenta) in 1993. Mesotrione (Fig. 16.2c) 2-nitro-4-methanesulphonylbenzoyl-1,3-cyclohexanedione was subsequently launched by Syngenta in 2001 into the US and European maize markets, under the brand name CALLISTO, after the Latin name of the bottle brush plant. When applied pre- and post-emergence, mesotrione provides control of all the important broad-leaved weeds in maize together with suppression/control of some of the annual grass weeds. Typical use rates for mesotrione range from 100 to 225 g/ha when applied pre-emergence, and 70–150 g/ha for post-emergence applications (Mitchell et al. 2001).

Since the introduction of sulcotrione and mesotrione a number of other agrochemical companies have introduced products with the same mode of action including, Topramezone, [3-(4, 5-dihydro-isoxazol-3-yl)-4-methylsulfonyl-2-methylphenyl](5-hydroxy-1-methyl-1*H*-pyrazol-4-yl)methanone (Grossmann and Ehrhardt 2007); Tembotrione, 2-[2-chloro-4-(methylsulfonyl)-3-[(2,2,2-trifluoroethoxy)methyl]benzoyl]-1,3-cyclohexanedione (Santel 2009); Isoxaflutole, 5-cyclopropyl-1,2-isoxazol-4-yl ααα-trifluoro-2-mesyl-p-tolyl ketone (Pallett et al. 2001); Bicyclopyrone, 4-hydroxy-3-[2-(2-methoxy, ethoxymethyl)-trifluoro-methylpyridine-3-carbonyl]-bicyclo[3.2.1]oct-3-en-2-one (Syngenta Crop Protection); Pyrasulfotole, (5-hydroxy-1,3-dimethyl-1*H*-pyrazol-4-yl) [2-(methylsulfonyl)-4-trifluoromethyl)phenyl] methanone (Reddy et al. 2012) and Tolpyralate (Ishihara Sangyo Kaisha Ltd).

The mode of action of this series of benzoyl syncarpic acids and benzoyl cylcohexane-1,3-dione for killing plants was not understood. Initially it was thought that they might act by inhibiting phytoene desaturase, a step in carotenoid biosynthesis (Mayonado et al. 1989; Sandmann et al. 1990). Clues to the mode of action came from toxicological studies where it was shown that rats dosed with 2-(2-nitro-4-trifluoromethylbenzoyl)-cyclohexane-1,3-dione (NTBC/ Nitisinone, Fig. 16.2d) exhibited increased concentrations of tyrosine in their blood and excreted increased amounts of 4-hydroxyphenylpyruvate in their urine (Ellis et al. 1995; Lock et al. 1996 Lee et al. 1997). This suggested a block in the catabolic degradation of tyrosine and further investigative work went on to show that NTBC is a potent inhibitor of mammalian 4-hydroxyphenylpyruvate dioxygenase (HPPD)(EC 1.13.11.27) (Ellis et al. 1995). This enzyme catalyses the oxidative decarboxylation and rearrangement of 4-hydroxyphenyl pyruvate to homogentisate in tyrosine degradation (Fig. 16.3) (Lindblad et al. 1970). For detailed information on the interaction of NTBC with HPPD, see Kavana and Moran (2003) and Moran (2005).

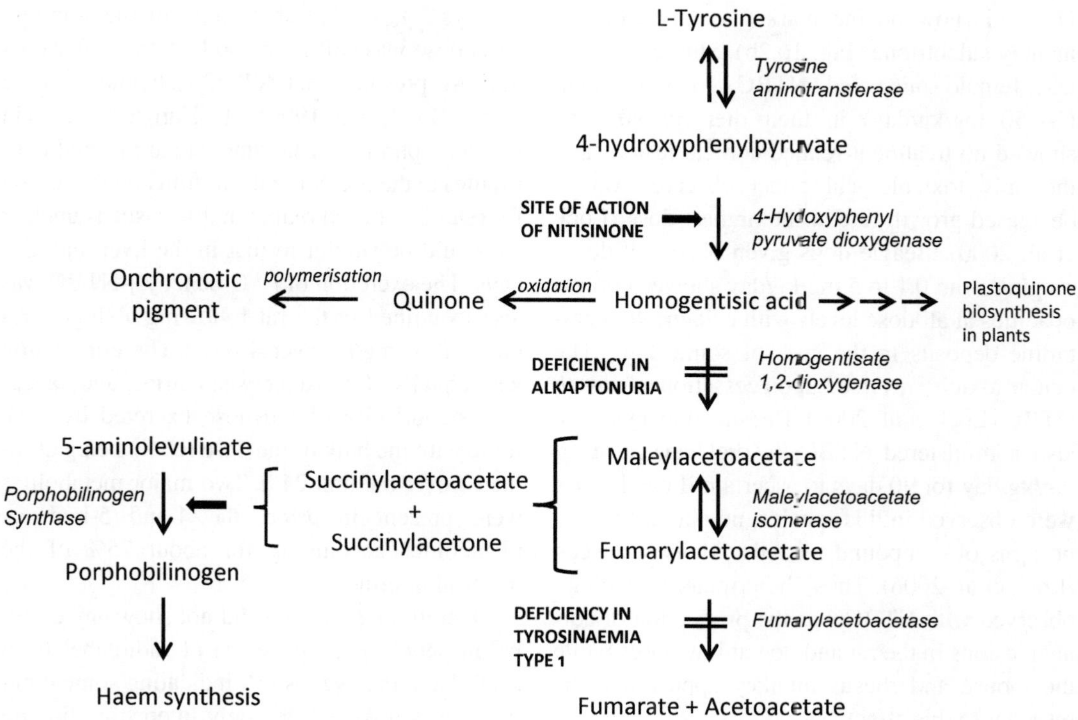

Fig. 16.3 The tyrosine degradation pathway and it relevance to tyrosinaemia type 1, alkaptonuria and the site of action of Nitisinone

At the time of discovery of the mode of action, it was unclear if plants had HPPD but subsequent work by Prisbylla et al. (1993) showed the enzyme was present in corn and was sensitive to NTBC. Treatment of *Ipomeanol Hederacea* with NTBC produced an increase in plant tyrosine concentration and a corresponding decrease in plastoquinone that accounts for the bleaching. Schulz et al. (1993) who reported that suclotrione was a potent inhibitor of HPPD confirmed this. It is important to note that in plants, HPPD is part of the synthesis pathway for essential components such as plastoquinone and tocopherols while in mammals HPPD is involved in the catabolism of tyrosine.

16.4 Toxicology of NTBC/Nitisinone to Mammals

Routine toxicology studies on NTBC and related structures at the Environmental Health Center Laboratories in Farmington, CT, USA, showed NTBC had a moderate to low acute oral toxicity. However, when fed in the diet to male and female rats at 1–800 ppm (0.05–40 mg/kg/day) for 90 days, NTBC produced corneal opacities with and without accompanying vascularisation, with the opacities being either bilateral or unilateral (Robinson 1995). The earliest response occurred about 1 week after starting the diet, with the maximal response seen towards the end of the study, typically with an incidence of about 80% at the highest doses (Lock et al. 1996). Hence, some rats fed NTBC for 90 days did not respond, the reason for this is currently unclear. Upon cessation of feeding the diet containing NTBC the corneal lesions recovered, all eyes being either normal or those with vascularisation just showing residual "ghost" blood vessels (Robinson 1995). Chronic exposure of rats to NTBC at 0.05–40 mg/kg/day for 2 years to assess its carcinogenic potential stopped after 1 year as NTBC was withdrawn from development. No additional NTBC-related toxicity was seen when compared to that after 90 days exposure. A structural analogue of

NTBC is now on the market as an herbicide, namely sulcotrione (Fig. 16.2b). Studies in male and female mice fed NTBC 10–3500 ppm, (1–350 mg/kg/day) in their diet for 90 days showed no treatment-related corneal lesions and the only toxicological effect observed was a decreased growth rate at the highest dose (Lock et al. 2000). Beagle dogs given NTBC at doses ranging from 0.1 to 5 mg/kg/day showed corneal opacities at all dose levels with evidence of crystalline deposits in the eyes of some dogs. The ocular toxicity reversed upon cessation of dosing NTBC (Lock et al. 2006). Rhesus monkeys were also administered NTBC by oral gavage at 10 mg/kg/day for 90 days to ascertain if eye lesions were observed in this species, no ocular toxicity or signs of compound related toxicity was seen (Lock et al. 2006). Thus, the primary toxicology observed with NTBC was the production of corneal lesions in the rat and dog at low doses while the mouse and rhesus monkey appeared to be resistant to this effect.

16.5 Tissue Distribution, Metabolism and Excretion of NTBC in the Rat

Research studies aimed to try to understand the basis for the ocular toxicity observed in the rat after dosing with NTBC. Initial studies focussed on the tissue distribution of [^{14} C-benzoyl]-NTBC using whole body autoradiography, following a single oral dose at 10 mg/kg. This showed that radioactivity spread throughout the body at early times after dosing, while after 24 h and 48 h radiolabel was still present in the liver, kidneys and Harderian gland with no selective retention in the cornea or eye. Following a low dose of 0.1 mg/kg [^{14} C-benzoyl]-NTBC to rats the plasma concentration peaked 2 h after dosing at about 70 pmol/ml, remained elevated for up to 8 h but was then cleared to below the limit of detection <1 pmol/ml by 24 h. Any radiolabel detected in the eye was always at a much lower concentration than in the plasma. The radiolabel associated with the liver and kidneys changed little over the ensuing 3 days, the liver retained 46% and the kidneys 3%, so that about 50% of the administered dose was still in the body 4 days after dosing. We presume that NTBC is bound in these organs (Lock et al. 1996). The Harderian gland in the rat is part of the lacrimal apparatus and contributes to the pre-corneal tear film and the reason for retention of radiolabel in this tissue is unclear but could be similar to that in the liver and kidneys. The excretion of [^{14} C-benzoyl]-NTBC was also examined in the rat following a single oral dose of 10 mg/kg over 4 days. The elimination was equally divided between urine and faeces with about 45% of the dose excreted by each route with the bulk of the dose (63%) being eliminated over the first 24 h. Two major metabolites were present in urine, the 4-and 5-hydroxy metabolites accounting for about 75% of the material in urine.

Overall, these studies did not show any correlation between the presence of radiolabel from NTBC and the eye lesions indicating some other factor was involved, the only interesting finding was the retention of a large part of the dose in the liver and to a lesser extent the kidneys.

16.6 Mechanism of Action of NTBC

The breakthrough in our understanding came from two pieces of information (1) speculation that the structure of NTBC resembled that of structures in a pharmaceutical patent for tyrosine hydroxylase inhibitors and (2) metabonomic studies on dog and rat urine from NTBC-treated animals using ^{1}H–NMR spectroscopy. In the first urine from NTBC treated rats was analysed for tyrosine and its metabolites using ferric chloride (which detects α-keto acids) and nitrosonaphthol (which detects hydroxylated aromatic compounds) and in both cases, the tests were positive. The study then looked at the plasma amino acid content of NTBC treated rats, they showed a marked elevation in plasma tyrosine concentration of about 2000 nmol/ml compared to 100 nmol/ml in untreated rat plasma (Lee et al. 1997). The second approach using ^{1}H–NMR on urine from dogs exposed to NTBC showed a marked increase in peaks in the aromatic range of the spectrum (Ian Wilson, ICI Pharmaceuticals) that were later identified as 4-hydroxyphenylpyruvate and 4-hydroxy-phenyllactate metabolites of tyrosine in NTBC treated rats (Ellis et al. 1995). The reason for the accumulation of tyrosine in rats (Ellis et al. 1995) and plants (Prisbylla et al. 1993; Schulz et al. 1993) is due to inhibition of the enzyme HPPD. 4-Hydroxyphenylpyruvate is the substrate for HPPD, which catalyses its conversion to homogentisic acid (Fig. 16.3), involving oxidative decarboxylation and ring migration of an enzymatically derived pyruvate moiety during catalysis. We consulted and had discussions with Sven Lindstedt and Elizabeth Holme of Gothenburg University as Sven was the expert on this enzyme, having isolated and purified it from human liver (Lindblad et al. 1977; Rüetschi et al. 1993) and worked with this enzyme for many years due to his interest in tyrosinaemia. Using the method of Lindstedt and Odelhög (1987) enzyme kinetic studies with rat liver cytosol showed that the inhibition was both time- and dose-dependent, the rate constant for the formation of the enzyme-inhibitor complex being about $9.9 \times 10^{-5} s^{-1} nmol/L^{-1}$ (Ellis et al. 1995). However, NTBC is not covalently bound to the enzyme; as the enzyme-inhibitor complex will dissociate with an estimated half-life in vitro at 25 °C of 63 h. Thus, the interaction of NTBC with the enzyme can be characterised by a rapid inactivation step to form the complex that can dissociate slowly with recovery of enzyme activity. This suggests that NTBC binds to the same part of the active site as the substrate. More recent studies have examined the interaction of NTBC with HPPD from *Streptomyces avermitilis* in greater detail showing that NTBC preferentially binds to the complex of HPPD when the non-haem Fe is in the Fe(II) ferrous state (Kavana and Moran 2003). They determined this process occurs in two phases and comprises three steps: in the first step NTBC binds initially reversibly proximal to the active site metal ion followed by a bidentate association with the Fe(II) atom producing an initial charge transfer absorbance. The Lewis-acidic metal ion then assists in the irreversible deprotonation of the enol to form the enolate complex, producing a second and final charge transfer band with a slightly increased intensity (Kanavan and Moran 2003). The three-dimensional structure of the Fe(II) form of HPPD from *Streptomyces avermitilis* in complex with NTBC has been determined at a resolution of 2.5 A (Brownlee et al. 2004) also see review on HPPD by Moran (2005). In these studies with *Streptomyces avermitilis* the reaction with NTBC is irreversible, while studies on the interaction of NTBC with HPPD in rat liver cytosol showed some reversibility (Ellis et al. 1995). One possible explanation for this difference is that HPPD exists as a tetramer in rat and human liver while in bacteria it is a dimer.

In the rat a single dose of NTBC as little as 100 μg/kg given orally completely inhibited the activity of HPPD in the liver within a couple of hours, resulting in an increase in the concentration of tyrosine in the plasma (2000 nmol/ml) and ocular fluid (3500 nmol/ml). The tyrosine concentration peaked 24 h after dosing when the HPPD activity in the liver was still 95% inhibited, but by 48 h after dosing plasma, tyrosine was back within the normal range, while the return to normal in the eye took about 72 h. The recovery of HPPD activity was only 40% of control 48 h after dosing and then 50% of control 72 h and 96 h after dosing. These finding are consistent with those in vitro with rapid inhibition followed by a very slow recovery and tell us that >5–40% recovery of the enzyme is enough to enable tyrosine values to return to normal. Later times after expose recovery of HPPD activity could be due to synthesis of new enzyme, however with 46% of the dose of the radiolabel from NTBC still in the liver at this time, any free NTBC would inactivate HPPD (Lock et al. 1996). The rate of loss of NTBC from the liver after 0.1 mg/kg declines over 4 days giving an approximate half-life of 5.5 days, much longer than in vitro. At higher doses of NTBC, e.g. 10 mg/kg removal from the liver 24 h after exposure is slower with little loss between 4 and 7 days after exposure suggesting a much longer half-life (Lock et al. 1996). The implication of this are that following a second and third daily dose HPPD is completely inhibited leading ultimately to a steady state concentration of NTBC and tyrosine in the plasma following multiple dosing which leads to ocular injury. That the cause of the eye lesion was due to tyrosinaemia was confirmed by feeding rats on a low protein diet containing 5% w/w tyrosine which resulted in a similar ocular toxicity to that seen with NTBC, although with tyrosine alone, there was a 100% incidence of eye lesions which were more severe than those seen with NTBC.

Mice in contrast were resistant to ocular lesions produced by NTBC even at doses up to 160 mg/kg/day for 6 weeks (Lock et al. 2000). Mice given a single oral dose of 10 mg/kg NTBC had HPPD activity inhibited in their liver >95% which only started to recover about 4 days after dosing. In these mice, the plasma tyrosine concentration peaked at around 700 nmol/ml 8 h after dosing and then slowly declined to approach normal over 7 days. The liver content of radiolabelled NTBC peaked 2 h after dosing and then rapidly cleared with an initial half-life of about 7 h followed by a very slow phase with little elimination over the ensuing 7 days (Lock et al. 2000). Thus in the mouse, despite almost complete inhibition of hepatic HPPD, tyrosine was cleared from the plasma more rapidly than in the rat. One explanation for the species difference is that hepatic activity of tyrosine aminotransferase is much higher in the mouse and hence able to clear tyrosine as 4-hydroxyphenylpyruvate and lactate and excrete it in urine. These tyrosine findings in the rat and mouse led us suggest that there was a threshold for plasma tyrosine above which corneal injury was observed, which was 1000 nmol/ml (Lock and Smith 2003).

Further studies showed that NTBC treatment of beagle dogs and rabbits resulted in tyrosinaemia with marked inhibition of hepatic HPPD; rhesus monkeys also developed tyrosinaemia after NTBC (Lock et al. 2006). As mentioned earlier dogs, developed corneal injury at the lowest dose of 0.1 mg/kg while both rabbits and rhesus monkey given 10 mg/kg for 6 weeks and 12 weeks respectively did not develop corneal injury. The findings in the dog where plasma tyrosine concentrations were above 1000 nmol/ml plasma is consistent with the threshold concept, but as with rats some dogs had high plasma tyrosine levels but did not show ocular impairment. However, findings in the rabbit and rhesus monkey where plasma tyrosine concentrations were above 1000 nmol/ml without any ocular injury indicated that there are other factors that we do not currently understand that lead to corneal injury. Tyrosine is a rather insoluble amino acid with a solubility of 4000 nmol/ml in water at 37 °C this is close to the concentration found in the aqueous humour and would seem to support the findings in rats (Gipson and Anderson 1997) and dogs of crystalline structures in the eye.

Thus, all HPPD inhibitors will cause an increase in plasma and ocular tyrosine concentration in mammals which if marked and sustained will result in corneal lesions in rats and dogs.

From the toxicology viewpoint a key question was how would humans respond to NTBC with regard to tyrosinaemia and corneal injury?

16.7 Development of NTBC/Nitisinone to Treat Tyrosinaemia Type 1

It was during the first visit by Sven Lindstedt and Elizabeth Holme that Sven told us he had been searching for inhibitors of HPPD for many years to treat a small number of children in Sweden with tyrosinaemia type 1 and he asked if he could have NTBC for this purpose. The concept was that by blocking HPPD in these patients you would prevent the formation of toxic metabolites downstream of homogentisic acid thereby improving their clinical condition. Zeneca consisted of two major businesses, one in Agrochemicals and the other in Pharmaceuticals, with the toxicology testing for Agrochemicals conducted at the Central Toxicology Laboratories (CTL) where I was working. Internal discussion took place between CTL and Agrochemicals, as NTBC was no longer in development; they were supportive for its clinical use if it could improve the quality of life for these children. However, agreement of the Pharmaceutical business was required. In October 1989, Sven formally requested the use of NTBC to treat children with tyrosinaemia type 1. As an initial step, it was important to establish that NTBC inhibited human HPPD and Sven was happy for Martin Ellis and I to visit his laboratory and conduct studies with NTBC using the purified human liver enzyme. These studies showed NTBC was a potent inhibitor of HPPD with an IC_{50} of 5 nM and with this information the director of CTL Dr. Iain Purchase and I started a dialogue with the Pharmaceuticals business. This took time but the outcome was supportive on the understanding that NTBC is used according to good clinical practice, with the approval of University of Gothenburg Ethical Committee and the consent of the children's parents. A small project team under my leadership with a clinician, patent experts, lawyer, and regulatory product advisor to move the project forward. Various documents were prepared outlining the case for treating these patients including the toxicology database. Professor Lindstedt then made a submission for the proposed clinical use of NTBC to treat a small number of children with tyrosinaemia type 1 to the Ethical Committee at Gothenburg University and to the Medical Products Agency (MPA) in Uppsala, Sweden. In February 1991, following approval by the MPA Zeneca supplied a quantity of NTBC for clinical use to treat a seriously ill 2-month-old baby with tyrosinaemia type 1. The outcome was dramatic and the initial studies on five patients was reported in the Lancet (Lindstedt et al. 1992). As expected plasma tyrosine was elevated in these patients but no adverse ocular effects were seen. Eventually NTBC was patented for clinical use worldwide by Zeneca Pharmaceuticals and then sub-licensed to Swedish Orphan International, that undertook the necessary studies on stability, formulation, and gathered additional clinical data to support drug registration. In May 1995, Nitisinone (Orfadin) was designated an orphan drug by the US Office for Orphan Product Development. The US Food & Drug Administration gave approval in January 2002 and European approval followed in February 2005. The worldwide rights are currently with Swedish Orphan Biovitrum AB (Sobi). Aspects of the discovery of NTBC have been previously reported Lock et al. (1998). Clinical studies conducted by CTL on ten human volunteers given a single dose of 1 mg/kg Nitisinone orally in solution or as a capsule showed it was slowly eliminated from the plasma with a half-life of 54 h. Plasma tyrosine gradually increased reaching a plateau of 1100 nmol/ml after 3 days and remained elevated for the 5 days examined. There was then a 14-day rest period followed by a second dose of 1 mg/kg orally. Prior to the second dose a blood sample was taken that showed plasma tyrosine was still elevated at 800 nmol/ml, i.e. a drop of only 300 nmol/ml over 14 days indicating a half-life in humans of >14 days, indicative of prolonged inhibition of hepatic HPPD (Hall et al. 2001).

This study also examined the structurally related HPPD inhibitor mesotrione (Fig. 16.2c), used as an herbicide, in human volunteers at doses 0.1, 0.5 and 4 mg/kg and showed that mesotrione was rapidly eliminated from the plasma with a half-life of 1 h. As a consequence there was only a very small increase in plasma tyrosine after the two lower doses (towards the upper end of normal) over a few hours after dosing, while with 4 mg/kg mesotrione plasma tyrosine was increased to about 300 mnol/ml 8 h after dosing and was back within the normal range after 48 h. This indicates that mesotrione inhibits human liver HPPD but compared to NTBC is rapidly cleared from the body with 72% of the dose excreted unchanged in the urine (Hall et al. 2001).

16.8 Use of NTBC/Nitisinone to Treat Alkaptonuria

Nitisinone is also being used off-label to treat people with the inborn error of metabolism alkaptonuria (AKU) which is due to a deficiency of the enzyme homogentisic acid dioxygenase (HGD) (EC. 1.13.11.5) (Introne et al. 2011; Olsson et al. 2015). AKU is an autosomal recessive disorder where the deficiency of HGD, prevents the metabolism of homogentisic acid (HGA) to maleylacetoacetic acid (Fig. 16.3) (O'Brien et al. 1963). The result is the accumulation of HGA in body fluids despite extraordinarily efficient renal excretion, both by glomerular filtration and tubular secretion. However, despite the ability of the kidney to eliminate HGA, an increase in circulating HGA occurs. Conversion of circulating HGA to a pigmented polymer that can be visible in the ears and eyes, a process termed ochronosis (Fig. 16.3), which is first observed around the age of 30 years. This ochronotic pigment is deposited in tissues, in a highly selective manner that is still not fully understood; for example, pigmentation is marked in the aortic valve and aortic root but is virtually absent in the pulmonary valve and pulmonary trunk. The ochronotic process consistently targets cartilage, so much, so that articular tissues endure the most damage caused by AKU. Studies have shown a marked reduction in the excretion of HGA following Nitisinone treatment in these patients, indicating the likelihood of reduced polymer formation, but it will take time to demonstrate clinical efficacy (Introne et al. 2011; Olsson et al. 2015; Milan et al. 2017). As expected plasma tyrosine concentrations are elevated, to date the only clinical adverse effect has been ocular keratopathy (Ranaganath, personal communication) in a number of patients.

In summary, this example with NTBC demonstrates the importance of understanding the mechanism whereby chemicals cause adverse toxic reactions and illustrates the application of this knowledge to help design effective herbicides and to treat people with rare inborn errors of tyrosine metabolism.

Acknowledgements I would like to thank all those who contributed to the toxicology and mode of action studies with NTBC and in particular the late Dr. Iain Purchase who was the Director of Zeneca CTL who strongly supported the mechanistic studies and the case for NTBC's use as a drug. In addition, my thanks to those in Zeneca Agrochemicals and Zeneca Pharmaceuticals who helped guide NTBC over the various hurdles to enable its clinical use. Finally, without an excellent working relationship with Professor Sven Lindstedt and Professor Elizabeth Holme and others at Sahlgrenska Hospital, NTBC would never have reached clinical use.

References

Beaudegnies R, Andrew JF, Edmunds AFJ, Fraser TEM, Hall G, Hawkes TR, Mitchell G, Schaetzer J, Wendeborn S, Wibley J (2009) Bioorg Med Chem 17:4134–4152

Brownlee JM, Johnson-Winters K, Harrison DH, Moran GR (2004) Biochemistry 43:6370–6377

Ellis MK, Whitfield AC, Gowans LA, Auton TR, Provan WM, Lock EA, Smith LL (1995) Toxicol Appl Pharmacol 133:12–19

Gipson IK, Anderson RA (1997) J Histochem Cytochem 125:1351–1362

Grossmann K, Ehrhardt T (2007) Pest Manag Sci 63:429–439

Hall MG, Wilks MF, Provan WM, Eksborg S, Lumholtz B (2001) Br J Clin Pharmacol 52:169–177

Hellyer R (1968) Aust J Chem 21:2825–2828

Introne WJ, Perry MB, Troendle J, Tsilou E, Kayser MA, Suwannarat P, O'Brien KE, Bryant J, Sachdev V, Reynolds JC, Moylan E, Bernardini I, Gahl WA (2011) Mol Genet Metab 103:307–314

Kavana M, Moran GR (2003) Biochemistry 42:10238–10245

Kollberg G, Clayton P (2015) Elisabeth Holme. J Inherit Metab Dis 38:1159–1160

Lee DL, Prisbylla MP, Cromartie TH, Dagarin DP, Howard ST, Provan WM, Ellis MK, Fraser T, Mutter CM (1997) Weed Sci 45:601–609

Lindblad B, Lindstedt G, Lindstedt S (1970) J Am Chem Soc 92:7446–7449

Lindblad B, Lindstedt G, Lindstedt S, Rundgren M (1977) J Biol Chem 252:5073–5084

Lindstedt S, Odelhög B (1987) Methods Enzymol 142:139–142

Lindstedt S, Holme E, Lock EA, Hjalmarson O, Strandvik B (1992) Lancet 340:813–817

Lock EA, Smith LL (2003) Toxicol Lett 140–141:317–322

Lock EA, Gaskin P, Ellis MK, Provan WM, Robinson M, Smith LL, Prisbylla MP, Mutter LC (1996) Toxicol Appl Pharmacol 141:439–447

Lock EA, Ellis MK, Gaskin P, Robinson M, Auton TR, Provan WM, Smith LL, Prisbylla MP, Mutter LC, Lee DL (1998) J Inherit Metab Dis 21:498–506

Lock EA, Gaskin P, Ellis MK, Provan WM, Robinson M, Smith LL (2000) Toxicology 114:179–187

Lock EA, Gaskin P, Ellis MK, Provan WM, Smith LL (2006) Toxicol Appl Pharmacol 215:9–16

Mayonado DJ, Hatzios KK, Orcutt DM, Wilson HP (1989) Pestic. Biochem Physiol 35:138

Milan AM, Hughes A, Davison AS Devine JM Usher JL Curtis S, Khedr M, Gallagher JA, Ranganath L (2017) ANNALS EXPRESS: The effect of nitisinone on homogentisic acid and tyrosine: a two-year survey of patients attending the National Alkaptonuria Centre, Liverpool. Ann Clin Biochem Jan 1:4563217691065

Mitchell G, Bartlett DW, Fraser TE, Hawkes TR, Holt DC, Townson JK, Wichert RA (2001) Pest Manag Sci 57:120–128

Moran GR (2005) Arch Biochem Biophys 433:117–128
O'Brien WM, La Du BN, Bunim JJ (1963) Biochemical, pathologic and clinical aspects of alcaptonuria, ochronosis and ochronotic arthropathy: review of world literature (1584–1962). Am J Med 34:813–838
Olsson B, Cox TF, Psarelli EE, Szamosi J, Hughes AT, Milan AM, Hall AK, Rovensky J, Ranganath LR (2015) J Inherit Metab Dis Rep 24:21–27
Pallett KE, Cramp SM, Little JP, Veerasekaran P, Crudace AJ, Slater AE (2001) Pest Manag Sci 57:133–142
Prysbilla MP, Onisko BC, Shribbs JM, Ellis MK, Hawkes TR, Mutter LC (1993) Proc Brighton Crop Protect Conf Weeds 2:731–738
Reddy SS, Stahlman PW, Geier PW, Peterson DE (2012) Am J Plant Sci 3:1613–1618
Robinson M (1995) In: Weise I, Hockwin O, Green K, Tropathi RC (eds) Ocular Toxicology. Plenum, New York, pp 327–334
Romagni JG, Meazza G, Nanayakkara NP, Dayan FE (2000) FEBS Lett 480:301–305
Rüetschi U, Dellsén A, Sahlin P, Stenman G, Rymo L, Lindstedt S (1993) Eur J Biochem 213:1081–1089
Sandmann G, Boger P, Kumita I (1990) Pestic Sci 30:353–355
Santel HJ (2009) Bayer Crop Sci J 69:95–108
Schulz A, Oswald O, Beyer P, Kleinig H (1993) FEBS Lett 318:162–166

The Québec NTBC Study

17

The Québec NTBC Study Group*,
Fernando Alvarez, Suzanne Atkinson,
Manon Bouchard, Catherine Brunel-Guitton,
Daniela Buhas, Jean-François Bussières,
Josée Dubois, Daphna Fenyves, Paul Goodyer,
Martyne Gosselin, Ugur Halac, Patrick Labbé,
Rachel Laframboise, Bruno Maranda,
Serge Melançon, Aicha Merouani,
Grant A. Mitchell, John Mitchell, Guy Parizeault,
Luc Pelletier, Véronique Phan,
and Jean-François Turcotte

Abstract

In this chapter we describe the current Quebec NTBC Study protocol. Quebec's unique characteristics have influenced the development of the protocol, including a high prevalence of hepatorenal tyrosinemia (HT1), universal newborn screening for HT1, availability of treatment with nitisinone (NTBC) and special diet, a large territory, where HT1 treatment is coordinated by a small number of centers. Screened newborns are seen

Submitted by Grant Mitchell for the Québec NTBC Study Group* including a section on diet therapy by Manon Bouchard.

The Québec NTBC Study Group*

F. Alvarez
Service de gastroentérologie, hépatologie et nutrition, CHU Sainte-Justine, Université de Montréal, 3175 chemin Côte-Sainte-Catherine, Montréal, Québec H3T 1C5, Canada

S. Atkinson • M. Bouchard
C. Brunel-Guitton • J.-F. Bussières • J. Dubois
M. Gosselin • A. Merouani G.A. Mitchell (✉)
V. Phan
Departments of Nutrition (MB) and Radiology (JD), Divisions of Gastroenterology, Hepatology and Nutrition (FA, UH), Medical Genetics (CBG, MG, GAM) and Nephrology (AM, VP), Department of Pediatrics and Department of Pharmacy (SA, JFB), CHU Sainte-Justine and Université de Montréal, Montreal, Canada
e-mail: grant.mitchell@recherche-ste-justine.qc.ca

U. Halac
Division of Gastroenterology, Hepatology and Nutrition, Department of Pediatrics, CHU Sainte-Justine and Research Center, Université de Montréal, Montreal, QC, Canada

D. Buhas • P. Goodyer • S. Melançon • J. Mitchell
Department of Genetics (DB, SM) and Divisions of Endocrinology (JM) and Nephrology (PG), Department of Pediatrics, Montreal Children's Hospital, McGill University, Montreal, Canada

D. Fenyves
Division of Hepatology, Department of Internal Medicine, CHUM, Université de Montréal (DF), Montreal, Canada

P. Labbé • G. Parizeault
Departments of Pediatrics (GP) and Internal Medicine (PL), Hopital de Chicoutimi, Chicoutimi, Québec, Canada

© Springer International Publishing AG 2017

R.M. Tanguay (ed.), *Hereditary Tyrosinemia*, Advances in Experimental Medicine and Biology 959,
DOI 10.1007/978-3-319-55780-9_17

within 3 weeks of birth. Patients with liver dysfunction (prolonged prothrombin time and/or international normalized ratio (INR) provide sensitive, rapidly available indicators) are treated by NTBC and special diet. The specific diagnosis is confirmed by diagnostic testing for succinylacetone (SA) in plasma and urine samples obtained before treatment. After an initial period of frequent surveillance, stable patients are followed every 3 months by assay of plasma amino acids and NTBC and plasma and urine SA. Abdominal ultrasound is done every 6 months. Patients have an annual visit to the coordinating center that includes multidisciplinary evaluations in metabolic genetics, hepatology, imaging (for abdominal ultrasound and magnetic resonance imaging) and other specialties as necessary. If hepatocellular carcinoma is suspected by imaging and/or because of progressive elevation of alphafetoprotein, liver transplantation is discussed. To date, no patient in whom treatment was started before 1 month of age has developed hepatocellular carcinoma, after surveillance for up to 20 years in some. This patient group is the largest in the world that has been treated rapidly following newborn screening. The protocol continues to evolve to adapt to the challenges of long term surveillance.

Keywords
Tyrosinemia • Québec • Quebec • Screening • Diet • Nitisinone • Surveillance • Treatment

Abbreviations

DRI	Dietary reference intake
HT1	Hereditary tyrosinemia type 1
INR	International normalized ratio
NTBC	Nitisinone
SA	Succinylacetone

17.1 Introduction

This chapter describes the core protocol currently used in Quebec for the medical treatment of patients with hepatorenal tyrosinemia (HT1). The clinical and biologic aspects of HT1 prior to the availability of NTBC treatment are reviewed elsewhere (Mitchell et al. 2001). Other chapters in this volume discuss the treatment and surveillance of liver disease in the context of the Quebec NTBC Study (See Halac et al., Chap. 6), the indications and results of liver transplantation (See Alvarez and Mitchell, Chap. 5) and future clinical challenges for HT1 treatment (See Mitchell and Yang, Chap. 19) and treatment and surveillance protocols elsewhere.

17.2 Background: A Québec Perspective on HT1

Following the report of Lindstedt and Holme describing the use of nitisinone (NTBC) in human patients with hepatorenal tyrosinemia (tyrosinemia type I, HT1) (Lindstedt et al. 1992), it was clear that NTBC treatment could rapidly improve the clinical manifestations of HT1.

This report followed upon decades of work by clinicians and biochemists in Québec to improve the

R. Laframboise • L. Pelletier • J.-F. Turcotte
Divisions of Medical Genetics (RL), Gastroenterology (LP) and General Pediatrics (JFT), Department of Pediatrics, Université Laval, Québec, Canada

B. Maranda
Department of Medical Genetics, Centre Hospitalier Universitaire de Sherbrooke and Université de Sherbrooke, Sherbrooke, Québec (BM), Canada

diagnosis, treatment and prognosis of HT1. Because of a genetic founder effect, HT1 is common in Québec (Grompe et al. 1994). Province-wide neonatal screening for tyrosinemia has been practiced since 1970. Unlike most places, in Québec, HT1 patients are first identified as clinically-asymptomatic newborns (Mitchell et al. 2001).

Geographically, Quebec covers a large territory. Provision of services to all patients, even those in remote locations, are coordinated by the four university hospital centers and by a regional center (Chicoutimi) in the area with the highest prevalence of HT1.

Clinicians in Québec had the strong impression that early initiation of a diet restricted in tyrosine and phenylalanine, with medical surveillance for the development of complications, resulted in a milder phenotype than if the disease was not detected by screening and instead came to medical attention when complications arose. However, the clinical effect of diet treatment alone was modest even if the diet was started within days of birth. This was seen biochemically as well. Succinylacetone (SA) is highly increased in HT1 and is closely linked to the pathophysiology of this condition. A marked elevation of SA is thought to be pathognomonic for HT1. SA levels in HT1 patients remained highly elevated despite dietary treatment. Even if they had excellent dietary adherence, HT1 patients treated by diet alone remained at risk for the life-threatening acute and chronic complications of HT1.

The development of a pediatric hepatic transplantation program in Québec, reviewed elsewhere in this book, provided hope for survival for tyrosinemic patients. Even with the combination of newborn screening and early diet therapy, most patients required liver transplantation before the age of 3 years. The two principal indications for liver transplantation were aggressive liver disease (episodes of hepatic failure or development of liver nodules) and the development of neurologic crises. Neurologic crises of HT1 are porphyria-like acute episodes of sensorimotor neuropathy that occurred in about half of HT1 patients in Québec before the availability of NTBC (Mitchell et al. 1990). Neurologic crises often caused severe neuropathic pain in the legs or paralysis of the legs. Diaphragmatic paralysis occurred more rarely but sometimes required prolonged mechanical ventilation.

Therefore, Québec was distinct from other places with respect to HT1. HT1 was more prevalent and patients were identified and treated from an early stage of the disease. In this patient population, early treatment with NTBC might possibly generate results that could not be directly compared with those obtained elsewhere, because of the differences in the age at which treatment could begin. Presumably, if patients were not screened and if they presented clinically with liver failure, cirrhosis and/or renal tubulopathy, permanent visceral damage might have occurred before diagnosis and the effect of treatment might not be as marked as if it was started shortly after newborn screening and before the development of clinical signs. Furthermore, because of the high level of clinical heterogeneity among HT1 patients, even within single families, it would be necessary to obtain the largest number possible of patients in order to draw conclusions of statistical significance. Finally, the toxicity of NTBC in humans was unknown at that time, which aroused major concern about initiating treatment in a large number of patients. If any side effect occurred, it was felt to be important that all of the Québec community be informed. In view of these considerations, all of the physicians who cared for tyrosinemic patients in Québec agreed to administer NTBC as part of a common protocol.

The first patients to whom NTBC was administered were older children who were several months or years of age. In some of them, NTBC was started during an acute decompensation. All patients showed rapid improvement of liver disease. No side effects were identified in these patients. The age at first treatment was progressively reduced to include babies immediately following their identification by newborn screening. Currently, after a positive screen for HT1, newborns are evaluated within days, often within hours.

17.3 Initial Diagnostic Evaluation

17.3.1 The Initial Visit

Currently, screen-positive babies are usually first seen in the second or third week of life. During the first month of life, HT1 patients are usually clinically asymptomatic (Mitchell et al. 2001), although neurologic crises have rarely been reported at 1 month of age (Mitchell et al. 1990).

Physical examination is usually normal but may reveal slight hepatomegaly. The initial samples in screen-positive newborns are obtained in the outpatient clinic. It is important to validate that the child has evidence of hepatic dysfunction. Even when evaluated following newborn screening, all HT1 patients have in our experience had evidence of coagulopathy. In practice this is detected as a prolonged prothrombin time and/or a high International Normalized Ratio (INR). These tests are available rapidly in all participating centers. Screen-positive patients who have normal liver function are not treated but are followed closely without treatment while awaiting the results of specific tests. Patients who have prolonged coagulation tests are offered treatment with NTBC and special diet.

Once the initial sampling is performed (Table 17.1) and the samples are confirmed to be adequate, and if the patient has prolonged coagulation tests, NTBC is prescribed. Prior to administration of NTBC, it is important to ensure that the samples, particularly those for assay of succinylacetone, are of adequate amount and quality. This is important because future metabolite levels will be compared with this value, and because SA decreases rapidly following the first dose of NTBC. Therefore, samples taken after treatment has begun, even rapidly after treatment has begun, may not reflect the levels of the patient before treatment.

When evidence of hepatic dysfunction is obtained in a screen-positive patient, the protocol coordinator authorizes the shipment of NTBC to the center. Usually, delivery requires less than 24 hours, approximately the time for the initial urine collection and sampling. Typically, multiples of the smallest pills (2 mg) are prescribed for newborn babies and a sufficient supply is available at the pharmacy of the coordination site to ship immediately for the first days of treatment.

Table 17.1 Pretreatment evaluation of HT1 patients

Blood/plasma
Complete blood count
Plasma electrolytes
Blood gases
Coagulation evaluation: international normalized ratio (INR) and prothrombin time (PT)
Aspartate aminotransferase
Bilirubin
Alphafoetoprotein
Succinylacetone
Amino acid chromatography
Urine
Urinalysis
Succinylacetone
Amino acid chromatography
Imaging
Abdominal ultrasound (often performed after the initial sampling)

NOTE: Treatment with NTBC may be indicated on an emergency basis, especially in patients who are not screened and who present clinically in liver failure or neurologic crisis. Access to NTBC varies greatly among different places. Physicians outside of Québec responsible for HT1 diagnosis and management are encouraged to establish procedures in advance for rapidly obtaining diagnostic results and rapidly obtaining NTBC if necessary.

17.3.2 First Hospitalization

In our experience a brief hospitalization is useful to ensure adequate diagnosis and to initiate treatment, and for the parents to feel confident about the treatment. Families are overloaded with information and emotion after the diagnosis is announced. Therefore, patients with abnormal coagulation testing at the first outpatient visit are hospitalized to start treatment.

The hospitalisation typically lasts about 48 h after the first dose of NTBC. By then, the parents know how to prepare the tyrosine and phenylalanine-restricted formula. The parents will be confident with NTBC administration, and understand how to obtain help in case of uncertainty. The hospitalization is useful in order to clear up potential misconceptions of the parents and to ensure strong links between the parents and the health care team. The family meets several times with members of the core treatment team (nurse, dietician, pharmacist and metabolic physician) who will be their primary contacts in the future.

17.4 Administration of NTBC

17.4.1 The First Dose of NTBC

We have never seen an acute deleterious effect following the first dose of NTBC. However, because NTBC has only been given to a comparatively small number of patients, we have continued to give the first dose under medical supervision, with monitoring of vital signs and of the general state of the patient for several hours afterward.

For a screened newborn, the first dose given is a single capsule of 2 mg, opened and mixed in a small volume (2–3 mL) of formula. On a practical note, it is important not to dilute the dose in a large volume because some might deposit on the walls of the container and the child may not drink the entire volume. A small volume of formula can be given orally to the child by syringe. The syringe can then be rinsed with an equal volume that is also given to the baby, followed by the rest of the feed.

The currently-available formulation of NTBC (Orphadin) is available as capsules of 2, 5 or 10 mg. After the first dose, NTBC is prescribed. The initial dose is the smallest multiple of 2 mg that will provide a dose of at least 1 mg/kg/day and at most 2 mg/g/day. For instance a 3.5 kg infant will initially receive 4 mg/day (2 mg twice daily) whereas a baby weighing 4.2 kg will receive a total of 6 mg/day (four in the morning and two at night). The prescription will subsequently be adjusted according to plasma levels of NTBC.

> NOTE: the doses of NTBC mentioned in this chapter are provided only as a starting guide. The dose should be adjusted according to the individual response of the patient.

17.4.2 Further Administration of NTBC

The goal of NTBC treatment is to suppress the level of SA, which is the measurable metabolite that is most closely related to the pathogenesis of HT1. We use plasma NTBC level to adjust the prescription of NTBC. We aim for plasma NTBC levels above 50 μmol/L. In our experience these levels produce a maximum suppression of the levels of SA. An alternate approach used by some groups is to adjust the dose of NTBC directly according to the level of SA in blood or urine, aiming to obtain a maximum suppression of SA levels. Often SA levels can be reduced to the normal range.

NTBC has a long half-life (~54 h in adults) (Hall et al. 2001). For this reason, some patients now receive NTBC as a single daily dose. We tend to reserve single daily dosing for older children because the pharmacokinetic parameters in infants are not well established.

> NOTE: It is important to clearly state the different roles of NTBC and of diet in the treatment of HT1. NTBC is given to produce a complete pharmacological block of 4-hydroxyphenylpyruvate dioxygenase. The prescription of NTBC is adjusted so as to minimize SA. In contrast, the dietary tolerance of the patient is determined based on plasma tyrosine level. Natural protein is given in amounts that provide a plasma tyrosine level of 200– 400 μmol/L. Therefore, the determination of the dose of NTBC is based on levels of SA and/or NTBC and is not determined by the levels of plasma tyrosine. The prescription of NTBC is adjusted to minimize SA levels, not to adjust blood tyrosine level.

As another consequence of the long half-life of NTBC, we wait at least 1 week to re-assess the plasma level of NTBC after changing the prescribed dose.

17.5 Dietary Treatment (Contributed by Manon Bouchard, DTP)

The goal of the dietary restriction is to prevent hypertyrosinemia and to achieve normal intake and levels of other nutrients while assuring a maximum level of dietary palatability.

We aim for plasma levels of tyrosine between 200 and 400 μmol/L and for plasma level of phenylalanine above the lower reference range (above ~25–30 μmol/L). Methionine intake is not restricted. Adequate intake of all other nutrients is another goal.

The main elements of the HT1 diet are restriction of natural protein, with additional phenylalanine and tyrosine-free formulas to provide sufficient protein and of low protein foods to meet energy requirements. Restriction of natural protein is made by calculation of the phenylalanine and tyrosine content of natural foods. This can be calculated with an exchange system (one exchange is equal to 25 mg of phenylalanine plus tyrosine combined (~0.3–0.5 g of natural protein)) or by counting grams of total natural proteins.

17.5.1 Initial Diet

For mothers wishing to breast feed, the initial prescription calls for breast feeds to alternate with feeds of special formula that contain no tyrosine or phenylalanine. For mothers who do not breast feed, the formula is a mixture of normal baby formula with a special formula that is depleted in phenylalanine and tyrosine. Several such formulas are available for HT1. A typical starting amount for total daily phenylalanine plus tyrosine combined is about 200 mg/day. This will be checked and adjusted frequently in the first weeks and months of treatment, according to the baby's growth and amino acid profile.

17.5.2 Restriction of Tyrosine and Phenylalanine

Phenylalanine and tyrosine are present in all natural proteins. To restrict these two amino acids, a restriction of all natural protein is needed. In general, this diet relies heavily on fruits, vegetables and cereals. Meat, nuts, seeds, legumes and dairy products are avoided.

The diet prescription is individualized for each patient. Frequent adjustments are needed in the first year of life because of the rapid growth and frequent intercurrent illnesses. The tolerance for tyrosine and phenylalanine can vary depending in part on growth rate, energy and protein intake. The dietary prescription is adjusted according to plasma levels of tyrosine and phenylalanine.

17.5.3 Protein and Other Nutrients

Protein requirements are calculated according to age and body weight. In the diet of these children, most protein intake (75–80 %) comes from free amino acids in the phenylalanine- and tyrosine-depleted formula. Free amino acids are rapidly absorbed and metabolized. Therefore, a higher protein intake than the dietary reference intake (DRI) for age is prescribed (DRI +25 %). Protein intake (the total of natural protein plus the proteins in the amino acid mixture) is distributed throughout the day. Energy, vitamins and minerals must meet the DRI for age.

Plasma amino acids, growth curves and plasma albumin are monitored in all patients, with other markers of protein nutrition as judged necessary. In addition, micro and macronutrients are evaluated, including levels of vitamin D and of iron. Bone mineral density is measured regularly.

NTBC-treated HT1 patients are generally stable clinically. Mild intercurrent illnesses are treated at home. They often induce a catabolic state that is associated with elevated levels of tyrosine and phenylalanine. During such episodes, phenylalanine and tyrosine intake is decreased by 50 % for 24–48 h and a high intake of energy is provided using metabolic formula that contains no phenylalanine and no tyrosine with and low protein foods.

17.5.4 Phenylalanine Supplementation

Some HT1 infants develop low levels of phenylalanine when treated with NTBC and a phenylalanine- and tyrosine-restricted diet (See Chap. 19 by Mitchell and Yang). The reason for this is not clear. Patients with low plasma phenylalanine are clinically asymptomatic. However, if patients have values below the reference range, we provide supplementation with phenylalanine, usually starting at 15 mg or 30 mg four times a day.

The other components of the diet, particularly the phenylalanine- and tyrosine-free formula, are important for ensuring adequate general nutrition. We have the impression that hypophenylalaninemia occurs less frequently in well-nourished patients. Many such patients do not require phenylalanine supplementation.

17.5.5 Sampling Frequency

The first weeks of life are a period of rapid growth and require frequent dietary adjustment of phenylalanine plus tyrosine intake. Amino acid chromatography is performed after 3 days, 1 week, 2 weeks, 3 weeks and once a month until one year of age. After one year, the protocol calls for sampling every 3 months although some physicians perform this more frequently depending on the patient's level of control. In patients who are growing rapidly, non-adherent to dietary treatment or otherwise unstable, samples are often obtained more frequently at the discretion of the treating physicians.

17.5.6 Older Children and Adults

Patients are encouraged to continue diet therapy for life. Several appetizing low protein food products are now available that can easily be integrated into school and work settings. This facilitates long term diet adherence.

17.6 Imaging Studies in HT1

Imaging studies are usually not urgent in screened patients and are usually not performed prior to NTBC treatment. Imaging studies are usually postponed until after the initiation of NTBC and collection of diagnostic blood and urine samples. If imaging is necessary before NTBC treatment, it is important to avoid sedation with barbiturates. The neurologic crises of tyrosinemia resemble acute episodes of hepatic porphyrias and can be precipitated by barbiturate administration. After NTBC treatment is established, treatment with barbiturates is allowed.

Abdominal ultrasound is more sensitive than magnetic resonance imaging (MRI) for the detection of small nodules (<10 mm). We have recently started to use Gd-EOB-DTPA (Primovist), which is reported to have excellent specificity and sensitivity for the detection of hepatocarcinomas, as discussed in the chapter by Halac (See Halac et al., Chap. 6).

The imaging timetable in Table 17.2 applies to patients with normal imaging results. If imaging abnormalities are discovered, further investigations are individualized in collaboration with the

Table 17.2 Baseline surveillance in the Quebec NTBC Study

Interval (months)	Test
3	Plasma SA, amino acids, AFP, NTBC; urine SA[a]
6	Abdominal ultrasound (exams alternate between the regional and coordinating centers)
12	Abdominal MRI[b] and annual multidisciplinary evaluation[c] at the coordinating center
48	GFR, echocardiography

[a]The results of tests in each participating center are shared with the coordinating center (CHU Sainte-Justine)
[b]The details of contrast media and timing of MRI images are discussed in the chapter in this volume about medical management and surveillance of the liver in tyrosinemia (see Halac et al., Chap. 6.)
[c]The multidisciplinary evaluation entails an annual visit to the coordination center for evaluation in genetics, hepatology, imaging and other specialities according to the above table

radiologists, hepatologists and other specialists affiliated with the QNS.

17.7 Other Evaluations

Other evaluations are performed as necessary. Nephrology, psychiatry, ophthalmology and cardiology consultants affiliated with the study, provide expert advice as required. An initial evaluation of psychomotor performance of the cohort was normal (P Robaey, personal communication). A second psychomotor evaluation after long-term treatment is currently underway. To date, renal function has been stable clinically and these data are also currently being evaluated and prepared for publication.

17.8 Results

The results of the first phase of the Québec NTBC protocol have been published (Larochelle et al. 2012). From 1988 to the time of this writing (September, 2016), no patient who has adhered to this protocol from the time of neonatal screening has developed a hepatocellular carcinoma, detectable renal disease or a neurologic crisis.

Two major prognostic variables are the age at which NTBC treatment is started and adherence to treatment. Both are addressed in Chap. 19 of this volume. If treatment is delayed, permanent damage can occur and such patients are at risk. As described in Chap. 5 (Alvarez and Mitchell), some patients in whom NTBC treatment was started after 1 month of age (because of non-detection by screening or because of birth outside of Québec in a place where neonatal screening for HT1 is not performed) required liver transplantation for the development of hepatic nodules.

Adherence to recommendations for diet and NTBC treatment is an important consideration. We have seen complications if the diet is not followed (corneal crystals resulting from hypertyrosinemia, malnutrition from non-compliance with the formula). Patients with partial adherence or non-adherence to NTBC can be clinically asymptomatic for prolonged periods. However, some develop severe complications. We have observed painful neurological crises and crises with paralysis requiring prolonged rehabilitation. Liver disease is presumably also active to some extent during such periods when NTBC levels are low. No liver cancer has yet been detected in any patient in the protocol who was treated from the time of newborn screening, even patients with suboptimal adherence to the prescribed therapy. However, HT1 patients with prolonged or recurrent sub-therapeutic levels of NTBC must be considered to be at increased risk for chronic liver pathology including hepatocellular carcinoma.

17.9 The Future

Life span, health and quality of life improved radically for HT1 patients who have benefited from the combination of newborn screening, NTBC and diet therapy that currently form the core of medical treatment of tyrosinemia in Québec. This combination is a major step towards the goal of providing normal health and normal quality of life to people affected by HT1.

NTBC-treated HT1 patients are at risk for a set of complications different from those of patients with untreated HT1. Although to date these complications have been minor compared with the high mortality and severe morbidity of HT1 prior to the availability of NTBC, it is important to be alert to the possible development of long-term complications and to new therapeutic opportunities. The number of patients and the overall follow-up time are still relatively small. Therefore, continued collection of data from each patient is important for understanding the long-term course of the disease, and to have high quality information to share with patients and families. Future clinical challenges in the diagnosis and treatment of HT1 are discussed elsewhere in this volume (see Mitchell and Yang, Chap. 19). Conversely, because of the good results obtained with the current protocol, new therapies will need to meet high standards if they are to be considered to replace the current treatment approach.

Acknowledgements We thank Martyne Gosselin and Yolande Lefèvre for organizing the NTBC study since its inception and André Imbeau for financial support. Yves Théoret, Department of Pharmacology, CHU Sainte-Justine, developed and supervises the clinical NTBC assay. Denis Cyr, Robert Giguère and Paula Waters of CHUS in Sherbrooke, developed sensitive clinical assays for succinylacetone. Shupei Wang coordinated the reception and distribution of clinical samples. Hao Yang contributed to the preparation of this manuscript. We thank the Groupe d'Aide aux Enfants Tyrosinémiques du Québec (GAETQ) for its collaboration and for its support of families and patients affected by HT1. Jean Larochelle's pioneering work with tyrosinemia was essential for establishing NTBC treatment for patients with HT1 in Québec.

Author Contributions

All authors revised the manuscript and provided comments as necessary. Manon Bouchard wrote the section about diet therapy. Grant Mitchell accepts responsibility for the accuracy of the text.

References

Grompe M, St-Louis M, Demers SI, Al-Dhalimy M, Leclerc B, Tanguay RM (1994) A single mutation of the fumarylacetoacetate hydrolase gene in French Canadians with hereditary tyrosinemia type I. N Engl J Med 331(6):353–357. doi:10.1056/NEJM199408113310603

Hall MG, Wilks MF, Provan WM, Eksborg S, Lumholtz B (2001) Pharmacokinetics and pharmacodynamics of NTBC (2-(2-nitro-4-fluoromethylbenzoyl)-1,3-cyclohexanedione) and mesotrione, inhibitors of 4-hydroxyphenyl pyruvate dioxygenase (HPPD) following a single dose to healthy male volunteers. Br J Clin Pharmacol 52(2):169–177

Larochelle J, Alvarez F, Bussieres JF, Chevalier I, Dallaire L, Dubois J, Faucher F, Fenyves D, Goodyer P, Grenier A, Holme E, Laframboise R, Lambert M, Lindstedt S, Maranda B, Melancon S, Merouani A, Mitchell J, Parizeault G, Pelletier L, Phan V, Rinaldo P, Scott CR, Scriver C, Mitchell GA (2012) Effect of nitisinone (NTBC) treatment on the clinical course of hepatorenal tyrosinemia in Quebec. Mol Genet Metab 107(1–2):49–54. doi:10.1016/j.ymgme.2012.05.022

Lindstedt S, Holme E, Lock EA, Hjalmarson O, Strandvik B (1992) Treatment of hereditary tyrosinaemia type I by inhibition of 4-hydroxyphenylpyruvate dioxygenase. Lancet 340(8823):813–817

Mitchell G, Larochelle J, Lambert M, Michaud J, Grenier A, Ogier H, Gauthier M, Lacroix J, Vanasse M, Larbrisseau A et al (1990) Neurologic crises in hereditary tyrosinemia. N Engl J Med 322(7):432–437. doi:10.1056/NEJM199002153220704

Mitchell GA, Grompe M, Lambert M, Tanguay RM (2001) Hypertyrosinemia. In: The Online Metabolic & Molecular Bases of Inherited Disease (OMMBID), 8th edn. McGraw Hill, New York

Dietary Considerations in Tyrosinemia Type I

18

Francjan J. van Spronsen, Margreet van Rijn, Uta Meyer, and Anibh M. Das

Abstract

Since the introduction of 2-(2 nitro-4-3 trifluoro-methylbenzoyl)-1, 3-cyclohexanedione (NTBC), life expectancy of HT1 patients greatly improved. However, due to treatment with NTBC, tyrosine concentrations greatly increase. As a consequence to possible neurocognitive problems, the main objective of dietary therapy in HT1 is to provide adequate nutrition allowing normal growth and development while strictly controlling tyrosine levels in blood (and tissues). Although no well-defined target levels exist, tyrosine concentrations below 400 μmol/L are considered to be safe. To achieve this aim a diet restricted in natural protein and supplemented with a special tyrosine and phenylalanine-free amino acid mixture is necessary.

Dietary management could be strenuous at diagnosis due to several different problems. If vomiting and diarrhea are a major issue at diagnosis, frequent feeding with additional energy from low protein food is needed

F.J. van Spronsen (✉)
Beatrix Children's Hospital, Division of Metabolic Diseases, University of Groningen, University Medical Center of Groningen,
HPC CA33, Antwoordnummer 333,
9700 VB Groningen, The Netherlands
e-mail: f.j.van.spronsen@umcg.nl

M. van Rijn
Department of Radiology, University of Groningen, University Medical Center of Groningen, Groningen, The Netherlands

U. Meyer
Department of Dietetics, Hannover Medical School, Hannover, Germany

A.M. Das
Department of Paediatrics, Hannover Medical School, Hannover, Germany

© Springer International Publishing AG 2017
R.M. Tanguay (ed.), *Hereditary Tyrosinemia*, Advances in Experimental Medicine and Biology 959,
DOI 10.1007/978-3-319-55780-9_18

for catch-up growth. Initiation of dietary treatment is usually easier if diagnosis is directly after birth. Based on newborn screening when infants are still reasonable healthy. If presenting clinically infants may experience serious difficulties in taking the amino acid mixtures probably due to feeding problems while when presenting after some 2–3 months taste development and the difference in the taste of amino acid mixtures compared to regular formula and breast milk increase difficulties with the treatment.

Following a dietary treatment is even harder than taking some medicine. Older children and adolescents often relax the diet and at some age become reluctant to stick to a strict regimen. Therefore, adequate training and information should be given to the patients and the family at regular intervals. To achieve this, a multidisciplinary approach involving pediatricians/physicians, dieticians, psychologists and social workers is an asset for the care of patients with HT1.

Keywords

Tyrosinemia type I • Diet • Amino acid • Phenylalanine • Tyrosine • Protein substitute

Abbreviations

HPPD	4-hydroxyphenylpyruvate dioxygenase
HT1	Hereditary Tyrosinemia type 1
NTBC	2-(2 nitro-4-3 trifluoro-methylbenzoyl)-1, 3-cyclohexanedione
PKU	Phenylketonuria

Until 1992 treatment for Hereditary Tyrosinemia type 1 (HT1) only consisted of a diet that was restricted in phenylalanine and tyrosine. Exclusively restricting dietary tyrosine is not enough as large parts of the precursor phenylalanine are converted to tyrosine. The diet sometimes led to temporary clinical stabilization but did not prevent development of liver cancer, renal tubulopathy and other serious sequelae. Thus, before the era of NTBC liver transplantation was necessary in every HT1-patient sooner or later (van Spronsen et al. 1994).

NTBC has led to a revolution of treatment in HT1. NTBC inhibits the activity of the enzyme 4-hydroxyphenylpyruvate dioxygenase (HPPD) upstream from the primary enzyme defect and thereby prevents production of toxic compounds (Lindstedt et al. 1992). NTBC gives rise to high tyrosine concentrations that might compromise neurocognitive development. Therefore, a diet low in phenylalanine and tyrosine is essential in all patients treated by NTBC.

It is essential to meet energy requirements to strictly prevent catabolism, which may give rise to elevated tyrosine levels. There are no well-defined target levels for tyrosine in blood, concentrations below 400 μmol/L are regarded as reasonably safe (de Laet et al. 2013; Mayorandan et al. 2014). To achieve this aim a diet restricted in natural protein and supplemented with a special tyrosine and phenylalanine-free amino acid mixture is necessary. The age at diagnosis (and thus the age at start of treatment) is variable (Mayorandan et al. 2014) and critically depends on the institution of a newborn screening programme for HT1.

18.1 Modes of Dietary Intervention

While the need for tyrosine (and phenylalanine) restriction is generally accepted, different modes of dietary intervention are used. In a recent multinational survey we found that there are differences in dietary treatment: 24% of centres used tyrosine and phenylalanine calculation, 38% of centres used protein restriction or tyrosine and phenylalanine calculation (depending on age), 19% of centres used protein restriction or tyrosine calculation (depending on age) and 19% of centres only recommended protein restriction (Mayorandan et al. 2014).

18.2 Dietary Management

18.2.1 Routine Dietary Management in HT1

For patients presenting in the first few months of life, severe vomiting and diarrhea may be a major issue. These children need frequent feeding with special food as well as additional energy for catch-up growth. The additional energy comes from low –protein food, and special protein-free modules with high energy content that are commercially available and provide energy without tyrosine/phenylalanine.

The intake of special amino acid mixtures is relatively easy when the patient can start with that regimen in the first few weeks of life. If newborn screening is not operative, most patients start treatment beyond the newborn period. Most of these babies experience serious difficulties in taking the amino acid mixtures. This probably relates to the taste development and considerable differences in taste of the amino acid mixtures compared to regular formula and breast milk.

The basic dietary principle is a restricted intake of phenylalanine and tyrosine. This regimen is comparable to phenylketonuria (PKU) where only phenylalanine intake is restricted (however, the intake of tyrosine is increased). In principle, the intake of phenylalanine in PKU and tyrosine in HT 1 can be calculated in milligrams per day or as protein equivalent in gram/day. Counting phenylalanine and tyrosine (in an exchange system) as well as counting natural protein is a strategy that works in HT1 as well as in PKU noticing that the content of phenylalanine in natural protein is generally 4.5–5% while the tyrosine content usually is somewhat lower (Table 18.1).

The food exclusively consists of mother's milk or formula supplemented with special tyrosine- and phenylalanine-free infant amino acid mixtures during the first 5–6 months of life which makes management by parents easy. Special tyrosine- and phenylalanine-free infant formula can be consumed ad libitum. For the first year of life, the dietitian should calculate nutrient intake at regular intervals. Supply of energy and natural protein/tyrosine intake is commensurate with weight development and tyrosine levels in blood. The total protein intake required is calculated based on the recommendations for the general population and "some" extra to compensate for the use of 70–90% of the recommendation of protein as amino acids mixture. Worldwide we see differences in recommendations in HT1 for the total protein based on different insights how much to calculate as com-

Table 18.1 Additional practical information for the dietary treatment

Estimation[a] of the amino acid content of foods (mg/g protein)		
Food	PHE	TYR
Fruits	25–30	25
Vegetables	35–40	25
Bread, crackers, pasta, cereals	50–55	33
Potatoes	40–49	59
Meat, poultry, fish	44–48	34–42
Dairy products	50	50
Egg	53	39

[a]USDA Agriculture Research Service, Nutrient Data Laboratory (http://www.nal.usda.gov/fnic/foodcomp/search/)

Table 18.2 Protein recommendations based on several age groups

Age	Protein [g] per kg body weight Natural protein + protein via AAM	Protein [g] per kg bodyweight via AAM Adapted according to DACH 2000 recommendations
0–2 months	2.5	2.0–2.5 (to be checked by dietary protocol)
2–12 months	2.0	2.0
1–4 years	1.5	1.0
4–10 years	1.2	0.9
10–14 years	1.1	0.9
>14 years	1.0	0.9

The total protein recommendation is on the left side. On the right side is the protein recommendation only for the AAM. This corresponds to our DACH reference values. The difference between the two is the estimated protein intake of natural foods that you have specified in the supplementary data "Expected natural protein tolerance in HT1" (supplementary data 1)

pensation. In the Table 18.2 we show 'DACH 2000' for the German-speaking countries.

A whole range of different commercial products with mixtures of l-amino acids devoid of phenylalanine and tyrosine, supplemented with vitamins and minerals, are available on the market providing essential amino acids and micronutrients (Table 18.3). For older children the amino acid mixture can be calculated according to 'DACH recommendation 2000'. The minimum protein requirement is completely supplied by special amino acid mixtures (see Table 18.3)

The phenylalanine requirements for healthy individuals are subject to discussion as reviewed by Pencharz et al. (2007). Such studies are necessary for all age groups and are very laborious. Initial data suggested that phenylalanine and tyrosine intake (in total for both) should be somewhere between 15 and 39 mg/kg per day with a ratio of 60:40. However, later studies have shown that phenylalanine intake by itself should be between 38 and 48 mg/kg per day. Therefore, the maximum phenylalanine requirement (if no tyrosine is given) is considered to be 42 mg/kg per day, while about 9 mg/kg per day is needed when enough tyrosine is provided.

Some children with HT1 on a phenylalanine/tyrosine -restricted diet have been found to have

Table 18.3 Special dietary products (metabolic formula) for tyrosinemia[a]

Product	Protein equivalent (g)	Energy (kcal)	Age indication	Flavored	Micronutrients added
Powdered (400–500 g cans); details shown for 100 g					
TYR Anamix Early Years	13.5	473	Infants, children <3 year	–	+
TYR Anamix Next	28	385	>1 year	–	+
XPhe, XTyr Maxamaid	25	324	Young children	+	+
XPhe, XTyr Analog	13	475	Infants, children <3 year	–	+
XPHEN, TYR Maxamaid	25	309	Young children	–	+
XPHEN, TYR Maxamum	39	297	Older children and adults	–	+
Tyrex-1 (Abbott)	15	480	0–12 month	–	+
Tyrex-2 (Abbott)	30	410	>1 year	–	+
Comida-TYRo A	11.8	506	0–12 month	–	+
Comida-TYRo B	31.1	420	1–14 year	–	+
Tyros 1 (Enfamil)	16.7	500	Infants, toddlers	–	+
Tyros 2 (Enfamil)	22	410	Children, adults	+	+
Zero TP Infant mix LCP	11	508	Infants	–	+
Zero TP Kid	72	289	4 month's-6 year	–S	+
Zero TP junior	75	300	7–14 year	–S	+
Zero TP Advance	77	309	>15 year	–S	+
Prepacked portions					
Tyr Gel (Vitaflo)	10	81	>1 year	–S	+
Tyr Express 15 (Vitaflo)	15	74	>3 year	–S	+
Tyr Express 20 (Vitaflo)	20	99	>3 year	–S	+
Zero TP minis (MetaX)	5	24	>3 year		+
Ready to use					
Cooler 10 (Vitaflo)	10	62	>3 year	+	+
Cooler 15 (Vitaflo)	15	92	>3 year	+	+
Cooler 20 (Vitaflo)	20	124	>3 year	+	+
Tylactin RTD (Cambrooke)	15	200	Children, adults	+	+
Tylactin restore (Cambrooke)	10	170	Children, adults	+	+

(continued)

Table 18.3 (continued)

Product	Protein equivalent (g)	Energy (kcal)	Age indication	Flavored	Micronutrients added
TYR Lophlex LQ	20	120	>4 year	+	+
PT-am infant	13	466	0–12 month	–	+
Milupa Tyr 1	50	290	> 0 month	–	+
Milupa Tyr 2 prima	60	289	1–8 year	–	+
Milupa Tyr 2 secunda	70	291	9–14 year	–	+
Milupa Tyr 3 advanta	70	297	>14 year	–	+
PT-am Anamix Frucht Vanille	29	374	>3 year	–	+
Tyr Anamix Junior LQ orange	8	95	<3 year	–	+
Tyr lophlex LQ 20 Juicy Berries	16	96	>4 year	+	+

[a]Worldwide, different products and other manufacturers are available; see country details on websites
+ = Yes; – = No; S = separate flavoring

low phenylalanine levels (Wilson et al. 2000; van Vliet et al. 2015). There has been concern that these low levels could be detrimental to the child's development, and it was thought that the addition of phenylalanine to the diet would increase levels in blood and tissues without increasing the tyrosine level. An initial study found that supplementing the diet with an additional 20–40 mg/kg per day phenylalanine increased the blood phenylalanine concentration while maintaining the tyrosine level (Wilson et al. 2000). Therefore, supplementation with phenylalanine may be beneficial, but the risk of high tyrosine should always be kept in mind. Further long-term studies are required before a recommendation for standard supplementation can be given. In this context, it should be considered whether normal bread instead of phenylalanine should be provided. This is based on the fact that normal bread contains more phenylalanine than tyrosine (Table 18.1). This relaxation is likely to improve dietary compliance.

18.2.2 Special Issues Regarding Diet in HT1

Diagnosis and initiation of dietary and pharmacological treatment are a big challenge for affected families. In "*Additional practical information for the dietary treatment*" (Table 18.1) we give practical recommendations to start dietary treatment.

A new period begins when the child can open the cupboard or fridge himself. When patients grow-up and become independent of their parents, new issues may arise: going to school and eating out, acceptance of dietary requirements by their peers, having parties etc.. Special attention has to be paid to new responsibilities and creative introduction of new roles to parents and the patient. Adolescence is a very difficult period for the patient, aspects like self-empowerment are important issues. The family and the child must be well trained and informed at regular intervals starting in early childhood and continuing during adulthood.

Following a dietary treatment is even harder than taking some medicine. A simple food selection is useful to improve dietary compliance (see Table 18.4). Young people often relax the diet and are reluctant to stick to a strict regimen. Before they stop sticking to their diet, a wider food selection should be considered. This should be discussed with physician and dietician.

Other issues may include the reimbursement of amino acid mixtures and low protein foods that varies around Europe already (Belanger-Quintana et al. 2012). From Mayorandan et al. (2014) we know that reimbursement of HT1 treatment varies considerably, similar to PKU (Ahring et al. 2009).

While age-specific target levels for phenylalanine exist in PKU, the situation in HT1 is less

Table 18.4 Simple recommendations for the older patient when choosing food

Very low protein food	Moderate protein food	
< 2–3 g protein per 100 g	< 6 g protein per 100 g	High protein food
Fruit	Normal bread and	Meat
Vegetables	Bakery products	Sausage
Potatoes, rice	(under 6 g E/100 g)	Egg
Butter, oil, mayonnaise, tartar sauce	Pasta	Poultry
Cream, creme fraiche, sour cream	Milk	Fish
Sugar,	Nuts	Seafood
Sugar confectionery		
Juice drinks and lemonade		
Rice milk, oat milk		
Special low protein foods		
You can eat as much as you like!	Be careful!	Not allowed
	May be taken depending on your tyrosine level	

clear. In PKU we know that diet for life is required in adults and women who aim to become pregnant have to stick to a stricter diet, we cannot give recommendations for HT1 yet.

Like in PKU, glycomacropeptide may be a dietary option, their role in the treatment of HT1 still has to be elucidated. In its pure form, glycomacropeptide does not contain phenylalanine nor tyrosine. Due to the fact that the glycomacropeptide has to be removed out of the whey protein, the product will contain some phenylalanine and tyrosine.

18.2.3 Dietary Management and Follow-Up

The objective of dietary therapy in HT1 is to provide adequate nutrition allowing normal growth and development while strictly controlling tyrosine levels in blood (and tissues). Changes in growth velocity, intercurrent illness and dietary indiscretions may result in large fluctuations of tyrosine concentrations. Therefore, especially during the early years, we need a very frequent follow-up by (home) blood sampling and frequent clinical and laboratory checks at the outpatient clinic with dietary assessment.

To conclude, life-long adherence to the tyrosine and phenylalanine restricted diet is necessary to prevent tyrosine concentrations to increase. Regular follow-up of growth, development and metabolic control needs to be done to ensure sufficient intake of all amino acids, especially including phenylalanine. Adherence to the dietary treatment is however a great challenge when the patient gets older. This, in combination with other issues associated with the disease, makes multidisciplinary treatment necessary. A multiprofessional team with expertise in the treatment of HT1 patients including paediatricians/physicans, dieticians, psychologists and social workers is an asset for the care of patients with HT1. To this aim Centers of Expertise are being formed throughout Europe to guarantee optimal care. These Centers of Expertise are members of a specific European Reference Network.

Supplementary Data

Supplementary data 1: Expected natural protein tolerance in HT1

	Protein	Protein
Age years	g/d	g/kg/d
<2	2–6	0.4–0.5
2–9	5–10	0.2–0.5
10–14	9–20	0.3–0.4
>15	11–25	0.2–0.4

Around 70–90% from total protein must be covered by the amino acid mixture

Notice: natural protein tolerance individually to be defined. Intake of tyrosine/natural protein is altered according to plasma tyrosine concentrations (influenced by residual enzyme activity, growth velocity

Supplementary data 2: Contact information for manufacturers of tyrosine- and phenylalanine-free medical foods

Manufacturer and website address	Contact details
Abbott Nutrition	3300 Stelzer Road, Columbus, OH 43219
www.abbottnutrition.com	Office Phone: 614–624-4416, Toll Free: 800–986-8755
	Cell: 614–264-4388, Fax: 614–727-4416
Vitaflo	Distributed by Vitaflo USA
www.vitaflousa.com	211 N. Union St. Suite 100, Alexandria, VA 22314
	Phone: 1–888-VITAFLO (1–888–848-2356), Fax: 1–631–693-2002
	Email: vitaflo@vitaflousa.com
Vitaflo Europe	Vitaflo International Ltd
www.vitaflo.co.uk	Suite 1.11, South Harrington Building, 182 Sefton Street Brunswick Business Park
	Liverpool L3 4BQ UNITED KINGDOM
	Metabolic product queries: +44 (0)151702 4938
	Email: vitaflo@vitaflo.co.uk
Mead Johnson (Enfamil)	Phone: 1–812–429-6399
www.meadjohnson.com/pediatrics/us-en/product-information/products	Email: mjmedicalaffairs@mjn.com
Cambrooke	4 Copeland Drive, Ayer, MA 01432
www.cambrooke.com	Phone: 866,456 9776, Fax: 978,443 1318
	Mail: info@cambrooke.com
Nutricia-NA USA	Phone: 301–795-2300 or 1–800–365-7354
www.medicalfood.com	Fax: 301–795-2301
Nutricia-NA Canada	Phone: 877–636-2283
www.Nutricia-NA.com	Fax: 514–745-6625
Nutricia	Nutricia Advanced Medical Nutrition
http://www.nutricia.com	Danone Place Schiphol, Tower E, World Trade Centre
	1118BG Schiphol, The Netherlands.
	Phone: +31 20,456 9000
	Email: medicalnutrition@nutricia.com
Dr Schär	In various countries with various phone and
http://drschaer.com	e-mail contacts around the world as in UK,
	Germany, Italy, Spain and with a translated
	website in various languages.
MetaX	Email: info@metax-gmbh.de
www.metax.org/EN/Products/Metabolics.aspx	
Nutricia GmbH	Email: info-metabolc@nutricia.com
www.nutricia-metabolics.info	Phone: 00800–74773799

Supplementary data 3: Monitoring the diet

Age		Frequency
0–6	Months	Fortnightly
6–12	Months	Fortnightly to monthly
1–2	Years	Quarterly
		(when applicable monthly Tyr-Phe in bloodspot)
> 3	Years	Every 3–6 months
		(when applicable monthly Tyr-Phe in bloodspot)

As soon as nutritional support is well established for the patient, the prescription should be routinely and frequently fine-tuned to growth and lab results. Changes in intake take time to be reflected in lab results. The time needed to evaluate effects increases with age

Monitoring of Nutritional Status

- Plasma amino acids
- Nutrient intake
- Growth (linear and weight)
- Condition of hair and skin
- Biochemistry

can verify the adequacy of intake.

Introduction of Solids: Follow General Advice When Possible

- At about 4–5 months of age
- Developmental readiness
- Important source of unidentified nutrients such as fiber
- Contributing to a child's acceptance of a variety of tastes and textures
- Development of jaw muscles important for speech

References

Ahring K, Bélanger-Quintana A, Dokoupil K, Gokmen Ozel H, Lammardo AM, MacDonald A, Motzfeldt K, Nowacka M, Robert M, van Rijn M (2009) Dietary management practices in phenylketonuria across European centres. Clin Nutr 28(3):231–236

Belanger-Quintana A, Dokoupil K, Gokmen-Ozel H, Lammardo AM, MacDonald A, Motzfeldt K, Nowacka M, Robert M, van Rijn M, Ahring K (2012) Diet in phenylketonuria: a snapshot of special dietary costs and reimbursement systems in 10 international centers. Mol Genet Metab 105(3):390–394

de Laet C, Dionisi-Vici C, Leonard JV et al (2013) Recommendations for the management of tyrosinaemia type 1. Orphan J Rare Dis 8:8

Ekvall SW, Ekvall VK (eds) (2016, April 27) Pediatric and adult nutrition in chronic diseases, developmental disabilities, and hereditary metabolic disorders prevention, assessment, and treatment, 3rd edn. Oxford University Press, ISBN: 9780199398911

Lindstedt S, Holme E, Lock EA, Hjalmarson O, Strandvik B (1992) Treatment of hereditary tyrosinemia type I by inhibition of 4-hydroxyphenylpyruvate dioxygenase. Lancet 340:813

Mayorandan S, Meyer U, Gokcay G, Segarra NG, de Baulny HO, van Spronsen F, Zeman J, de Laet C, Spiekerkoetter U, Thimm E, Maiorana A, Dionisi-Vici C, Moeslinger D, Brunner-Krainz M, Lotz-Havla AS, Cocho de Juan JA, Couce Pico ML, Santer R, Scholl-Bürgi S, Mandel H, Bliksrud YT, Freisinger P, Aldamiz-Echevarria LJ, Hochuli M, Gautschi M, Endig J, Jordan J, McKiernan P, Ernst S, Morlot S, Vogel A, Sander J, Das AM (2014) Cross-sectional study of 168 patients with hepatorenal tyrosinaemia and implications for clinical practice. Orphanet J Rare Dis 9:107. doi:10.1186/s13023-014-0107-7

Pencharz PB, Hsu JW, Ball RO (2007) Aromatic amino acid requirements in healthy human subjects. J Nutr 137(Suppl 1):1576S

van Spronsen FJ, Thomasse Y, Smit GPA et al (1994) Hereditary tyrosinaemia type I: a new classification with difference in prognosis on dietary management. Hepatology 20:1187

van Vliet D, van Dam E, van Rijn M et al (2015) Infants with tyrosinemia type 1: should phenylalanine be supplemented? JIMD Rep 18:117

Wilson CJ, Van Wyk KG, Leonard JV, Clayton PT (2000) Phenylalanine supplementation improves the phenylalanine profile in tyrosinemia. J Inherit Metab Dis 23:677

19

Remaining Challenges in the Treatment of Tyrosinemia from the Clinician's Viewpoint

Grant A. Mitchell and Hao Yang

Abstract

This chapter provides a clinical perspective on the challenges that stand between current clinical practice and a cure for hepatorenal tyrosinemia (HT1). HT1 has been transformed in the last 50 years from an aggressive often undiagnosed childhood disease causing liver failure or liver cancer, with infant death in most patients, to a condition that is detectable at birth, and for which treatment with nitisinone (NTBC) and diet can prevent detectable liver or kidney abnormalities. What challenges remain? The properties of the affected metabolic pathway and the broad spectrum of severity seen in untreated patients are incompletely understood but potentially important for patients. Available treatments have potential complications, including liver transplantation (risks of surgery and of immunosuppression to prevent rejection), nitisinone and diet therapy (hypertyrosinemia, corneal opacities, nutritional imbalances and possibly developmental delay). The detection of liver cancer is imperfect and laborious. The effects of tyrosinemia during pregnancy are little-known. Although animal models of HT1 are becoming standard research tools in cell replacement and gene modification therapy, these techniques are not currently applicable to HT1 itself. Treatment adherence is variable, causing concern about long term outcome for some patients. Around the world, there are great disparities in the diagnosis and treatment of HT1. Most affected individuals are born in places where newborn screening for HT1 is not performed and where appropriate treatment is not available. We hope that this list will help to focus on some of these remaining obstacles to a cure for HT1.

Both authors participated in writing the chapter. Grant Mitchell accepts responsibility for the accuracy of the text.

G.A. Mitchell (✉)
Departments of Nutrition (MB) and Radiology (JD), Divisions of Gastroenterology, Hepatology and Nutrition (FA, UH), Medical Genetics (CBG, MG, GAM) and Nephrology (AM, VP), Department of Pediatrics and Department of Pharmacy (SA, JFB), CHU Sainte-Justine and Université de Montréal, Montreal, Canada
e-mail: grant.mitchell@recherche-ste-justine.qc.ca

H. Yang
Division of Medical Genetics, Department of Pediatrics, CHU Sainte-Justine and Université de Montréal, Montréal, QC, Canada

© Springer International Publishing AG 2017

R.M. Tanguay (ed.), *Hereditary Tyrosinemia*, Advances in Experimental Medicine and Biology 959,
DOI 10.1007/978-3-319-55780-9_19

Keywords
Tyrosinemia • Cancer • Adherence • Development • Cornea • Eye

Abbreviations

4HPPD	4-hydroxyphenylpyruvate dioxygenase
AFP	Alpha fetoprotein
FAA	Fumarylacetoacetate
FAH	Fumarylacetoacetate hydrolase
HCC	Hepatocellular carcinoma
HT1	Tyrosinemia type 1
NTBC	Nitisinone (2-(2-N-4-trifluoromethylbenzoyl)-1,3-cyclohexanedione)
SA	Succinylacetone.

19.1 Introduction

There are few medical conditions for which treatment has a greater beneficial effect than hepatorenal tyrosinemia (HT1). Newborn screening, liver transplantation and treatment with nitisinone and diet have transformed the outcome of HT1. Early reports of HT1 describe a course of suffering. Most patients had progressive liver disease leading to death from acute liver failure in infancy. Painful neurologic crises occurred in some patients, sometimes with fatal respiratory paralysis or with limb weakness that required months to recuperate. Some patients showed a more prolonged clinical course, developing cirrhosis, chronic liver failure and liver cancer, often with renal tubulopathy and rickets, and death during childhood or sometimes in young adulthood. The availability of liver transplantation revolutionized the surgical treatment of HT1, allowing normal liver function and removing the high risk of developing liver cancer. The use of nitisinone plus a phenylalanine- and tyrosine-restricted diet was a radical improvement in medical treatment. Within the first hours of treatment with nitisinone, there is a marked reduction of the levels of toxic metabolites, which persists through the course of treatment (Larochelle et al. 2012). Current data indicate that a patient with HT1 who is detected by neonatal screening, who receives early treatment with diet and nitisinone, and who adheres to this treatment, can expect to have normal liver and kidney function and freedom from neurological crises, and can avoid liver transplantation. At first glance, one might speak of a "cure" and relegate outstanding questions to a category of minor curiosities.

Patients remind us that for most people, "cure" means the enjoyment of normal health, preferably with no limitations upon the choice of lifestyle. From this perspective, it is well to consider some of the clinically important obstacles to cure for HT1 that remain at every level from cell physiology to health care policy.

19.2 The Biology of Tyrosinemia

Much is still unknown about the biology of tyrosinemia. HT1 is caused by genetic deficiency of the last step of the degradation pathway of phenylalanine and tyrosine. This classic metabolic pathway is a staple of introductory courses in genetics and biochemistry. Other diseases related to this pathway include phenylketonuria (the first inborn error of amino acid metabolism for which dietary treatment was shown to be efficacious and the first aminoacidopathy for which newborn screening was introduced (Scriver 2007)) and alkaptonuria (one of the four conditions that inspired Garrod to develop his concept of biochemical individuality (Scriver 2008)). These six enzymatic steps still can surprise us.

To cite one example, many HT1 patients treated by diet and NTBC have low phenylalanine levels. This is the opposite of what was expected. Plasma phenylalanine levels were predicted to increase, because phenylalanine is situated before the metabolic block induced by

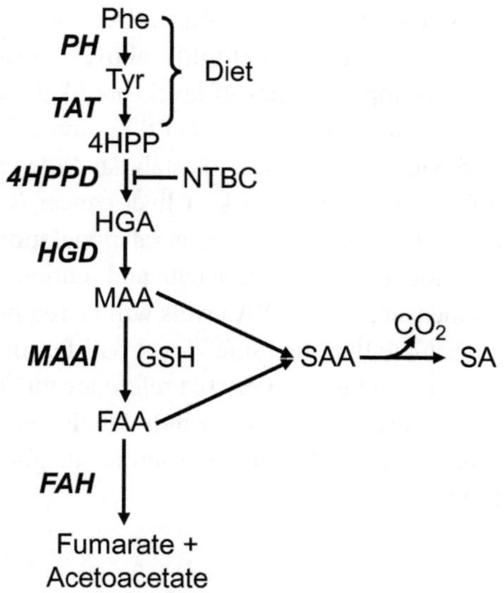

Fig. 19.1 The tyrosine degradation pathway. Understanding the structure of this pathway is useful for understanding the unique clinical challenges in HT1. Fumarylacetoacetate (*FAA*) and possibly maleylacetoacetate (*MAA*) are reduced to succinylacetoacetate (*SAA*) (Lindblad et al. 1977). *SAA* can spontaneously decarboxylate to form succinylacetone (*SA*) (Lindblad and Steen 1982), which is used to screen and to diagnose HT1. *FAA* and *MAA* are toxic and can react with other molecules, causing mutations in *DNA* (Jorquera and Tanguay 1997), succinylacetone adducts in proteins (Manabe et al. 1985) and oxidative stress with depletion of glutathione (*GSH*) (Stoner et al. 1984; Langlois et al. 2008). Other abbreviations not used in the text: *HGA* homogentisic acid, *HGD* homogentisic acid 1,2-dioxygenase, *MAAI* maleylacetoactate isomerase

NTBC (Fig. 19.1). The reason why some NTBC-treated HT1 patients have low phenylalanine levels is not clear, but it affects patients, many of whom are supplemented with phenylalanine to maintain their plasma level of phenylalanine in the normal reference range.

19.3 Natural Phenotypic Diversity in HT1

Before the availability of NTBC, HT1 patients from opposite ends of the clinical spectrum were scarcely recognizable as having the same disease. Early articles distinguished between "acute" and "chronic" patients. The former developed liver failure in the first few months of life. "Chronic" patients developed cirrhosis, liver cancer and renal Fanconi syndrome. Patients with intermediate clinical courses were reported, and occasional patients survived to adulthood with mild growth retardation and apparently normal liver function. The neurological phenotype was also incompletely penetrant (Mitchell et al. 1990). This is reminiscent of acute intermittent porphyria, a biochemically-related condition in which a high fraction of affected individuals are clinically asymptomatic (Whatley and Badminton 1993).

One author (GM) recalls an illustrative family with three affected siblings. The first died in infancy of liver failure, before the availability of screening. The second was treated from birth with diet and had an uneventful pediatric course, but had progressive renal disease with declining glomerular filtration rate during adolescence. This patient was considered for combined liver-kidney transplantation before the availability of NTBC. With NTBC treatment however, glomerular function improved somewhat then had remained stable for nearly twenty years. Transplantation has been avoided. Genetic studies of the patient and the parents later showed that the patient was homozygous for the Québec founder mutation. Encouraged by the initial clinical course of the second affected child, the parents had another pregnancy. The child was also affected. Although treated by diet from the neonatal period, the third affected sibling developed signs compatible with a severe neurologic crisis during which she died. This example shows that, even within a single family, affected siblings that are homozygous for a classical disease-causing FAH allele can have markedly different clinical courses dominated by each of the principal classical target organs of tyrosinemia.

The classical clinical observations show clearly that protective factors exist. Presumably these involve genetic variants in genes other than *FAH* and also dietary and other environmental factors. Today, this broad natural clinical spectrum is concealed by NTBC treatment. However, it shows that naturally-occurring factors exist which, if they could be defined, might identify important variables that might direct the development of new treatments for tyrosinemia.

19.4 Medical Treatment with Diet and NTBC

It is important to establish the levels of tyrosine that indicate optimal control. Patients are clinically asymptomatic over a broad range of concentrations of plasma tyrosine. It is tempting to counsel perfection, and to recommend that plasma levels of tyrosine should be maintained in the normal reference range. In practice this is impossible. In NTBC-treated patients who receive enough phenylalanine plus tyrosine to allow normal growth and maintenance, the level of plasma tyrosine is elevated.

It is clear that uncontrolled hypertyrosinemia is to be avoided. One complication that is directly related to high plasma tyrosine is the occurrence of corneal crystals. These are irritating, painful and reversible by adherence to a diet with low phenylalanine and tyrosine content. There are reports of impaired psychomotor function in some HT1 patients (see below). Whether this is related to hypertyrosinemia is unproven but developmental delay has been reported in other types of tyrosinemia (reviewed in Mitchell et al. 2001). Given these considerations, it is reasonable to maintain a level of tyrosine as close as possible to normal and that permits normal growth. This is becoming easier, because of the development of many palatable low protein foods and of nutrient mixtures devoid of tyrosine and phenylalanine for which the packaging and taste are increasingly attractive to patients. Current dietary treatment practices are reviewed elsewhere in this book (See Chap. 18 by van Spronsen et al. and Chap. 17 by the Québec NTBC study group).

What is an acceptable level of SA? Many articles speak of patients under NTBC treatment as being "negative" for SA. This gives the false impression that SA is completely absent. In fact, when assayed by sensitive methods, SA is measureable even in nontyrosinemic individuals (Jakobs et al. 1985). The levels of SA in untreated HT1 are about 1,000-fold above normal. Currently-employed cut-off levels are likely adequate for the diagnosis of HT1 in non-NTBC treated patients. However, to establish "safe" levels of SA at which the risk of liver cancer is not different from that of the general population, a large amount of prospective data and routine sensitive measurement of SA levels will be required. We speculate that the "safe" level will fall in the grey zone that lies between the reference range of SA concentration in normal individuals and the threshold of detection of routine diagnostic techniques.

19.5 HT1 and Neurodevelopment

Delayed development and attention deficit disorders have been reported in patients with HT1 treated with diet and nitisinone (Mayorandan et al. 2014; Thimm et al. 2012). Most patients in the Québec NTBC protocol who are treated with NTBC and diet attend age-appropriate classes in school and some young adult patients are now in post-secondary training or in the workforce. Some are academically talented; others require special help. An study of psychomotor development of HT1 patients in Québec was performed in the first years of the Quebec NTBC Study. No significant differences were seen with the control group (in preparation). However, the question of psychomotor development in HT1 patients is very important. A prospective study of psychomotor evaluation of HT1 patients is underway as part of the Québec NTBC treatment protocol.

19.6 Liver Cancer in Tyrosinemia

Early detection of hepatocellular carcinoma (HCC) is a priority for the clinical management of HT1. Nontransplanted HT1 patients, even if they are treated by NTBC, are considered to be at risk and are followed for the development of HCC. Current protocols are discussed in the chapters in this volume by the

Quebec NTBC Study group and medical management of liver complications (See Chap. 6 by Halac et al.). Patients can be reassured that current surveillance protocols detect the great majority of liver cancers at an operable stage. Conversely, it is important to recognize that there is no perfectly reliable marker for the early detection of HCC in tyrosinemia. Some HCCs are associated with normal or only mildly elevated plasma levels of alpha fetoprotein (Paradis et al. 1990). Some HCCs have subtle or atypical imaging characteristics. The development of reliable noninvasive methods for early detection of HCC in HT1 deserves a high priority in tyrosinemia research.

A fundamental unanswered question is whether the behavior of HCC that develops in tyrosinemic patients resembles that of HCC resulting from other causes. Histologically, liver cancers in HT1 patients resemble those occuring in other conditions. However, the metabolic environment inside HT1 hepatocytes is different from that of other liver diseases. In HT1 patients that are not treated with NTBC, hepatocytes continually produce toxic compounds that are found only in HT1 (Fig. 19.1) and HCC often develops in childhood or adolescence. In non-tyrosinemic patients with HCC, the precipitating factors, when known, are usually exogenous, such as infections or environmental toxins, and patients are usually adults. This is discussed further in this volume, in the chapter dealing with hepatic transplantation. In clinical practice, it cannot be assumed that HCCs in HT1 and in other conditions arise by the same steps or that the cancers will behave in the same way. Further understanding of the distinct features of HCC in tyrosinemia may yield important information about the clinical management of HT1.

Liver transplantation revolutionized the treatment of HT1 but it carries the risks of operative complications including death, and of medical complications such as graft rejection. If in the future major improvements occur in surgical safety, organ availability or host tolerance of liver transplants, these developments could change the approach to treatment.

19.7 HT1 and Pregnancy

Pregnancy in women affected by HT1 is a largely unexplored domain. Three situations can occur: pregnancies in which the fetus is affected by HT1, in which the mother is affected by HT1, and in which both are affected.

The first situation is the commonest. In most pregnancies with a HT1 fetus, the mother is a healthy heterozygote who does not herself require NTBC treatment. In most instances the couple does not know that they are at risk and learns this only after the birth of an affected newborn. In some instances, such as for couples from regions of high risk in Québec, carrier screening for prevalent mutations is available. If both spouses are known to be carriers, preconception genetic counselling and preimplantation or prenatal molecular diagnoses are discussed. If such couples decide upon a pregnancy, and wish to continue the pregnancy even if the fetus is affected, invasive, prenatal diagnosis is not necessary, because it currently does not change medical or obstetrical management of the pregnancy. At birth, diagnosis can be performed rapidly. Treatment can be started shortly after birth if the child is affected.

Only a small number of pregnancies are recorded in mothers who are under medical treatment for HT1 (Kassel et al. 2015; Vanclooster et al. 2012; Garcia Segarra et al. 2010). To our knowledge there is no evidence that NTBC is teratogenic in humans, but currently-available data are insufficient to conclude on this point. Current recommendations, based on general considerations, are that nontransplanted tyrosinemic women who wish to become pregnant should continue treatment with NTBC and diet. We recommend that strict dietary and pharmacologic monitoring be performed starting before conception, and that contraception be stopped only when sustained adequate biochemical and pharmacological control is attained. It would be dangerous to stop NTBC treatment before or during pregnancy. This is because it is predicted that hepatic or neurological decompensation could occur within days after NTBC is stopped, and such complications could endanger the health of the mother and the fetus. Of note, liver transplantation is not a contraindication to pregnancy, and women with HT1 who were treated by liver transplantation have had normal pregnancies.

One pregnancy has been described in a woman with HT1 whose fetus was also affected (Garcia Segarra et al. 2010). NTBC and diet treatment were prescribed, although adhesion to treatment recommendations was suboptimal. Of note, the level of AFP in cord blood was normal. In HT1 patients, alpha fetoprotein is increased at birth (Hostetter et al. 1983). Therefore it is reasonable to suspect that prior to birth, the liver of HT1 fetus is subject to a period of toxicity. Based on this admittedly anecdotal report of a single affected fetus exposed to NTBC before birth, it appears likely that prenatal treatment of the mother with NTBC protected the fetal liver. At present, NTBC treatment during pregnancy of non-tyrosinemic women bearing an affected fetus cannot be recommended, because it exposes the fetus and the mother to NTBC and to nonphysiologically high maternal plasma levels of tyrosine and to the potential risks a suboptimal artificial diet during pregnancy. Prenatal treatment does not seem necessary because newborns treated within the first month of life rapidly gain normal liver function (see Chap. 17 about the Quebec NTBC study protocol). This important observation does however hold exceptional interest and must be noted as more information is gathered about HT1 and NTBC in pregnancy.

Because of the scarcity of available data, recording and compiling data on all pregnancies is a useful goal. Firm evidence-based recommendations are lacking for nearly all aspects of pregnancy and HT1. Preliminary data are promising as regards teratogenicity and fetal and maternal health during pregnancy. However, compiling clinical information about all pregnancies of women with HT1 is an important and feasible clinical priority.

19.8 The Psychological Paradigm Shift

Families that have not experienced the complications of untreated HT1 on a first-hand basis may have a different relationship to the disease than those in which an affected person has been gravely ill. The treatment of HT1 involves a palatable but lifelong special diet and taking NTBC every day. Some families and some adolescent and young adult patients become focused on the negative aspects of this situation. Some patients who have never been symptomatic question whether HT1 is really a severe condition. Some adopt a non-adherent lifestyle, and risk the complications of untreated tyrosinemia. The problem is compounded by the efficiency of treatment. Patients are asymptomatic when adherent to treatment and periods of non-adherence are often not accompanied by symptoms. We have repeatedly seen hypertyrosinemia with corneal lesions and neurological crises in patients who do not adhere to recommendations for diet and NTBC treatment.

It is unknown to what extent chronic non-adherence to NTBC administration influences the accumulation of mutations in the liver and the risk of liver cancer. Clinical assessment of liver function is usually normal in non-adherent patients, although within a few days after completely stopping NTBC, AFP levels can start to increase. SA levels increase if the levels of NTBC are subtherapeutic. The concentration of SA is presumably related to the intracellular concentration of fumarylacetoacetate, the substrate of the enzyme deficient in HT1 (Fig. 19.1). Fumarylacetoacetate is a reactive compound. It is felt to deplete antioxidants and may act directly as a mutagen. No liver cancer has yet been detected in any patient in the protocol who was treated from the time of newborn screening, even patients with suboptimal adherence to the prescribed therapy. Basic considerations however strongly suggest that periods of non-adherence to NTBC will be accompanied by reduced liver protection and perhaps by increased mutagenesis, which is predicted to increase the lifetime risk for the development of liver cancer.

Communication with patients and families about the importance of long term treatment adherence and a better understanding of the psychology underlying patient decisions about treatment adherence are important in all medical situations and this is especially true in tyrosinemia.

19.9 New Genetic Therapies

For three decades, it has been proposed to genetically modify liver cells in order to treat tyrosinemia. Progress has been made, most recently with the popular CRISPR technology of gene modification (Sander and Joung 2014; Yin et al. 2014, 2016). Because of the widespread media attention that focusses on such promising and spectacular techniques, it is not rare for parents to directly question physicians about these techniques. The physician must be prepared to provide a reasoned answer to such anxious and understandable questioning.

It is to be hoped that efficient and specific modification of the FAH gene will be applicable to HT1 at some point in the future. However, one important consideration suggests strongly that the use of cell or gene modification treatments will be more challenging in tyrosinemia than in many other genetic liver diseases. In most of these conditions, the ability to introduce normal function of the deficient gene in most cells of the liver would suffice to produce a clinical cure. In HT1, available evidence strongly suggests that *all* HT1 cells would have to be replaced.

All indications suggest that the pathophysiology of HT1 in the liver and kidneys is cell-autonomous. This is an important concept for clinicians and patients. It means that one or more hepatocytes with FAH deficiency, even if it is surrounded by cells with normal FAH activity and normal function, and even if it is exposed to a normal extracellular concentration of tyrosine, will produce the toxic compounds at the end of the tyrosine degradation pathway just as if it were surrounded by FAH-deficient cells. Such a cell could accumulate mutations and would be at high risk of becoming a cancer. The cell autonomy concept is strongly supported by clinical observations and basic reasoning. Clinically, before the availability of NTBC, many HT1 patients had normal plasma levels of tyrosine and phenylalanine (reviewed in Mitchell et al. 2001). Also, some patients were demonstrated to have revertant nodules that contained normal FAH activity (Kvittingen et al. 1994). However, despite this, these HT1 patients developed cirrhosis and sometimes cancer.

Biochemically, FAA and also its precursor, maleylacetoacetate (Figure), are unstable. They are thought to react with other intracellular molecules before they can escape from cells, causing mutations and other damage. Even if the surrounding cells are normal, only a fraction of FAA and MAA would escape from the HT1 cell for degradation by the surrounding normal cells. Therefore, cell(s) deficient in FAH would continue to produce and to be damaged by these toxic compounds. Such cells might become a cancer.

The presence of cell autonomous damages in HT1 has not been explicitly proven, but it is supported by available evidence. It must be considered in evaluating any treatment strategy for HT1 involving cell replacement or gene therapy. A treatment that proposes the genetic or metabolic correction of some but not all HT1 hepatocytes would have to be combined with a satisfactory additional strategy to eradicate or otherwise deal with any remaining FAH-deficient hepatocytes. If techniques become available to identify and efficiently destroy FAH-deficient hepatocytes, they could change the treatment paradigm for HT1.

19.10 Universality

Methods of newborn screening and treatment described in this volume allow affected individuals and families to enjoy a high level of health, but we suspect that a large fraction of children born today with HT1, perhaps the majority, still endure the natural course of HT1 as described in the first paragraph of this chapter. The greatest disease burden of tyrosinemia is now in places without screening. There, patients present with clinical signs and may already have permanent damage to their liver and other organs. In the developing world, the treatments discussed above for tyrosinemia are often not available. In many cases, the diagnosis of HT1 may not be considered at presentation or during the course of the disease.

The diagnosis and treatment of tyrosinemia and of inborn errors in general in the third world is complex, and is dominated by the extreme penury of resources. We cannot propose a solution, but in the quest to cure tyrosinemia, it is important to recall that the majority of the world's HT1 patients are probably not treated, and to be alert to opportunities to improve access to modern care for the largest possible number of these individuals.

The course of tyrosinemia has been repeatedly transformed by innovative ideas in biochemistry, genetics, medicine, surgery and public health. Hopefully the list of remaining challenges provided in this chapter may help to identify some milestones on the route to a cure.

Acknowledgements Thanks to colleagues of the Quebec NTBC Study Group, including Martyne Gosselin and Yolande Lefèvre, who have coordinated the project from its inception, to Manon Bouchard, dietician and to the many clinicians and researchers worldwide who share the dream of curing tyrosinemia.

References

Garcia Segarra N, Roche S, Imbard A, Benoist JF, Greneche MO, Davit-Spraul A, Ogier de Baulny H (2010) Maternal and fetal tyrosinemia type I. J Inherit Metab Dis 33(Suppl 3):S507–S510. doi:10.1007/s10545-012-9569-8

Hostetter MK, Levy HL, Winter HS, Knight GJ, Haddow JE (1983) Evidence for liver disease preceding amino acid abnormalities in hereditary tyrosinemia. N Engl J Med 308(21):1265–1267. doi:10.1056/NEJM198305263082105

Jakobs C, Kvittingen EA, Berger R, Haagen A, Kleijer W, Niermeijer M (1985) Prenatal diagnosis of tyrosinaemia type I by use of stable isotope dilution mass spectrometry. Eur J Pediatr 144(2):209–210

Jorquera R, Tanguay RM (1997) The mutagenicity of the tyrosine metabolite, fumarylacetoacetate, is enhanced by glutathione depletion. Biochem Biophys Res Commun 232(1):42–48. doi:10.1006/bbrc.1997.6220

Kassel R, Sprietsma L, Rudnick DA (2015) Pregnancy in an NTBC-treated patient with hereditary tyrosinemia type I. J Pediatr Gastroenterol Nutr 60(1):e5–e7. doi:10.1097/MPG.0b013e3182a27463

Kvittingen EA, Rootwelt H, Berger R, Brandtzaeg P (1994) Self-induced correction of the genetic defect in tyrosinemia type I. J Clin Invest 94(4):1657–1661. doi:10.1172/JCI117509

Langlois C, Jorquera R, Orejuela D, Bergeron A, Finegold MJ, Rhead WJ, Tanguay RM (2008) Rescue from neonatal death in the murine model of hereditary tyrosinemia by glutathione monoethylester and vitamin C treatment. Mol Genet Metab 93(3):306–313. doi:10.1016/j.ymgme.2007.09.018

Larochelle J, Alvarez F, Bussieres JF, Chevalier I, Dallaire L, Dubois J, Faucher F, Fenyves D, Goodyer P, Grenier A, Holme E, Laframboise R, Lambert M, Lindstedt S, Maranda B, Melancon S, Merouani A, Mitchell J, Parizeault G, Pelletier L, Phan V, Rinaldo P, Scott CR, Scriver C, Mitchell GA (2012) Effect of nitisinone (NTBC) treatment on the clinical course of hepatorenal tyrosinemia in Quebec. Mol Genet Metab 107(1–2):49–54. doi:10.1016/j.ymgme.2012.05.022

Lindblad B, Lindstedt S, Steen G (1977) On the enzymic defects in hereditary tyrosinemia. Proc Natl Acad Sci U S A 74(10):4641–4645

Lindblad B, Steen G (1982) Identification of 4,6-dioxoheptanoic acid (succinylacetone), 3,5-dioxooctanedioic acid (succinylacetoacetate) and 4-Oxo-6-hydroxyheptanoic acid in the urine from patients with hereditary tyrosinemia. Biomed Mass Spectrom 9(10):419–424. doi:10.1002/bms.1200091003

Manabe S, Sassa S, Kappas A (1985) Hereditary tyrosinemia. Formation of succinylacetone-amino acid adducts. J Exp Med 162(3):1060–1074

Mayorandan S, Meyer U, Gokcay G, Segarra NG, de Baulny HO, van Spronsen F, Zeman J, de Laet C, Spiekerkoetter U, Thimm E, Maiorana A, Dionisi-Vici C, Moeslinger D, Brunner-Krainz M, Lotz-Havla AS, Cocho de Juan JA, Couce Pico ML, Santer R, Scholl-Burgi S, Mandel H, Bliksrud YT, Freisinger P, Aldamiz-Echevarria LJ, Hochuli M, Gautschi M, Endig J, Jordan J, McKiernan P, Ernst S, Morlot S,

Vogel A, Sander J, Das AM (2014) Cross-sectional study of 168 patients with hepatorenal tyrosinaemia and implications for clinical practice. Orphanet J Rare Dis 9:107. doi:10.1186/s13023-014-0107-7
Mitchell G, Larochelle J, Lambert M, Michaud J, Grenier A, Ogier H, Gauthier M, Lacroix J, Vanasse M, Larbrisseau A et al (1990) Neurologic crises in hereditary tyrosinemia. N Engl J Med 322(7):432–437. doi:10.1056/NEJM199002153220704
Mitchell GA, Grompe M, Lambert M, Tanguay RM (2001) Hypertyrosinemia. The online metabolic & molecular bases of inherited disease (OMMBID), 8th edn. McGraw Hill, New York
Paradis K, Weber A, Seidman EG, Larochelle J, Garel L, Lenaerts C, Roy CC (1990) Liver transplantation for hereditary tyrosinemia: the Quebec experience. Am J Hum Genet 47(2):338–342
Sander JD, Joung JK (2014) CRISPR-Cas systems for editing, regulating and targeting genomes. Nat Biotechnol 32(4):347–355. doi:10.1038/nbt.2842
Scriver CR (2007) The PAH gene, phenylketonuria, and a paradigm shift. Hum Mutat 28(9):831–845. doi:10.1002/humu.20526
Scriver CR (2008) Garrod's croonian lectures (1908) and the charter 'Inborn Errors of Metabolism': albinism, alkaptonuria, cystinuria, and pentosuria at age 100 in 2008. J Inherit Metab Dis 31(5):580–598. doi:10.1007/s10545-008-0984-9
Stoner E, Starkman H, Wellner D, Wellner VP, Sassa S, Rifkind AB, Grenier A, Steinherz PG, Meister A, New MI et al (1984) Biochemical studies of a patient with hereditary hepatorenal tyrosinemia: evidence of glutathione deficiency. Pediatr Res 18(12):1332–1336
Thimm E, Richter-Werkle R, Kamp G, Molke B, Herebian D, Klee D, Mayatepek E, Spiekerkoetter U (2012) Neurocognitive outcome in patients with hypertyrosinemia type I after long-term treatment with NTBC. J Inherit Metab Dis 35(2):263–268. doi:10.1007/s10545-011-9394-5
Vanclooster A, Devlieger R, Meersseman W, Spraul A, Kerckhove KV, Vermeersch P, Meulemans A, Allegaert K, Cassiman D (2012) Pregnancy during nitisinone treatment for tyrosinaemia type I: first human experience. JIMD Rep 5:27–33. doi:10.1007/8904_2011_88
Whatley SD, Badminton MN (1993) In: Pagon RA, Adam MP, Ardinger HH et al (eds) Acute intermittent porphyria. GeneReviews, Seattle
Yin H, Song CQ, Dorkin JR, Zhu LJ, Li Y, Wu Q, Park A, Yang J, Suresh S, Bizhanova A, Gupta A, Bolukbasi MF, Walsh S, Bogorad RL, Gao G, Weng Z, Dong Y, Koteliansky V, Wolfe SA, Langer R, Xue W, Anderson DG (2016) Therapeutic genome editing by combined viral and non-viral delivery of CRISPR system components in vivo. Nat Biotechnol 34(3):328–333. doi:10.1038/nbt.3471
Yin H, Xue W, Chen S, Bogorad RL, Benedetti E, Grompe M, Koteliansky V, Sharp PA, Jacks T, Anderson DG (2014) Genome editing with Cas9 in adult mice corrects a disease mutation and phenotype. Nat Biotechnol 32(6):551–553. doi:10.1038/nbt.2884

Fah Knockout Animals as Models for Therapeutic Liver Repopulation

20

Markus Grompe

Abstract

Several animal models of Fah deficiency have been developed, including mice, pigs and most recently rats. Initially, the murine models were developed with the intent to mirror the human disease for pathophysiologic and therapeutic studies. However, it soon became apparent that Fah-positive hepatocytes have a potent selective growth advantage in mutant liver and can extensively repopulate the diseased organ. For this reason, Fah mutant mice have become a workhorse for liver biology and are widely used in liver stem cell and hepatic gene therapy research. Immune deficient Fah-knockout mice can be repopulated with human hepatocytes, creating "mice with human livers". These chimeric animals have become an important preclinical model for infectious diseases, metabolism and gene therapy. The potent expansion of human hepatocytes in Fah knockout mice has given rise to the concept of using Fah mutants as living bioreactors to produce large quantities of fully mature hepatocytes. As a consequence, larger animal models of Fah deficiency have recently been developed.

Keywords

Animal model • Gene therapy • Selective advantage • Hepatocyte • Cell transplantation • Chimeric animals • Liver repopulation • Hepatocellular carcinoma

Abbreviations

Fah	Fumarylacetoacetate hydrolase
HCC	Hepatocellular carcinoma
Ipsc	Induced pluripotent stem cells
CRM	Cross-reacting material
ENU	Ethylnitrosurea
FAA	Fumarylacetoacetate
rAAV	Recombinant adeno-associated virus

M. Grompe (✉)
Oregon Stem Cell Center, Department of Pediatrics, Oregon Health and Science University, Portland, OR 97239-3098, USA
e-mail: grompem@ohsu.edu

© Springer International Publishing AG 2017

R.M. Tanguay (ed.), *Hereditary Tyrosinemia*, Advances in Experimental Medicine and Biology 959,
DOI 10.1007/978-3-319-55780-9_20

20.1 Mouse Models of Fah Deficiency

20.1.1 FAH Deficiency and the Hepatocyte Specific Developmental Regulation Locus (*hsdr-1*)

The c^{14CoS} lethal albino mouse is one of a series of mutants bearing large overlapping X-ray induced deletions of chromosome 7 (Gluecksohn-Waelsch 1979; Russell et al. 1979), discovered in the 1970s. Mice homozygous for the c^{14CoS} deletion die within a few hours after birth and display a striking hepatic phenotype. The mRNA levels of many cAMP inducible hepatic enzymes are absent or reduced. These include tyrosine aminotransferase (*Tat*) (Schmid et al. 1985), glucose-6 phosphatase (*G-6P*) (Gluecksohn-Waelsch 1979), glutamine synthetase (*Gs*) (Gluecksohn-Waelsch 1979), phosphoenolpyruvate carboxykinase (*Pepck*) (Loose et al. 1986), aldolase B and albumin (Ruppert et al. 1990; Sala-Trepat et al. 1985). In contrast, other liver mRNAs inducible by DNA or oxidative damage such as *Chop* and *Nmo-1* are increased (Fornace et al. 1989; Petersen et al. 1989). Histological abnormalities are also found in both liver and kidney in these mice. Because of this down-regulation of multiple liver enzymes in the homozygous deletion, it was hypothesized that a liver transcription factor (hepatocyte specific developmental regulation locus = *hsdr-1*) is localized within the region (Niswander et al. 1991; Ruppert et al. 1990, 1992).

In 1992, however, two groups found that the mouse *Fah* gene was within the c^{14CoS} deletion and thus *Fah* became a candidate to represent the *hsdr-1* gene (Klebig et al. 1992; Ruppert et al. 1992). Subsequently, our work (Grompe et al. 1993) and that of the Schütz group (Kelsey et al. 1993) has clearly shown that FAH deficiency is responsible for all the major phenotypic effects of the c^{14CoS} deletion.

Hence, *Fah* and *hsdr-1* are actually identical and the profound gene expression changes found in the lethal albino mouse are not caused by a mutation in a transcription factor, but in a metabolic enzyme.

20.1.2 The *Fah*$^{\Delta exon5}$ Mouse

Before it was known that the lethal albino mouse was a model for Fah deficiency our lab decided to create a model of hereditary tyrosinemia by knocking out the gene in embryonic stem cells (Fig. 20.1). Exon 5 of the 14 exon gene was deleted using homologous recombination, creating a complete null mutation. Not surprisingly, mice homozygous for the deletion completely phenocopied the lethal albino mouse (Grompe et al. 1993). *Fah*$^{-/-}$ pups died as neonates with severe hepatic dysfunction. Luckily, NTBC was discovered around this time (Lindstedt et al. 1992; Lindstedt and Holme 1994) and it became possible to treat the disease even prenatally by adding it to the drinking water (Grompe et al. 1995). *Fah*$^{-/-}$ mutant mice treated with NTBC showed complete correction of the metabolic liver disease, including normalization of the abnormal gene expression (Grompe et al. 1995). Survival of *Fah*$^{-/-}$ mutants improved to 100 % and it even became possible to maintain breeding pairs of homozygous mutants. As in humans, NTBC-treated mice have very high blood tyrosine levels, but unlike in rats this does not cause keratitis of the eye even on a diet containing normal amounts of protein.

While the c14CoS mouse has Fah deficiency, its genomic deletion is one megabase in size and contains several other genes, including the tyrosinase locus which causes the albino (white coat color) phenotype. For this reason the cleaner *Fah*$^{\Delta exon5}$ mouse has become the standard animal model for human HT1 and has been used by many laboratories worldwide to study the pathophysiology and treatment of the disease. It was initially created on the 129Sv strain of mice, but also has been backbred to generate a congenic C57/BL6 strain.

20.1.3 Fah Mice with Point Mutations

ENU mutagenesis generated several point mutation alleles of Fah deficiency (Aponte et al. 2001). Our laboratory imported and

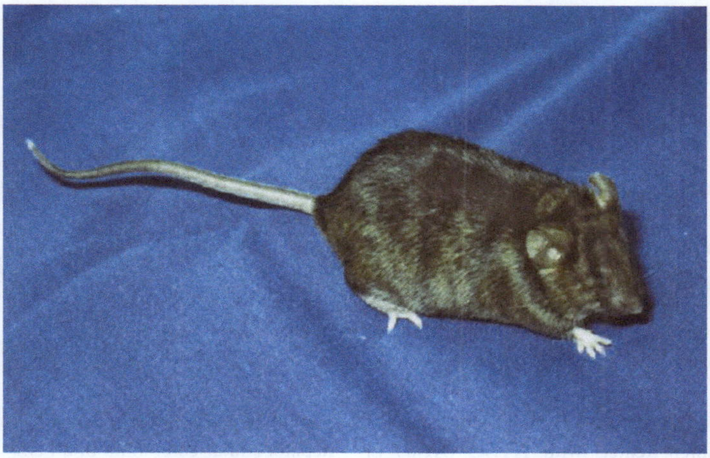

Fig. 20.1 The original $Fah^{\Delta exon5}$ founder. Image of the male germ line chimera that resulted from targeting exon 5 of the *Fah* gene in mouse embryonic stem cells. This animal gave rise to the thousands of Fah mutant mice used for liver research around the globe today

expanded one of these strains as a model for targeted gene correction in hepatocytes (Paulk et al. 2010, 2012). We chose a point mutation at the junction of exon 7 and intron 8, Fah^{5961SB} (Aponte et al. 2001). This single base change disrupts splicing and creates a protein null allele. Hence, the biochemical phenotype of these animals is the same as in other $Fah^{-/-}$ mutant animals.

20.2 Pathophysiologic Insights Obtained with Fah Mouse Models

Much of our current molecular understanding of HT1 disease pathophysiology has been obtained by studying the $Fah^{\Delta exon5}$ mouse. The typical experiment consists of taking NTBC-treated and fully healthy mutants and then withdrawing NTBC for 2 weeks or more. Complete NTBC withdrawal in combination with a normal protein content diet results in liver and renal disease, weight loss starting at about 2 weeks and typically results in death about 6 weeks later. It is possible to keep Fah mutants alive for extended periods without NTBC if the amount of phenylalanine and tyrosine in the diet is severely restricted (Grompe et al. 1998).

20.2.1 Cell Death Resistance

Acute FAA accumulation causes cell death by apoptosis (Jorquera and Tanguay 1999) and it was initially thought that apoptotic cell death is an important component of the in vivo liver disease. However, transaminase levels are curiously low in humans with HT1 and also not highly elevated in the $Fah^{-/-}$ mouse. Surprisingly, we were able to show that the hepatocytes of $Fah^{-/-}$ mice off NTBC are paradoxically apoptosis resistant (Vogel et al. 2004). Much higher doses of several liver toxins are needed to kill $Fah^{-/-}$ mice with liver disease than in wild-type controls (Vogel et al. 2006). Subsequent studies revealed that the oxidative damage inducible transcription factor Nrf2 is responsible for this adaptation, along with the cell cycle regulator p21 (Marhenke et al. 2008; Willenbring et al. 2008). FAA accumulation strongly induces both Nrf2 and p21, thereby producing both paradoxical apoptosis resistance and cell cycle arrest. A model emerged indicating that rapid accumulation of FAA can acutely cause hepatocyte death, but that a more gradual accumulation of the metabolite permits adaptive responses and the survival of dysfunctional cells. This phenomenon explains why synthetic liver functions are impaired in HT1 out of proportion to the relatively low transaminase levels.

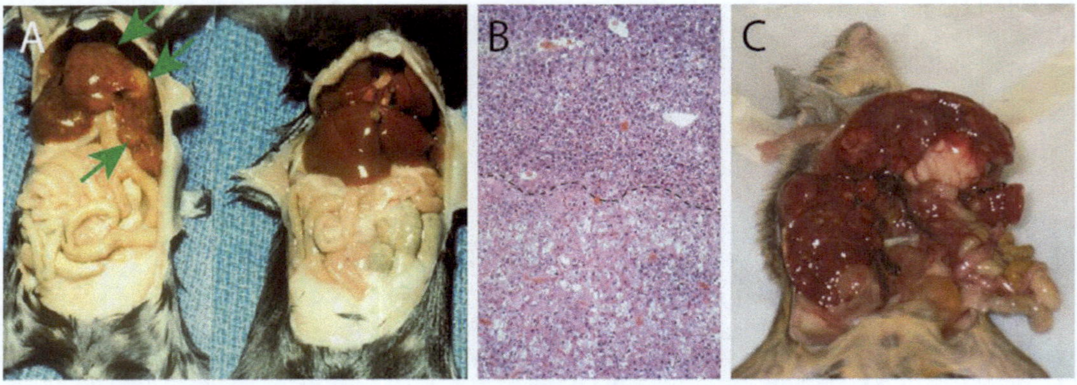

Fig. 20.2 Cancer in Fah$^{-/-}$ mice. (**a**) Macroscopic images of Fah$^{-/-}$ mice either treated with NTBC (*right*) or exposed to repeated cycles of NTBC withdrawals. The liver of the treated mouse looks completely normal, whereas the untreated liver is cholestatic (*orange color*) and full of tumor nodules (*green arrows*). (**b**) Histology of hepatocellular carcinoma in a Fah$^{-/-}$ mouse. Two tumors (*separated by dotted line*) with distinctive morphology are visible. (**c**) Massive, multifocal cancer in a Fah/p21 double mutant animal

20.2.2 Liver Cancer

Humans with Fah deficiency are very prone to developing HCC (Russo and O'Regan 1990). In the pre-NTBC era a large percentage of HT1 patients developed cancer before age 5 years. Fah mutant mice also develop liver cancer even when treated with NTBC and a low tyrosine diet (Overturf et al. 1999). p21 has emerged as key molecule in the evolution of HCC in Fah deficiency (Willenbring et al. 2008). Mice doubly mutant in both Fah and p21 developed multifocal liver cancer in only 6 weeks (Willenbring et al. 2008) (Fig. 20.2). These findings highlight the extreme mutagenicity of FAA (Jorquera and Tanguay 1997), a compound which can damage both DNA and DNA repair proteins.

20.3 Reversion and Liver Mosaicism in Fah Deficiency

In 1994, a very important clinical observation was made in human patients undergoing orthotopic liver transplantation for HT1: at the time of transplant, most children had multiple large and small hepatic nodules as is typical for cirrhosis. Surprisingly, some of these nodules were FAH+ (Kvittingen et al. 1994), had FAH enzyme activity and normal cell morphology. Skin fibroblasts obtained from the same patients were CRM- indicating that the original disease causing mutation lead to complete absence of the protein. Molecular studies then demonstrated that the FAH^{+} hepatic nodules arose by somatic reversion of an inherited *FAH* gene mutation generating clonal nodules of healthy hepatocytes (Kvittingen et al. 1993, 1994).

The co-existence of FAH deficient and FAH expressing tissue in the same liver illustrates an important pathophysiologic fact: the adverse effects of FAH deficiency are limited to those cells in which the hepatotoxic compounds are generated. MAA and FAA have such short half-lives that they do not leave the cells in which they are generated. FAH expressing hepatocytes are not adversely affected by being immediately adjacent to a deficient cell. This is also confirmed by the fact that non-hepatocyte liver cells (such as Kupffer cells) are healthy in FAH deficient mice (Kelsey et al. 1993). In other words: FAH acts cell autonomously.

It is attractive to speculate that the presence of both acute and chronic presentations of HT1 in the same sibship may be due to such somatic reversion events and partial correction of the biochemical defect. Several of the patients described by Tanguay were homozygous for the same French Canadian splice mutation (Demers et al. 2003) yet had variable biochemical phenotypes, with patients having partial enzyme activity and CRM+ material.

Somatic reversion followed by selection in hepatocytes is not without precedent. Transgenic mice which express urokinase tissue plasminogen activator show this phenomenon (Sandgren et al. 1991). Urokinase is toxic to hepatocytes and therefore cells, which inactivate the transgene have a selective advantage and repopulate the liver in a nodular fashion. Similarly, normal hepatocytes transplanted into the liver of transgenic mice expressing urokinase can replace the entire liver cell mass by selection (Rhim et al. 1994). Thus, the regenerative potential of both human and mouse hepatocytes is large enough to permit the generation by selection of macroscopic nodules consisting of millions of cells.

20.4 Therapeutic Liver Repopulation in the *Fah*$^{\Delta exon5}$ Mouse

The positive selection of FAH+ hepatocytes observed in human HT1 patients led us to test whether this phenomenon could be replicated in *Fah*$^{\Delta exon5}$ mice (Overturf et al. 1996). Wild-type hepatocytes were transplanted into FAH null mice and if these animals were maintained on NTBC after transplant only scattered small clusters of transplanted FAH+ hepatocytes were detected. However, if NTBC treatment was discontinued shortly after cell transplantation, liver injury occurred and the FAH+ cells proliferated extensively, forming large clusters within 3 weeks and replacing most of the liver mass within 6 weeks (Overturf et al. 1996). FAH null mice with repopulated livers remained healthy, had normal liver function tests and showed a relatively normal liver structure for many months. These studies provided proof of principle that liver repopulation can effectively cure a metabolic disease (Fig. 20.3).

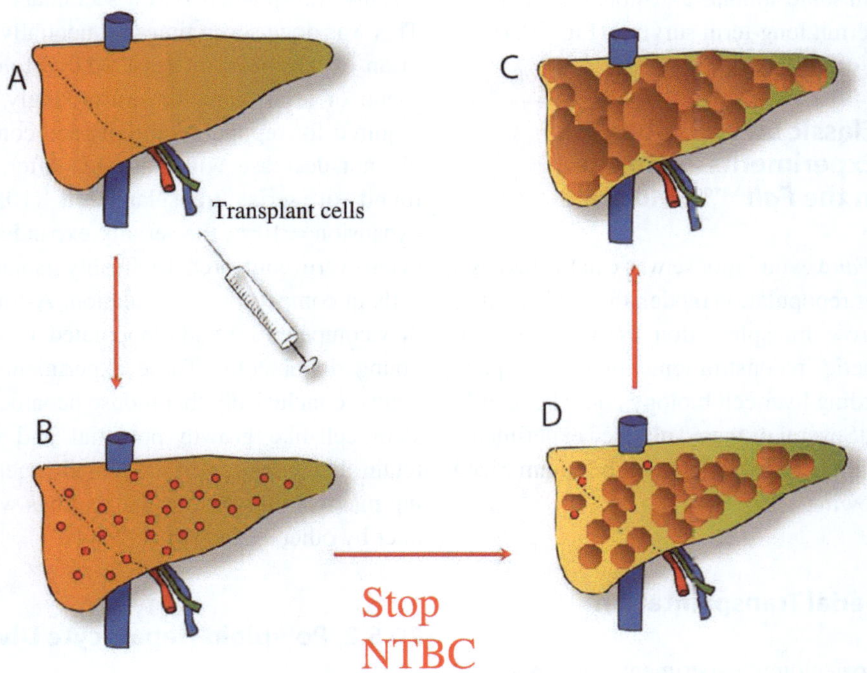

Fig. 20.3 Liver repopulation in the Fah knockout mouse. (**a**) Wild-type hepatocytes are injected into the portal vasculature of Fah$^{-/-}$ mice on NTBC; (**b**) individual wild-type hepatocytes (*red circles*) engraft at low density; (**c**) after NTBC withdrawal the FAH+ hepatocytes divide and form clonal nodules. (**d**) Eventually the FAH+ nodules become confluent and >95 % of the liver mass is replaced by donor hepatocytes

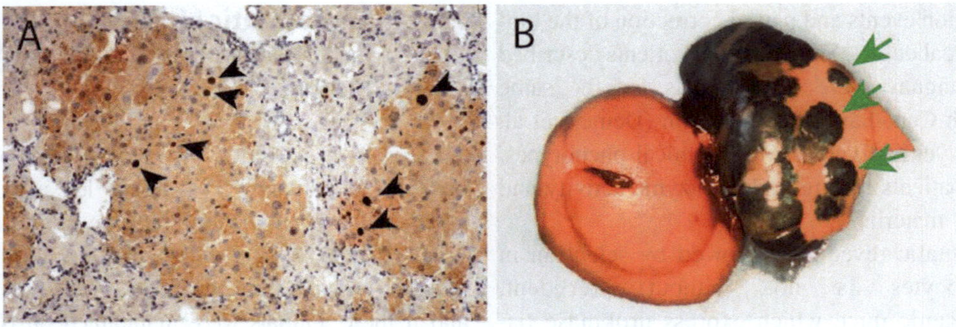

Fig. 20.4 Repopulation nodules. (**a**) Double-staining for FAH (*cytoplasmic brown*) and BrDU (*dark nuclei*) in a *Fah*$^{-/-}$ liver 4 weeks after transplantation with wild-type cells. The animal was pulsed with BrDU 2 h before harvest. Large FAH+ repopulation nodules with BrDU+ replicating hepatocytes (*arrows*) are visible. No BrDU incorporation wasseen in the FAH negative hepatocytes. (**b**) Macroscopic view of an *Fah*$^{-/-}$- liver transplanted with 200 Rosa26 transgenic wild-type hepatocytes. Clonal beta-galactosidase positive nodules of donor hepatocytes are clearly visible (*green arrows*)

After establishing the feasibility of therapeutic liver repopulation in this model, we explored quantitative aspects of this system. We found that transplantation of only 10,000 (1/5000 replacement ratio) hepatocytes reliably rescued adult Fah mutant mice. Transplantation of 1000 cells rescued 90 % of the animals and even 100 cells was sufficient in some animals to restore hepatic function and permit long-term survival (Fig. 20.4).

20.5 Classic Liver Biology Experiments Performed in the *Fah*$^{\Delta exon5}$ Mouse

Once the *Fah*Δexon5 mouse was established as a robust liver repopulation model, the equivalent of bone marrow transplantation as a read-out of hematopoietic reconstitution, important questions regarding liver cell biology could be readily addressed. Several of these "classic" experiments with high impact on the field will be summarized in the following.

20.5.1 Serial Transplantation

Partial hepatectomy experiments had revealed the extensive regenerative ability of mature hepatocytes (Stocker and Pfeifer 1965), but the upper limit of this potential was unknown. In order to explore this boundary, serial transplantation was performed in the *Fah*$^{-/-}$ mouse using limiting doses of donor hepatocytes (Overturf et al. 1997). Only 3000 FAH+ hepatocytes were transplanted into a mutant recipient (Fig. 20.5). After full repopulation and weight stabilization, hepatocytes from the repopulated animals were harvested and serially transplanted into a secondary recipient. This was done seven times sequentially resulting in an expansion of at least 10,000-fold in each round of transplantation. Surprisingly, the time required for repopulation remained constant and did not decrease with passage. After the sixth round of serial transplantation (10,000^{6}-fold expansion = 10^{24}), the serially expanded hepatocytes were compared to freshly isolated naïve cells in competitive repopulation. Astonishingly, they competed 1:1 and repopulated as well as the young donor cells. These experiments demonstrated conclusively that mouse hepatocytes have stem cell-like growth potential and that they retain their mature, differentiated functions during massive growth. Similar findings were made later by others (Wang et al. 2014).

20.5.2 Polyploid Hepatocyte Division

Liver hepatocytes are known to frequently be polyploid. In typical adult mice, the majority of hepatocytes are either tetraploid or octaploid

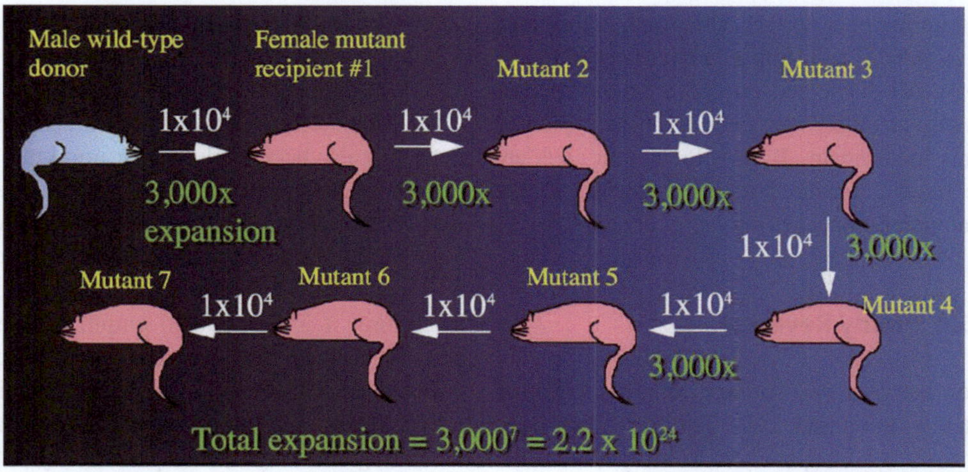

Fig. 20.5 Serial transplantation. Schematic of serial transplantation in the *Fah*$^{-/-}$ model system. 10,000 FAH+ hepatocytes from a wild-type male donor were serially transplanted into *Fah*$^{-/-}$ female mice. Complete liver repopulation was achieved in each round, representing an expansion of at least 3000-fold (an adult mouse liver contains 3×10^7 hepatocytes). After seven rounds of repopulation the initial donor cells had expanded a least 2.2×10^{24}-fold, requiring 80 cell divisions

(Guidotti et al. 2003). Polyploidization is known to occur by frequent failure of cytokinesis, resulting in binucleated cells. In the past, polyploid hepatocytes had been thought to be postmitotic (Faktor and Uryvaeva 1975), similar to other polyploid cell types such as skeletal myocytes. However, the Fah repopulation model made it possible for the first time to directly test the replicative ability of polyploid hepatocytes. Hepatocytes of different ploidy classes were purified and examined for their ability to repopulate Fah knockout mice (Overturf et al. 1999). Surprisingly, it was shown that polyploid hepatocytes were not postmitotic at all, but were able to divide extensively. Others confirmed this, using a different transplantation model (Weglarz et al. 2000).

20.5.3 Cell Fusion

In the late 1990s a series of publications suggested that hematopoietic stem cells could function as multipotential stem cells for many tissues, not only blood. Several reports indicated that hematopoietic stem cells could produce multiple endodermal epithelial lineages, including hepatocytes (Krause et al. 2001; Theise et al. 2000). To directly test this provocative hypothesis, bone marrow transplantation was carried out in Fah knockout mice. The donor cells came from a lacZ positive *Rosa26* mouse. Interestingly, animals with high percentages of lacZ positive hepatocytes were found and liver function was fully rescued. This experiment therefore confirmed that bone marrow could indeed generate hepatocytes (Lagasse et al. 2000). Bone marrow derived hepatocytes functioned like regular hepatocytes and completely rescued the metabolic liver disease of HT1. Initially, the mechanism underlying this finding was thought to be direct differentiation of hematopoietic stem cells into hepatocytes in the context of injury. An alternative mechanism, however, was considered after cell fusion was shown to underlie lineage transitions after co-culture of pluripotent stem cells with bone marrow (Terada et al. 2002; Ying et al. 2002). Targeted experiments revealed that cell fusion between bone marrow derived donor cells and *Fah*$^{-/-}$ hepatocytes was the actual mechanism by which bone marrow derived hepatocytes arise (Vassilopoulos et al. 2003; Wang et al. 2003). It was subsequently shown that myelomonocytic cells were the main hematopoietic lineage involved in the fusion phenomenon

(Willenbring et al. 2004). This work highlighted the existence and importance of cell fusion in vivo. All of the earlier reports on unexpected plasticity of blood stem cells could be explained by this mechanism. Today blood stem cells are once more considered to be precursors to the hematopoietic lineages only.

20.5.4 Ploidy Reversal

Fusion between two somatic cells by definition creates a polyploid cell. The fact that bone marrow-derived hepatocytes can repopulate the *Fah*$^{-/-}$ mouse confirmed the earlier observation demonstrating that polyploid hepatocytes are in fact not postmitotic. Although hybrid cells are by definition at least tetraploid when they are generated, karyotypic analysis of fusion hepatocytes showed that many of them had become diploid during liver repopulation (Wang et al. 2003). This observation was very unexpected because it suggested that ploidy reversal is possible in noncancerous somatic cells. Ploidy reduction was thought to occur only in meiosis. Studies to elucidate the mechanism of ploidy reversal in fusion hepatocytes were initiated. Single cell molecular analysis of diploid hepatocytes derived from polyploid hepatocyte precursors revealed random segregation of donor and host cell genomic loci, following Mendelian rules (Duncan et al. 2009). This detailed analysis confirmed the results obtained from cytogenetics and directly demonstrated that polyploid hepatocytes can become diploid in vivo. Liver cell imaging was then used to study the mitosis of polyploid hepatocytes and frequent multipolar mitoses were observed (Duncan et al. 2010). Multipolar mitoses were previously thought to occur only in cancer cells. We coined the term ploidy conveyor to describe the fact that hepatocytes can both polyploidize (by failed cytokinesis) and subsequently undergo reductive divisions (by multipolar mitosis). The consequence of this cycle is a very high incidence of aneuploidy in the normal liver of both mice and humans (Duncan et al. 2012).

20.6 Classic Gene Therapy Experiments

The strong selective advantage of FAH+ hepatocytes has also been exploited for important studies on liver-directed gene therapy. Once a Fah mutant hepatocyte has been transduced with a Fah-expressing transgene, the cell has a selective advantage and will form repopulation nodules. Therefore, it is possible to amplify these rare genetic correction events and generate clonal nodules derived from a transgene integration event. Many important studies have been carried out taking advantage of this powerful system.

20.6.1 Retroviral Vectors

The first gene therapy vector used in the *Fah*$^{-/-}$ mouse was an oncoretroviral vector. *Fah*$^{-/-}$ mice were infused with virus through the portal vein and then subjected to NTBC withdrawal. Gene therapy corrected FAH+ hepatocytes repopulated the liver and completely restored function (Overturf et al. 1996) (Fig. 20.6). Importantly, this experiment also revealed that gene therapy was not a good clinical option for the treatment of HT1 despite the selective advantage. Although >80% of the liver mass was corrected, the residual pockets of *Fah*$^{-/-}$ hepatocytes eventually gave rise to hepatocellular carcinoma. Hence, HT1 is a disorder that requires replacement of 100 % of the mutant hepatocytes with healthy cells. This cannot be achieved by either gene therapy or hepatocyte transplantation.

20.6.2 Transposons and Recombinases

Non-viral methods for gene therapy of the liver are currently inefficient, but hold significant promise for the future. The *Fah*$^{\Delta exon5}$ model was used to study several of these approaches, including transgene integration by sleeping beauty

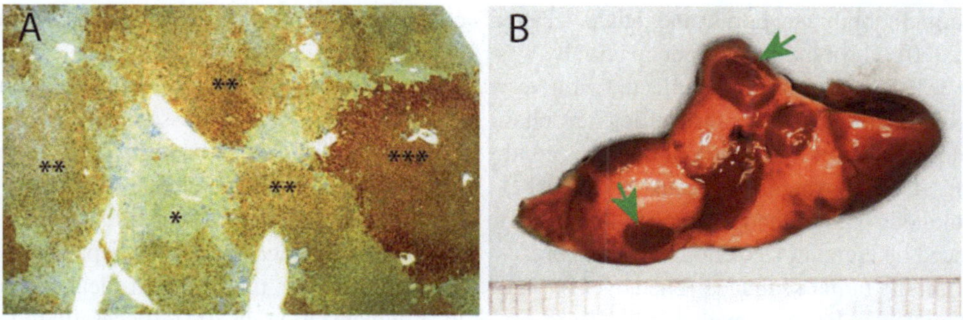

Fig. 20.6 Fah mutant liver corrected by gene therapy. A MoMLV retroviral vector containing a Fah minigene was infused into the portal vein of a $Fah^{-/-}$ mouse, followed by NTBC withdrawal for 8 weeks. (**a**) Gene corrected hepatocytes expanded clonally and repopulated the mutant liver. Different genomic integration sites generated nodules with different levels of FAH expression, ranging from high (***), to medium (**) to low (*). (**b**) The gene corrected FAH+ tissue can be distinguished from mutant, diseased tissue by its healthy red color (*green arrows*)

transposon (Montini et al. 2002) as well as phiC31 recombinase (Held et al. 2003). The Fah model permitted an accurate assessment of the integration efficiencies of these systems and also revealed morphologic changes in hepatocytes caused by the recombinase method.

20.6.3 AAV and Adenoviral Integrations

Several important "episomal" gene therapy vectors have the ability to integrate into the host chromosome with low frequencies. These rare events are difficult to study, because of the high background of episomal, non-integrated vectors. Recombinant adeno-associated vectors (rAAV) and adenoviral vectors both are considered to be non-integrating. The $Fah^{\Delta exon5}$ mouse was used to determine the frequency and the chromosomal location of integrations for both systems. To do this, virally transduced hepatocytes were serially transplanted into secondary $Fah^{-/-}$ recipients to remove vector episomes and permit repopulation with hepatocytes bearing integrated transgenes. The integration frequency of adenoviral vectors was found to be very low, less than 1/100,000 transduced hepatocytes (Stephen et al. 2010). However, the frequency for the clinically important rAAV vectors was much higher, measuring between 1 % and 0.1 % (Nakai et al. 2003, 2005). In addition, integration site analysis showed the rAAV vectors integrate preferentially into actively transcribed genes. It is now clear that rAAV -while mostly episomal- does integrate with sufficient frequency to raise concerns about insertional mutagenesis. These data explain why rAAV is able to cause liver cancer in animals (Donsante et al. 2007; Wang et al. 2012) and why wild-type AAV2 can cause hepatocellular carcinoma in humans (Nault et al. 2015, 2016).

20.6.4 Gene Repair

Gene therapy for inherited diseases has traditionally consisted of gene addition, i.e. the introduction of cDNA containing minigenes, driven by artificial promoters. This approach is confounded by the lack of physiologic transgene regulation, lack of permanence in the case of episomal vectors and insertional mutagenesis with integrating vectors. All of these drawbacks can be potentially overcome by gene repair, a method where the disease causing mutation is replaced by wild-type sequence using homologous recombination Homologous recombination is a rare event in resting cells in vivo and therefore the Fah point-mutation model (Aponte et al. 2001) was ideally suited for the first proof-of-principal experiments. rAAV of serotypes 2 and 8 were used to correct the disease causing mutation in both newborns and adults (Paulk et al. 2010). The maximal efficiency initially was about 1/200 in newborn mice injected with very high vector doses. To enhance efficiency pharmacological inhibition of non-homologous end-joining (NHEJ) was used (Paulk et al. 2012). Homologous recombination can be significantly enhanced by creating DNA double-strand breaks (DSC) in the target sequence. The advent of targeted nucleases has now made it possible to create site-specific DSB, even in vivo. The Fah^{5961SB} mouse became the very first example of in vivo gene repair enhanced by CRISPR/cas9 (Yin et al. 2014). Very high gene repair frequencies of >1 % were achieved even in adult animals. Efficiency was then further enhanced by delivering the gene correction donor in the form of an rAAV along with CRISPR mRNA in a nanoparticle (Yin et al. 2016). Gene correction levels of ~5 % were obtained, a level that would be sufficient to cure most metabolic diseases.

20.7 Humanized FRG

Animal models are used to study many aspects of human disease and to test therapeutic interventions but many very important features of human biology cannot be replicated in animals, even in non-human primates or transgenic rodents engineered with human genes. Most human microbial pathogens do not infect animals and the metabolism of many xenobiotics is different between humans and animals. Therefore, chimeric mice harboring human cells have long been used in research. The first examples are xenografts of human cancer cells in nude mice, permitting the study of human tumors in vivo (Rygaard and Povlsen 1969). Human primary cell types could not be engrafted in rodents until more severely immune deficient recipients became available. Considerable success was achieved in generating human blood cell chimerism in mice deficient in T- and B-cells as well as NK-cells, such as the NOD-Scid strain (Ramirez et al. 1998). Additional deletion of the common γ-chain of the interleukin 2 receptor (Igc) yielded mice completely unable to reject human cells (Traggiai et al. 2004).

The liver plays a central role in many human-specific biological processes and mice with "humanized livers" can be used to model human metabolism, liver injury, gene regulation, drug toxicity and hepatotropic infections. The strong selective advantage for FAH+ hepatocytes suggested the $Fah^{-/-}$ mouse as an ideal system for liver humanization. In 2007, successful repopulation of immune deficient Fah mice was reported (Azuma et al. 2007). This FRG mouse (deficient in Fah, Rag2 and the interleukin receptor common γ-chain) has since become the mainstay for experiments involving liver repopulation by human cells (Grompe and Strom 2013). FRG mice have been used to evaluate human liver stem/progenitor cells, gene therapy vectors and microbes attacking the liver. Examples of important studies done with immune deficient $Fah^{-/-}$ mice will be discussed in the following.

20.7.1 Modeling of Human Genetic Diseases

FRG mice can be repopulated not only by healthy hepatocytes, but also with cells from patients with genetic liver diseases (Bissig-Choisat et al. 2016; Gramignoli et al. 2013). If the disease is cell autonomous at the level of hepatocytes, this created the opportunity to explore therapeutic approaches such as gene therapy in a human context (Bissig-Choisat et al. 2016).

20.7.2 Hepatocytes from Stem Cells

Embryonic stem cells from the mouse were long known to be able to produce all of the somatic lineages of the adult organism after blastocyst injection or morula aggregation (Ramirez-Solis and Bradley 1994). Therefore the discovery of human embryonic stem cells in 1998 (Thomson et al. 1998) raised the hope that these extensively expandable cells would be a source of many transplantable cell types, including hepatocytes.

Several groups developed protocols to generate hepatocyte-like cells by first generating definitive endoderm and then subsequent maturation steps (Basma et al. 2009; Gouon-Evans et al. 2006; Si-Tayeb et al. 2010). The advent of induced pluripotent stem cells (Ipsc) (Takahashi et al. 2007) further fueled this line for research, permitting for the first time the production of hepatocyte-like cells from defined human genotypes, including genetic diseases. While the cells generated from pluripotent precursors appeared to have many features of mature hepatocytes, their utility for cell transplantation needed to be tested by in vivo experiments. The Fah knockout mouse has been used extensively for this purpose.

Definitive proof-of-principle for the hepatocytic potential of Ipsc was provided by Willenbring's group in 2010 (Espejel et al. 2010). They injected wild-type mouse Ipsc into *Fah*$^{-/-}$ mutant blastocysts. Importantly, all hepatocytes as well as proximal tubular renal cells of live-born chimeras originated from the wild-type FAH+ Ipsc (Espejel et al. 2010). This experiment conclusively demonstrated that pluripotent stem cells have the potential to produce fully mature and differentiated hepatocytes. Despite this inherent potential, hepatocyte-like cells generated from stem cells in vitro have not shown good transplantability. Compared to cadaveric hepatocytes from human donors, the engraftment efficiency has been very poor and to date no protocol has yielded full repopulation. Differentiation protocols will have to be significantly improved before hepatocyte-like cells derived from pluripotent precursors have the potential for cell therapy as primary human hepatocytes.

In addition to pluripotent stem cells, adult liver progenitors have also been studied as potential hepatocyte precursors. Lgr5+ liver organoids have been generated from both mouse and human liver tissue (Huch et al. 2013, 2015). "Stem cells" derived from EPCAM+ human ducts have been used as well (Schmelzer et al. 2007). While cells derived from these precursors express some markers hepatocytes upon in vitro differentiation, their transplantability is quite poor, yielding only minimal liver repopulation in the *Fah*$^{-/-}$ model (Dorrell et al. 2014).

20.7.3 Hepatocytes Generated by Direct Reprogramming of Somatic Cells

Direct reprogramming of somatic cells with hepatic transcription factors has been slightly more successful in producing transplantable hepatocyte-like cells. Several groups have reported the conversion of human fibroblasts into hepatocyte-like cells in vitro. Modest levels of liver repopulation in the FRG mouse were seen after transplantation of large numbers of these iHeps (induced hepatocyte like cells) (Du et al. 2014; Huang et al. 2014; Yu et al. 2013; Zhu et al. 2014).

20.7.4 Infectious Diseases

Infectious diseases are an important application of humanized FRG mice. Most human hepatic pathogens cannot infect laboratory animals, but readily infect liver chimeric mice (Bissig et al. 2010). Multiple reports have shown that both hepatitis B and C virus can infect humanized FRG mice (Bissig et al. 2010). In addition, the effects of antiviral drugs can be observed in this model (Bissig et al. 2010). Malaria is a disease that affects millions of humans and causes about one million deaths annually worldwide. The life cycles of both *Plasmodium falciparum* and *Plasmodium vivax* begin with the infection of hepatocytes by sporozoites after a mosquito bite. After massive replication in the hepatocyte, the cell lyses, releasing the parasite into the circulation and resulting in a red blood cell infection. Until recently, the hepatic phase of the malaria life cycle was not amenable to direct study and the entire life cycle (liver + blood) had not been modeled in an animal in vivo. However, highly humanized FRG mice can be infected with Plasmodium falciparum, complete their entire replication cycle in the human hepatocytes, lyse and then infect human erythrocytes injected into the infected animals (Vaughan et al. 2012, 2015). The ability to model the entire disease process now opens up the possibility of targeting the hepatic phase of the life cycle with novel therapeutics. Fully double-humanized mice harboring not only human erythrocytes and hepatocytes, but also human immune cells will enable even more relevant modeling of malaria in the future.

20.7.5 Gene Therapy

Although related to infectious disease research, the questions that can be answered in humanized FRG mice for gene therapy applications are different. The biology of viral and non-viral vector delivery systems can differ substantially between species. Capsids and envelopes that work well for the transduction of mouse hepatocytes do not necessarily work equally well on human cells. The DNA repair machinery and propensity for transgene integration is different as well. siRNA therapeutics are highly sequence specific and the optimization of gene knockdown therapies in the liver would benefit from having a humanized system. For all of these reasons, liver humanized mice represent an ideal platform to optimize and test human liver gene therapy vectors. For example, liver-humanized mice were used to evolve a novel rAAV capsid capable of efficiently transducing human hepatocytes (Lisowski et al. 2013). This novel LK03 capsid is completely inactive on mouse hepatocytes.

20.8 Other Animal Models of Fah Deficiency

20.8.1 Fah Knockout Pigs

Fah knockout mice have been instrumental in many areas of liver biology and therapeutic research. However, there are limitations to small animal models and pigs are the large animal model of choice for many applications (Suzuki et al. 2011). Surgical manipulations are much easier in this species and the anatomy is much more akin to humans than rodents. In addition, the size of pigs makes them a potentially attractive choice for expanding human cells or even entire organs. For this reason, our group decided to create a porcine model of Fah deficiency. A rAAV vector was used to target exon 5 of the porcine FAH locus in fetal fibroblasts (Hickey et al. 2011). Successfully targeted fibroblasts were then used for somatic cell nuclear transfer (cloning) and a herd of heterozygous FAH mutant pigs was generated. These heterozygotes were subsequently bred to homozygosity and faithfully recapitulate the human disease phenotype (Hickey et al. 2014). Mutant animals require NTBC therapy to survive and untreated animals develop severe liver disease. Recently, the $FAH^{-/-}$ pig was used to show proof-of-principle for liver repopulation with FAH+ hepatocytes. Immune deficient FAH pigs may be useful for the in vivo expansion of human hepatocytes. They also will be useful as a large animal model for hepatocellular carcinoma and end-stage liver failure.

20.8.2 Fah Knockout Rats

Very recently Fah knockout rats have also been generated using targeted nucleases (Kuijk et al. 2016). The Fah deficiency allele has also been combined with immune deficiency mutations (Kuijk et al. 2016). Rats are about 10× larger than mice and are an important animal model in preclinical pharmaceutical research, especially in toxicology (Mashimo and Serikawa 2009). It seems likely that immune deficient rats will be amenable to humanization, as in mice and that liver chimeric rats will be an important resource in the future.

20.9 Conclusions

Fah deficiency creates a very potent selective environment for FAH+ hepatocytes in the liver of multiple mammalian species including humans, mice, pigs and rats. Because Fah deficiency acts cell-autonomously the selection is very tight and permits the expansion of even single events leading to FAH expression in hepatocytes. The system therefore is ideal as a read-out of stem cell activity, correction by any method of gene therapy and for cell fusion. Immune deficient Fah mutant animals can be extensively reconstituted with human hepatocytes and hence the model can be used to study many human-specific liver functionalities, including infection by liver pathogens, drug transport and metabolism and modeling of human genetic diseases. It appears likely that Fah mutant animals will remain an important tool for the study of liver biology for many years to come.

References

Aponte JL, Sega GA, Hauser LJ, Dhar MS, Withrow CM, Carpenter DA, Rinchik EM, Culiat CT, Johnson DK (2001) Point mutations in the murine fumarylacetoacetate hydrolase gene: animal models for the human genetic disorder hereditary tyrosinemia type 1. Proc Natl Acad Sci U S A 98:641–645

Azuma H, Paulk N, Ranade A, Dorrell C, Al-Dhalimy M, Ellis E, Strom S, Kay MA, Finegold M, Grompe M (2007) Robust expansion of human hepatocytes in Fah-/-/Rag2-/-/Il2rg-/- mice. Nat Biotechnol 25:903–910

Basma H, Soto-Gutierrez A, Yannam GR, Liu L, Ito R, Yamamoto T, Ellis E, Carson SD, Sato S, Chen Y et al (2009) Differentiation and transplantation of human embryonic stem cell-derived hepatocytes. Gastroenterology 136:990–999

Bissig KD, Wieland SF, Tran P, Isogawa M, Le TT, Chisari FV, Verma IM (2010) Human liver chimeric mice provide a model for hepatitis B and C virus infection and treatment. J Clin Inv 120:924–930

Bissig-Choisat B, Kettlun-Leyton C, Legras XD, Zorman B, Barzi M, Chen LL, Amin MD, Huang YH, Pautler RG, Hampton OA et al (2016) Novel patient-derived xenograft and cell line models for therapeutic testing of pediatric liver cancer. J Hepatol 65:325–333

Demers SI, Russo P, Lettre F, Tanguay RM (2003) Frequent mutation reversion inversely correlates with clinical severity in a genetic liver disease, hereditary tyrosinemia. Hum Pathol 34:1313–1320

Donsante A, Miller DG, Li Y, Vogler C, Brunt EM, Russell DW, Sands MS (2007) AAV vector integration sites in mouse hepatocellular carcinoma. Science 317:477

Dorrell C, Tarlow B, Wang Y, Canaday PS, Haft A, Schug J, Streeter PR, Finegold MJ, Shenje LT, Kaestner KH et al (2014) The organoid-initiating cells in mouse pancreas and liver are phenotypically and functionally similar. Stem Cell Res 13:275–283

Du Y, Wang J, Jia J, Song N, Xiang C, Xu J, Hou Z, Su X, Liu B, Jiang T et al (2014) Human hepatocytes with drug metabolic function induced from fibroblasts by lineage reprogramming. Cell Stem Cell 14:394–403

Duncan AW, Hickey RD, Paulk NK, Culberson AJ, Olson SB, Finegold MJ, Grompe M (2009) Ploidy reductions in murine fusion-derived hepatocytes. PLoS Genet 5:e1000385

Duncan AW, Taylor MH, Hickey RD, Hanlon Newell AE, Lenzi ML, Olson SB, Finegold MJ, Grompe M (2010) The ploidy conveyor of mature hepatocytes as a source of genetic variation. Nature 467:707–710

Duncan AW, Hanlon Newell AE, Smith L, Wilson EM, Olson SB, Thayer MJ, Strom SC, Grompe M (2012) Frequent aneuploidy among normal human hepatocytes. Gastroenterology 142:25–28

Espejel S, Roll GR, McLaughlin KJ, Lee AY, Zhang JY, Laird DJ, Okita K, Yamanaka S, Willenbring H (2010) Induced pluripotent stem cell-derived hepatocytes have the functional and proliferative capabilities needed for liver regeneration in mice. J Clin Investig 120:3120–3126

Faktor VM, Uryvaeva IV (1975) Progressive polyploidy in mouse liver following repeated hepatectomy. Tsitologiia 17:909–916

Fornace AJ Jr, Nebert DW, Hollander MC, Luethy JD, Papathanasiou M, Fargnoli J, Holbrook NJ (1989) Mammalian genes coordinately regulated by growth arrest signals and DNA-damaging agents. Mol Cell Biol 9:4196–4203

Gluecksohn-Waelsch S (1979) Genetic control of morphogenetic and biochemical differentiation: lethal albino deletions in the mouse. Cell 16:225–237

Gouon-Evans V, Boussemart L, Gadue P, Nierhoff D, Koehler CI, Kubo A, Shafritz DA, Keller G (2006) BMP-4 is required for hepatic specification of mouse embryonic stem cell-derived definitive endoderm. Nat Biotechnol 24:1402–1411

Gramignoli R, Tahan V, Dorko K, Skvorak KJ, Hansel MC, Zhao W, Venkataramanan R, Ellis EC, Jorns C, Ericzon BG et al (2013) New potential cell source for hepatocyte transplantation: discarded livers from metabolic disease liver transplants. Stem Cell Res 11:563–573

Grompe M, Strom S (2013) Mice with human livers. Gastroenterology 145:1209–1214

Grompe M, Al-Dhalimy M, Finegold M, Ou CN, Burlingame T, Kennaway NG, Soriano P (1993) Loss of fumarylacetoacetate hydrolase is responsible for the neonatal hepatic dysfunction phenotype of lethal albino mice. Genes Dev 7:2298–2307

Grompe M, Lindstedt S, Al-Dhalimy M, Kennaway NG, Papaconstantinou J, Torres-Ramos CA, Ou CN, Finegold M (1995) Pharmacological correction of neonatal lethal hepatic dysfunction in a murine model of hereditary tyrosinaemia type I. Nat Genet 10:453–460

Grompe M, Overturf K, Al-Dhalimy M, Finegold M (1998) Therapeutic trials in the murine model of hereditary tyrosinaemia type I: a progress report. J Inherit Metab Dis 21:518–531

Guidotti JE, Bregerie O, Robert A, Debey P, Brechot C, Desdouets C (2003) Liver cell polyploidization: a pivotal role for binuclear hepatocytes. J Biol Chem 278:19095–19101

Held P, Olivares E, Montini E, Finegold M, Calos M, Grompe M (2003) Characterization of the Phage phiC31 integrase system for in vivo hepatic gene delivery. Mol Ther J Am Soc Gene Ther 7:S184

Hickey RD, Lillegard JB, Fisher JE, McKenzie TJ, Hofherr SE, Finegold MJ, Nyberg SL, Grompe M (2011) Efficient production of Fah-null heterozygote pigs by chimeric adeno-associated virus-mediated gene knockout and somatic cell nuclear transfer. Hepatology 54:1351–1359

Hickey RD, Mao SA, Glorioso J, Lillegard JB, Fisher JE, Amiot B, Rinaldo P, Harding CO, Marler R, Finegold MJ et al (2014) Fumarylacetoacetate hydrolase deficient pigs are a novel large animal model of metabolic liver disease. Stem Cell Res 13:144–153

Huang P, Zhang L, Gao Y, He Z, Yao D, Wu Z, Cen J, Chen X, Liu C, Hu Y et al (2014) Direct reprogramming of human fibroblasts to functional and expandable hepatocytes. Cell Stem Cell 14:370–384

Huch M, Dorrell C, Boj SF, van Es JH, Li VS, van de Wetering M, Sato T, Hamer K, Sasaki N, Finegold MJ et al (2013) In vitro expansion of single Lgr5+ liver stem cells induced by Wnt-driven regeneration. Nature 494:247–250

Huch M, Gehart H, van Boxtel R, Hamer K, Blokzijl F, Verstegen MM, Ellis E, van Wenum M, Fuchs SA, de Ligt J et al (2015) Long-term culture of genome-stable bipotent stem cells from adult human liver. Cell 160:299–312

Jorquera R, Tanguay RM (1997) The mutagenicity of the tyrosine metabolite, fumarylacetoacetate, is enhanced by glutathione depletion. Biochem Biophys Res Commun 232:42–48

Jorquera R, Tanguay RM (1999) Cyclin B-dependent kinase and caspase-1 activation precedes mitochondrial dysfunction in fumarylacetoacetate-induced apoptosis. FASEB J Faseb J 13:2284–2298

Kelsey G, Ruppert S, Beermann F, Grund C, Tanguay RM, Schütz G (1993) Rescue of mice homozygous for lethal albino deletions: implications for an animal model for the human liver disease tyrosinemia type I. Genes Dev 7:2285–2297

Klebig ML, Russell LB, Rinchik EM (1992) Murine fumarylacetoacetate hydrolase (Fah) gene is disrupted by a neonatally lethal albino deletion that defines the hepatocyte-specific developmental regulation 1 (hsdr-1) locus. Proc Natl Acad Sci U S A 89:1363–1367

Krause DS, Theise ND, Collector MI, Henegariu O, Hwang S, Gardner R, Neutzel S, Sharkis SJ (2001) Multi-organ, multi-lineage engraftment by a single bone marrow-derived stem cell. Cell 105:369–377

Kuijk EW, Rasmussen S, Blokzijl F, Huch M, Gehart H, Toonen P, Begthel H, Clevers H, Geurts AM, Cuppen E (2016) Generation and characterization of rat liver stem cell lines and their engraftment in a rat model of liver failure. Sci Rep 6:22154

Kvittingen EA, Rootwelt H, Brandtzaeg P, Bergan A, Berger R (1993) Hereditary tyrosinemia type I. Self-induced correction of the fumarylacetoacetase defect. J Clin Investig 91:1816–1821

Kvittingen EA, Rootwelt H, Berger R, Brandtzaeg P (1994) Self-induced correction of the genetic defect in tyrosinemia type I. J Clin Investig 94:1657–1661

Lagasse E, Connors H, Al-Dhalimy M, Reitsma M, Dohse M, Osborne L, Wang X, Finegold M, Weissman IL, Grompe M (2000) Purified hematopoietic stem cells can differentiate into hepatocytes in vivo. Nat Med 6:1229–1234

Lindstedt S, Holme H (1994) Interim report on the use of NTBC for treatment of hereditary tyrosinemia type 1., submitted

Lindstedt S, Holme E, Lock EA, Hjalmarson O, Strandvik B (1992) Treatment of hereditary tyrosinaemia type I by inhibition of 4-hydroxyphenylpyruvate dioxygenase. Lancet 340:813–817

Lisowski L, Chu K, Kay MA (2013) A species restricted AAV obtained from viral capsid shuffling. Mol Ther 21:S3

Loose DS, Shaw PA, Krauter KS, Robinson C, Englard S, Hanson RW, Gluecksohn-Waelsch S (1986) Trans regulation of the phosphoenolpyruvate carboxykinase (GTP) gene, identified by deletions in chromosome 7 of the mouse. Proc Natl Acad Sci U S A 83:5184–5188

Marhenke S, Lamle J, Buitrago-Molina LE, Canon JM, Geffers R, Finegold M, Sporn M, Yamamoto M, Manns MP, Grompe M et al (2008) Activation of nuclear factor E2-related factor 2 in hereditary tyrosinemia type 1 and its role in survival and tumor development. Hepatology 48:487–496

Mashimo T, Serikawa T (2009) Rat resources in biomedical research. Curr Pharm Biotechnol 10:214–220

Montini E, Held PK, Noll M, Morcinek N, Al-Dhalimy M, Finegold M, Yant SR, Kay MA, Grompe M (2002) In vivo correction of murine tyrosinemia type I by DNA-mediated transposition. Mol Ther J Am Soc Gene Ther 6:759–769

Nakai H, Montini E, Fuess S, Storm TA, Grompe M, Kay MA (2003) AAV serotype 2 vectors preferentially integrate into active genes in mice. Nat Genet 34:297–302

Nakai H, Wu X, Fuess S, Storm TA, Munroe D, Montini E, Burgess SM, Grompe M, Kay MA (2005) Large-scale molecular characterization of adeno-associated virus vector integration in mouse liver. J Virol 79:3606–3614

Nault JC, Datta S, Imbeaud S, Franconi A, Mallet M, Couchy G, Letouze E, Pilati C, Verret B, Blanc JF et al (2015) Recurrent AAV2-related insertional mutagenesis in human hepatocellular carcinomas. Nat Genet 47:1187–1193

Nault JC, Datta S, Imbeaud S, Franconi A, Mallet M, Couchy G, Letouze E, Pilati C, Verret B, Blanc JF et al

(2016) AAV2 and Hepatocellular Carcinoma. Hum Gene Ther 27:211–213

Niswander L, Kelsey G, Schedl A, Ruppert S, Sharan SK, Holdener KB, Rinchik EM, Edstrom JE, Magnuson T (1991) Molecular mapping of albino deletions associated with early embryonic lethality in the mouse. Genomics 9:162–169

Overturf K, Al-Dhalimy M, Tanguay R, Brantly M, Ou CN, Finegold M, Grompe M (1996) Hepatocytes corrected by gene therapy are selected in vivo in a murine model of hereditary tyrosinaemia type I. Nat Genet 12:266–273

Overturf K, al-Dhalimy M, Ou CN, Finegold M, Grompe M (1997) Serial transplantation reveals the stem-cell-like regenerative potential of adult mouse hepatocytes. Am J Pathol 151:1273–1280

Overturf K, Al-Dhalimy M, Finegold M, Grompe M (1999) The repopulation potential of hepatocyte populations differing in size and prior mitotic expansion. Am J Pathol 155:2135–2143

Paulk NK, Wursthorn K, Wang Z, Finegold MJ, Kay MA, Grompe M (2010) Adeno-associated virus gene repair corrects a mouse model of hereditary tyrosinemia in vivo. Hepatology 51:1200–1208

Paulk NK, Loza LM, Finegold MJ, Grompe M (2012) AAV-mediated gene targeting is significantly enhanced by transient inhibition of nonhomologous end joining or the proteasome in vivo. Hum Gene Ther 23:658–665

Petersen DD, Gonzalez FJ, Rapic V, Kozak CA, Lee JY, Jones JE, Nebert DW (1989) Marked increases in hepatic NAD(P)H:oxidoreductase gene transcription and mRNA levels correlated with a mouse chromosome 7 deletion. Proc Natl Acad Sci U S A 86:6699–6703

Ramirez M, Rottman GA, Shultz LD, Civin CI (1998) Mature human hematopoietic cells in donor bone marrow complicate interpretation of stem/progenitor cell assays in xenogeneic hematopoietic chimeras. Exp Hematol 26:332–344

Ramirez-Solis R, Bradley A (1994) Advances in the use of embryonic stem cell technology. Curr Opin Biotechnol 5:528–533

Rhim JA, Sandgren EP, Degen JL, Palmiter RD, Brinster RL (1994) Replacement of disease mouse liver by hepatic cell transplantation. Science 263:1149–1152

Ruppert S, Boshart M, Bosch FX, Schmid W, Fournier REK, Schütz G (1990) Two genetically defined trans-acting loci coordinately regulate overlapping sets of liver-specific genes. Cell 61:895–904

Ruppert S, Kelsey G, Schedl A, Schmid E, Thies E, Schütz G (1992) Deficiency of an enzyme of tyrosine metabolism underlies altered gene expression in newborn liver of lethal albino mice. Genes Dev 6:1430–1443

Russell LB, Russell WL, Kelly EM (1979) Analysis of the albino-locus region of the mouse. Origin and viability. Genetics 91:127–139

Russo P, O'Regan S (1990) Visceral pathology of hereditary tyrosinemia type I. Am J Hum Genet 47:317–324

Rygaard J, Povlsen CO (1969) Heterotransplantation of a human malignant tumour to "Nude" mice. Acta Pathol Microbiol Scand 77:758–760

Sala-Trepat JM, Poiret M, Sellem CH, Bessada R, Erdos T, Gluecksohn-Waelsch S (1985) A lethal deletion on mouse chromosome 7 affects regulation of liver cell-specific functions: posttranscriptional control of serum protein and transcriptional control of aldolase B synthesis. Proc Natl Acad Sci U S A 82:2442–2446

Sandgren EP, Palmiter RD, Heckel JL, Daugherty CC, Brinster RL, Degen JL (1991) Complete hepatic regeneration after somatic deletion of an albumin-plasminogen activator transgene. Cell 66:245–256

Schmelzer E, Zhang L, Bruce A, Wauthier E, Ludlow J, Yao HL, Moss N, Melhem A, McClelland R, Turner W et al (2007) Human hepatic stem cells from fetal and postnatal donors. J Exp Med 204:1973–1987

Schmid W, Müller G, Schütz G, Gluecksohn-Waelsch S (1985) Deletions near the albino locus on chromosome 7 of the mouse affect the level of tyrosine aminotransferase mRNA. Proc Natl Acad Sci U S A 82:2866–2869

Si-Tayeb K, Noto FK, Nagaoka M, Li J, Battle MA, Duris C, North PE, Dalton S, Duncan SA (2010) Highly efficient generation of human hepatocyte-like cells from induced pluripotent stem cells. Hepatology 51:297–305

Stephen SL, Montini E, Sivanandam VG, Al-Dhalimy M, Kestler HA, Finegold M, Grompe M, Kochanek S (2010) Chromosomal integration of adenoviral vector DNA in vivo. J Virol 84:9987–9994

Stocker E, Pfeifer U (1965) On the manner of proliferation of the liver parenchyma after partial hepatectomy. Autoradiography studies using 3H-thymidine. Naturwissenschaften 52:663

Suzuki Y, Yeung AC, Ikeno F (2011) The representative porcine model for human cardiovascular disease. J Biomed Biotechnol 2011:195483

Takahashi K, Tanabe K, Ohnuki M, Narita M, Ichisaka T, Tomoda K, Yamanaka S (2007) Induction of pluripotent stem cells from adult human fibroblasts by defined factors. Cell 131:861–872

Terada N, Hamazaki T, Oka M, Hoki M, Mastalerz DM, Nakano Y, Meyer EM, Morel L, Petersen BE, Scott EW (2002) Bone marrow cells adopt the phenotype of other cells by spontaneous cell fusion. Nature 416:542–545

Theise ND, Badve S, Saxena R, Henegariu O, Sell S, Crawford JM, Krause DS (2000) Derivation of hepatocytes from bone marrow cells in mice after radiation-induced myeloablation. Hepatology 31:235–240

Thomson JA, Itskovitz-Eldor J, Shapiro SS, Waknitz MA, Swiergiel JJ, Marshall VS, Jones JM (1998) Embryonic stem cell lines derived from human blastocysts. Science 282:1145–1147

Traggiai E, Chicha L, Mazzucchelli L, Bronz L, Piffaretti JC, Lanzavecchia A, Manz MG (2004) Development of a human adaptive immune system in cord blood cell-transplanted mice. Science 304:104–107

Vassilopoulos G, Wang PR, Russell DW (2003) Transplanted bone marrow regenerates liver by cell fusion. Nature 422:901–904

Vaughan AM, Mikolajczak SA, Wilson EM, Grompe M, Kaushansky A, Camargo N, Bial J, Ploss A, Kappe SH (2012) Complete Plasmodium falciparum liver-stage development in liver-chimeric mice. J Clin Investig 122:3618–3628

Vaughan AM, Pinapati RS, Cheeseman IH, Camargo N, Fishbaugher M, Checkley LA, Nair S, Hutyra CA, Nosten FH, Anderson TJ et al (2015) Plasmodium falciparum genetic crosses in a humanized mouse model. Nat Methods 12:631–633

Vogel A, van Den Berg IE, Al-Dhalimy M, Groopman J, Ou CN, Ryabinina O, Iordanov MS, Finegold M, Grompe M (2004) Chronic liver disease in murine hereditary tyrosinemia type 1 induces resistance to cell death. Hepatology 39:433–443

Vogel A, Aslan JE, Willenbring H, Klein C, Finegold M, Mount H, Thomas G, Grompe M (2006) Sustained phosphorylation of Bid is a marker for resistance to Fas-induced apoptosis during chronic liver diseases. Gastroenterology 130:104–119

Wang X, Willenbring H, Akkari Y, Torimaru Y, Foster M, Al-Dhalimy M, Lagasse E, Finegold M, Olson S, Grompe M (2003) Cell fusion is the principal source of bone-marrow-derived hepatocytes. Nature 422:897–901

Wang PR, Xu M, Toffanin S, Li Y, Llovet JM, Russell DW (2012) Induction of hepatocellular carcinoma by in vivo gene targeting. Proc Natl Acad Sci U S A 109:11264–11269

Wang MJ, Chen F, Li JX, Liu CC, Zhang HB, Xia Y, Yu B, You P, Xiang D, Lu L et al (2014) Reversal of hepatocyte senescence after continuous in vivo cell proliferation. Hepatology 60:349–361

Weglarz TC, Degen JL, Sandgren EP (2000) Hepatocyte transplantation into diseased mouse liver. Kinetics of parenchymal repopulation and identification of the proliferative capacity of tetraploid and octaploid hepatocytes. Am J Pathol 157:1963–1974

Willenbring H, Bailey AS, Foster M, Akkari Y, Dorrell C, Olson S, Finegold M, Fleming WH, Grompe M (2004) Myelomonocytic cells are sufficient for therapeutic cell fusion in liver. Nat Med 10:744–748

Willenbring H, Sharma AD, Vogel A, Lee AY, Rothfuss A, Wang Z, Finegold M, Grompe M (2008) Loss of p21 permits carcinogenesis from chronically damaged liver and kidney epithelial cells despite unchecked apoptosis. Cancer Cell 14:59–67

Yin H, Xue W, Chen S, Bogorad RL, Benedetti E, Grompe M, Koteliansky V, Sharp PA, Jacks T, Anderson DG (2014) Genome editing with Cas9 in adult mice corrects a disease mutation and phenotype. Nat Biotechnol 32:551–553

Yin H, Song CQ, Dorkin JR, Zhu LJ, Li Y, Wu Q, Park A, Yang J, Suresh S, Bizhanova A et al (2016) Therapeutic genome editing by combined viral and non-viral delivery of CRISPR system components in vivo. Nat Biotechnol 34:328–333

Ying QL, Nichols J, Evans EP, Smith AG (2002) Changing potency by spontaneous fusion. Nature 416:545–548

Yu B, He ZY, You P, Han QW, Xiang D, Chen F, Wang MJ, Liu CC, Lin XW, Borjigin U et al (2013) Reprogramming fibroblasts into bipotential hepatic stem cells by defined factors. Cell Stem Cell 13:328–340

Zhu S, Rezvani M, Harbell J, Mattis AN, Wolfe AR, Benet LZ, Willenbring H, Ding S (2014) Mouse liver repopulation with hepatocytes generated from human fibroblasts. Nature 508:93–97

Gene Therapy in Tyrosinemia: Potential and Pitfalls

21

Sophie Carter and Yannick Doyon

Abstract

In this chapter, we intend to review gene therapy concepts applied to the potential treatment of tyrosinemia for parents and pediatricians. Therefore, our main objective is to give general informations in a comprehensible manner. Considering the nature of tyrosinemia and the current state of technology, a particular focus will be put on strategies using viral delivery of DNA to the liver. In light of the recent development of the CRISPR technology and the revival of promises for previously unavailable therapeutical tools, the present chapter aims at presenting up to date facts and potential pitfalls towards an application for metabolic diseases, in particular tyrosinemia.

Keywords

Tyrosinemia • Liver disease • Gene therapy • Genome editing • Adeno-associated virus

Abbreviations

AAV	Adeno-associated virus
FIX	Factor IX
FAH	Fumarylacetoacetate hydrolase
HDR	Homology directed repair
HT1	Type 1 hypertyrosinemia
iPS	Induced pluripotent stem cells
LCA	Leber congenital amaurosis
NHEJ	Non-homologous end-joining
NTBC	[2-(2-nitro-4-trifluoromethylbenzoyl)-1–3-cyclohexanedione]
RGENS	RNA-guided endonucleases
TALENs	Transcription activator-like effector nuclease
ZFNs	Zinc finger nucleases

S. Carter • Y. Doyon (✉)
Centre Hospitalier Universitaire de Québec Research Center and Faculty of Medicine, Centre de recherche du CHU de Québec – Université Laval 2705, boulevard Laurier, T-3-67, Québec, QC G1V 4G2, Canada
e-mail: Sophie.Carter@crchudequebec.ulaval.ca; Yannick.Doyon@crchudequebec.ulaval.ca

© Springer International Publishing AG 2017

R.M. Tanguay (ed.), *Hereditary Tyrosinemia*, Advances in Experimental Medicine and Biology 959,
DOI 10.1007/978-3-319-55780-9_21

21.1 Gene Therapy for Inherited Diseases

The concept of gene therapy first emerged in the 1970s when Friedmann and Roblin (1972) proposed the use of exogenous *good* DNA to replace defective DNA. Gene therapy strategies comprise three critical elements: the gene to be transferred, the target tissue into which the gene will be introduced, and the gene delivery vehicle (vector) used to mediate entry of the gene to the target tissue. In order to achieve long-term expression of the transferred gene, one of two strategies can be employed: (i) ex vivo gene therapy, where cells (mostly from the hematopoietic system) are explanted, modified and re-implanted in patients, and (ii) in vivo gene therapy where DNA packaged in a virus is injected (intravenously or directly in the tissue) to infect a targeted organ. The latter approach is mainly used and allows for the targeting of organs or systems that cannot be explanted. The first generation of in vivo gene therapy mostly relied on modified viruses (retrovirus, lentivirus, adenovirus) to deliver foreign DNA. In some cases, a slight proportion of the transgene could be randomly incorporated in chromosomal DNA allowing its persistence, but raising concerns about potential oncogenicity (Ehrhardt et al. 2008). In vivo gene transfer strategies have now coalesced around the use of recombinant adeno-associated virus (AAV), a vector that has been successfully used in the clinic for disabling or fatal inherited metabolic disorders resulting from enzyme deficiencies (Bainbridge et al. 2008; Maguire et al. 2008, 2009; Nathwani et al. 2011b; Leone et al. 2012; Bryant et al. 2013; Smith et al. 2013; Testa et al. 2013). Recombinant AAV vectors are structurally simple and versatile, they are devoid of viral genes and instead contain an expression cassette for the gene of interest, which is limited to ~5 kb in length. Of particular interest, it is possible to transduce liver cells using a specific serotype of AAV in mice, primates and humans (Nathwani et al. 2011a, b; Markusic and Herzog 2012). This vector is ideally suited for liver gene transfer because of its ability to infect non-dividing cells such as hepatocytes, its low immunogenicity, and the persistence of vector genomes. Multiple long-term studies in mice, canines, non-human primates, and to date humans injected with AAV vectors show no indication of increased risk of hepatocellular carcinoma from AAV insertional mutagenesis (Li et al. 2011b; Nowrouzi et al. 2012; Kaeppel et al. 2013). Hence, the field of gene therapy is now focused on AAV mediated gene transfer.

21.2 Successes in AAV-Mediated Gene Therapy

A prototypical example of in vivo gene therapy is the treatment of Leber congenital amaurosis (LCA), a form of congenital blindness caused by a deficiency in the RPE65 gene (Bainbridge et al. 2008). In a pioneer clinical trial, young adults with early onset of the disease were treated by subretinal injection of AAV containing a working copy of RPE65 in one eye, the other serving as control (Maguire et al. 2008). Among three patients, one experienced an increase in visual mobility in low light that was significant enough to decrease the travel time to complete a predetermined course from 77 to 14 s (Bainbridge et al. 2008). Further clinical trials revealed that the best improvements were seen in children (Maguire et al. 2009). All pediatric subjects (age 8–11 years) developed the ability to navigate a standardized obstacle course and the effect has been persistent (7 years and counting) (Maguire et al. 2009). The earlier the treatment was initiated the better was the outcome, prompting regulatory agencies to allow researchers to treat younger children. Results obtained with LCA clinical trials were so encouraging that researchers went on to treat the contralateral-eye of patients involved in the initial trials, yielding similar results (Bennett et al. 2016).

Successful in vivo gene therapy has also been achieved in the liver to treat hemophilia B, a blood clotting disorder caused by a mutation of the factor IX (FIX) gene. Current therapy for hemophilia B is infusions of recombinant FIX but this process requires two to three weekly infusions, is expensive and is not curative. Most individuals suffering from hemophilia B have a severe form of the disease, with <1% normal circulating levels of FIX activity. Interestingly, increase of FIX levels to only 5% of normal is sufficient to prevent life-threatening bleeding episodes (Lofqvist et al. 1997; Ljung 1998). Gene therapy could offer hemophilia B patients the potential for a cure after a single injection of vector. Pre-clinical studies in monkeys have demonstrated the potential of using AAV to target the liver as a site of FIX production (Nathwani et al. 2002, 2006). In this model, the use of AAV yielded serum levels of FIX 3–8% of normal (Nathwani et al. 2002, 2006). The first successful transduction on human liver cells was accomplished in 2006, when AAV vectors were injected into the hepatic artery of seven subjects (Manno et al. 2006). Participants in the low and medium dose groups showed no signs of vector-related toxicity or efficacy, defined as circulating FIX >1% (Manno et al. 2006). The two subjects in the high dose group experienced a transient increase in FIX levels (11% and 3% respectively) that peaked 2 weeks following the infusion and started to decline after 5 weeks, reaching baseline levels after 10 weeks (Manno et al. 2006). Of note, during this period subjects reported an absence of bleeding episodes and FIX infusion despite trauma that would usually have required clotting factor infusion (Manno et al. 2006). Loss of FIX expression was attributed to the building of immunity against AAV capsid, which lead to the elimination of transduced hepatocytes by the immune system (Manno et al. 2006). These observations were the first to suggest that AAV delivery of DNA could elicit a therapeutic effect in humans. In 2011, Nathwani et al. infused AAV containing human FIX in six patients suffering from severe hemophilia B (Nathwani et al. 2011b). Subjects from the high dose group experienced increases in FIX levels from <1% to 3–12% of normal values, which was enough to confer a therapeutic effect (Nathwani et al. 2011b). It is noteworthy that in those participants, therapeutic effects are still ongoing (Nathwani et al. 2011b). Recently, Sparks therapeutics released the first results of their phase I-II clinical trial (NCT02484092) using a novel bio-engineered AAV capsid expressing a codon-optimized, high-activity human FIX variant. These results showed that four patients treated with the initial low dose experienced increases in FIX between 26% and 41% of normal values, and that after over 58 weeks of observation, none of the four subjects received regular infusions of FIX concentrates to prevent bleeding events (Dolgin 2016). These reports contributed to demonstrate that transgene delivery using AAV is safe and effective for liver targeting. Recent spectacular observations hint that AAV delivery of FIX to hepatocytes could be one of the next FDA approved treatment based on gene therapy. Fifteen years have passed since the first description of safe administration of FIX in humans, still no treatment is currently approved. This clearly demonstrates the amount of research necessary to develop a proper strategy for organ targeting and to allow the expression of an optimal recombinant protein yielding therapeutic effects. It is reasonable to believe that the strategies applied to hemophilia could be translated for the treatment of other hepatic diseases, making the development of further liver-based gene therapy easier. Only one gene therapy is currently approved and available for patients. Glybera is used to treat lipoprotein lipase deficiency by delivering in muscles a working copy of the gene using adenoviruses (Wierzbicki and Viljoen 2013). The approval of Glybera paved the way for further development of gene therapy, including approaches for treating tyrosinemia.

21.3 Potential Approaches for Gene Therapy in Tyrosinemia

Like hemophilia, type 1 tyrosinemia (HT1) is a monogenic disease affecting principally the liver (Grompe 2001). In HT1, patients lack the activity of the enzyme fumarylacetoacetate hydrolase (FAH). The development of gene therapy for these two diseases was conducted in parallel by using different strains of viruses. However, classical AAV-mediated therapy cannot be applied to tyrosinemia considering that episomal transgenes are lost because of injury-induced hepatocyte turnover and that the production of toxic metabolites increases risks of cancer (Nakai et al. 2001, 2003). For these reasons, current state of the art for in vivo gene therapy is unlikely going to work in FAH patients. A certain number of studies support these affirmations. Retroviral gene delivery was first tested in mice when human FAH cDNA was delivered via the portal vein (Overturf et al. 1996). In these animals, initial correction of hepatocytes was lower than 1%, but by combining gene therapy to in vivo selection, up to 90% of hepatocytes were corrected (Overturf et al. 1996). In vivo selection relies on the ability of corrected FAH hepatocytes to have a strong competitive growth advantage in contrast to diseased cells (Overturf et al. 1996). Considering the nature of the disease, tyrosinemic livers are in constant regeneration (Overturf et al. 1996; Hickey et al. 2015). Following viral injection, 2-(2-nitro-4trifluoromethylbenzoyl) (NTBC) is removed and corrected cells can expand, leading to the repopulation of a great proportion of the liver (up to 90%) (Overturf et al. 1996). However, only 52% of treated mice survived without NTBC, the majority of which had mosaic livers consisting of corrected and uncorrected tissues (Overturf et al. 1996). A single injection of retroviruses was not sufficient to normalize plasma levels of aspartate aminotransferase (AST), a marker of hepatocyte integrity, whereas five infusions of viruses have normalized AST and bilirubin (Overturf et al. 1996). Multiple virus injections also corrected plasma levels of succinylacetone a toxic metabolite, suggesting that the number of corrected and amplified hepatocytes arising from a single virus injection is not sufficient to exert therapeutic effects.

Considering that retroviruses can only infect dividing cells, researchers further studied the delivery of FAH using adenoviral vectors, which have the ability of infecting a wide variety of cell types, including non-dividing cells. Adenovirus-mediated gene therapy in tyrosinemic mice lead to FAH expression in hepatocytes, but was associated with high variability of the response between animals when compared to retroviral therapy (Overturf et al. 1996, 1997). Mice off NTBC treatment and injected at the same time had various degrees of liver inflammation and percent of hepatocytes positive for FAH (Overturf et al. 1997), suggesting that the use of adenoviral vectors might not be optimal. In line with previous observations made with a single injection of retroviruses, gene therapy using adenoviruses did not normalize circulating levels of succinylacetone and liver enzymes to those of wild-type controls (Overturf et al. 1997). Long-term experiments revealed a persistent expression of FAH transgene (5% of controls for at least 6 months), however this was not sufficient to protect from kidney damage. Kidneys of adenoviruses treated animals were enlarged and displayed extensive injuries (Overturf et al. 1997). Such injuries were not prevented in mice with the highest percentage of transduced hepatocytes (Overturf et al. 1997), implying that unique targeting of the liver might not be a viable approach in tyrosinemia. Long-term follow-up of treated mice also highlighted the risk of neoplasm in uncorrected hepatocytes. Of note, 77% of treated mice developed neoplasm, all of which were uniformly FAH negative, and there was no correlation between the percentages of FAH positive cells and prevention of hepatocellular neoplasm. This observation underscores the importance of correcting 100% of FAH -/- hepatocytes, which remains unachieved.

Further studies were conducted using self-inactivating lentiviral vectors, which are able to transduce non-dividing cells and integrate DNA. Mice were either submitted to in vivo gene therapy or were transplanted with corrected hepatocytes (Rittelmeyer et al. 2013). These animals showed patches of corrected hepatocytes, but 51% of all long-term observed animals showed nodular tissues consistent with liver tumor development (Rittelmeyer et al. 2013). This high incidence of cancer was attributed to endogenous non-corrected tissue, since tumorous tissue had low viral copy numbers (Rittelmeyer et al. 2013). Nonetheless, some of the top ten integration sites were located close to genes potentially implicated in hepatocellular carcinoma (Rittelmeyer et al. 2013). The concern for cancer is one of the major reasons why gene therapy has still not been tested in tyrosinemic patients. There are still underlying apprehensions in gene therapy about oncogenesis caused by random integration of the transgene. If hepatocellular carcinomas were to arise in tyrosinemic patients treated by gene therapy, it would be nearly impossible to determine whether the cancer was caused by therapy itself or by uncorrected cells. This could jeopardize future clinical trials for tyrosinemia as well as current trials based on similar delivery systems. From a different perspective, AAV were used in a mouse model of tyrosinemia to induce gene repair via homologous recombination (Paulk et al. 2010). This strategy, chosen because episomal vectors are not effective in the context of hepatocyte division (Nakai et al. 2001; Wang et al. 2012b), was aimed at inserting a corrected DNA fragment in the FAH locus by harnessing the ability of AAV vectors to integrate the DNA. Treated mice in which hepatocytes were positively selected using NTBC withdraw showed weight gain and improvements in liver function that were paralleled by the presence of >50% of FAH producing hepatocytes (Paulk et al. 2010). Both treatment of neonates and adult mice yielded integration frequencies of up to 10^{-3} that were confirmed by transplantation of hepatocytes from AAV treated mice to FAH −/− recipients (Paulk et al. 2010). Transplant recipients had successful engraftment and displayed clinical improvements, thus validating permanent integration (Paulk et al. 2010). This report highlighted the feasibility of in vivo gene correction in tyrosinemia. However, disadvantages of gene replacement include risks related to insertional mutagenesis, and loss of regulatory signals that control gene expression. Several strategies were used to explore the impact of in vivo gene therapy in tyrosinemia, however success is currently limited. Other approaches focusing on ex vivo cell therapy were also investigated.

21.4 Potential Approaches for Cell Therapy in Tyrosinemia

Recently, direct reprogramming of somatic cells into induced pluripotent stem cells (iPS) has been achieved in different contexts including tyrosinemia. A first report demonstrated how iPS cell-derived hepatocytes have the same functional output as adult hepatocytes and retain their ability to proliferate following injury (Espejel et al. 2010). In a second set of experiments, FAH −/− iPS cell lines were created and used to produce mice exhibiting tyrosinemia symptoms (Wu et al. 2011). Using a lentiviral vector, FAH-corrected iPS cell lines were established by insertion of FAH cDNA followed by selection of corrected cells (Wu et al. 2011). When aggregated with embryos, these corrected iPS repopulated the liver, leading to the expression of FAH in a portion of hepatocytes (Wu et al. 2011). Intriguingly, most of the transgene was silenced in liver cells; however in vivo selection using NTBC withdraw in these mice resulted in 50–70% of hepatocytes expressing FAH (Wu et al. 2011). This technology could eventually be used to create FAH corrected patient-derived iPS cells that could be transplanted to repopulate the liver. Further studies are needed to highlight the benefits of using iPS cells in tyrosinemia and to ascertain the efficacy of this technique. At the current state of this technology, chances for successful therapies are low.

Ex vivo hepatic gene therapy was also recently tested in different models of HT1 (Hickey et al. 2016; Zhang et al. 2016). In HT1 rats, infusion of WT hepatocytes yielded a 70–90% liver repopulation that was accompanied by weight gain and normalization of liver enzymes (Zhang et al. 2016). In FAH −/− mice and pigs, partial hepatectomy was conducted, followed by hepatocytes isolation. Isolated hepatocytes were then infected using lentiviruses expressing FAH under the control of a liver- specific promoter and reinjected through the spleen or portal vein for hepatic engraftment (Hickey et al. 2016). In some mice, transplant of corrected cells followed by NTBC removal was associated with a near complete liver repopulation and rescue of the phenotype. However, initial engraftment efficiency appeared to impact long-term repopulation (Hickey et al. 2016). Similar results were obtained when this technique was applied in pigs (Hickey et al. 2016). Twelve months following transplant, a near-complete repopulation of the liver was observed. Fibrosis was detected in some animals, but importantly no adenocarcinomas were detected (Hickey et al. 2016). Considering its ability to function in larger outbred mammals, this approach appears promising. Its most important advantage is the use of autologous cells, which alleviates the need for haplotype matching and potential immunosuppressive drugs. Additional larger studies are needed to determine the potential for this technology in humans and to assess integration profile.

21.5 Limits of Gene Therapy

The first gene therapy clinical trials highlighted the importance of sufficient expression of the transgene. In some occurrences, transgene was detectable by biochemical methods, but levels were insufficient to trigger therapeutical effects (Manno et al. 2006). Different factors can influence transgene expression including AAV serotype, dose and vector design. Optimal design will have to be determined for each condition, including tissue specific promoter, enhancers and DNA variants used. As it has been the case with hemophilia, numerous experiments in animal models and clinical trials will be necessary in order to identify the conditions required for therapeutic effects in different diseases, including tyrosinemia.

Development and validation of gene therapy in postmitotic cells is currently underway. Nonetheless, in a wide variety of genetic diseases, symptoms appear during the first months/years of life leading to mortality within few years. Unfortunately, as demonstrated in animal models, treatment of infants faces a major challenge because the loss of episomal vector genomes that accompanies hepatocellular proliferation during liver growth results in the rapid extinction of transgene expression (Cunningham et al. 2008, 2009; Weinstein et al. 2010; Cotugno et al. 2011; McKay et al. 2011; Wang et al. 2011; Lisowski et al. 2012; Kok et al. 2013). Thus, for genetic diseases that manifest at a young age with irreversible consequences, early treatment is currently unlikely to succeed. New approaches aimed at integrating a fragment of therapeutical DNA at a specific locus could allow permanent modification of the cells and circumvent transgene extinction.

Another limitation of gene therapy has been discussed above and is what others have identified as the fitness of edited cells (Cox et al. 2015), which is the ability of edited cells to have a selective advantage over unedited cells. This advantage confers to the initially low number of edited cells the ability to expand and reverse disease symptoms. For example in SCID-X1, restoration of IL2RG gene function in blood cells confers a selective advantage and allows the modified cells to expand relative to their diseased counterparts, leading to a therapeutic effect (Hacein-Bey-Abina et al. 2002; Gaspar et al. 2004). In some cases, the absence of fitness in modified cells can still lead to therapeutic effects. Most of these diseases, including hemophilia B, have a common denominator: even small changes in gene expression can alleviate symptoms or cure the disease. This suggests that for some diseases, modification of a limited number of liver cells could be clinically relevant. This, however does not appear to be the case in tyrosinemia.

As it could be anticipated, the administration of AAV has direct impacts on the immune response to foreign antigens. Immune responses can be mounted and directed against AAV capsid, the transgene or both (Mingozzi and High 2013). Animal models foretold many aspects of the human immune response to the transgene product but largely failed to predict response to AAV capsid (Mingozzi and High 2013). In a context where an increasing number of therapies are translated to clinical trials, viral load is an ever-growing concern. Although AAV itself is not associated with any known illness, most of the human population has been exposed to wild-type AAV since prevalence rate for antibody titers exceeds 60% among adults (Boutin et al. 2010). In clinical trials where AAV were delivered through the bloodstream, low titer of neutralizing antibody completely neutralized large doses of vector (Manno et al. 2006). Conversely, the presence of anti-AAV antibodies does not seem to affect efficiency when the vector is injected intraparenchymally or into the eye (Kaplitt et al. 2007; Bainbridge et al. 2008; Worgall et al. 2008; Brantly et al. 2009). Gene therapy in subjects with preexisting immunity to AAV remains a major challenge for systemic delivery through the bloodstream. The current approach excludes patients with antibodies against AAV, but since this involves up to 70% of the population, better solutions must be found. Different techniques including the use of immune suppressants are currently being tested to decrease the risks of immune reactions, but these are not free of potential side effects.

21.6 Genome Editing (Gene Therapy 2.0)

To circumvent certain limitations associated with AAV-based or in vivo gene therapy, permanent targeted integration of the transgene in DNA is required. The ability to modify the genomic sequence of living cells through a process called targeted homologous recombination has revolutionized biology. Unfortunately, the effectiveness of the conventional gene targeting tools has been limited to lesser organisms. However, the genomes of numerous species including mouse, fish, fly, cow, and human cells have now proven amenable to manipulation using a new class of tools termed engineered nucleases (Urnov et al. 2010; Joung and Sander 2013; Pennisi 2013). Specifically, three complementary classes of designer enzymes that cleave precise DNA sequences to introduce double-strand breaks have been described: zinc-finger nucleases (ZFNs) (Urnov et al. 2010), transcription activator–like effector nucleases (TALENs) (Joung and Sander 2013), and RNA-guided endonucleases (RGENs-CRISPR/Cas9 system) (Pennisi 2013). Independently of the platform, the action of engineered nucleases relies on the cells' ability to resolve the DNA break via evolutionary conserved pathways, either by a non-templated error-prone process called non-homologous end joining (NHEJ), or using an exogenous template to repair the break by homology directed repair (HDR) (Lieber 2010; Moynahan and Jasin 2010). Using this core technology, referred to as "genome editing", gene disruption (the targeted induction of minor insertions and deletions), gene correction (the introduction of discrete base substitutions specified by a homologous donor DNA construct) and targeted gene addition (aka. targeted integration – the transfer of entire transgenes into a native genomic locus) has been achieved in cancer, primary, and stem cells as well as in many model systems (Urnov et al. 2010; Joung and Sander 2013; Pennisi 2013). Of note, it is possible to design a donor molecule to take advantage of the active DNA repair pathway of the target cell. As a whole, this "genome editing" technology was named Science's breakthrough of the year in 2015 because it is highly efficient in all species tested and is now a standard experimental strategy for gene manipulation in pre-clinical models (Travis 2015).

Genome editing allows for the insertion of therapeutic DNA to a predetermined site in the genome. This represents a huge advantage for the treatment of children or the targeting of proliferative tissues, since it alleviates any issues related to dilution of the transgene. Considering that therapeutic DNA can be inserted downstream of strong promoters (Sharma et al. 2015), this technique could also circumvent low expression concerns. Neonatal gene therapy has the potential advantage of achieving therapeutic effects before disease manifestation. Furthermore viral vectors face a relatively immature immune system, which could limit the use of immunosuppressants and potentially adverse immune responses (Calcedo et al. 2011; McKay et al. 2011). Genome editing has been tested in a certain number of mouse models of pediatric diseases, where it has shown potential. ZFNs targeting FIX and a repair template were packaged in liver specific AAV and injected intraperitoneally in 2 days old neonatal mice (Li et al. 2011a). Eight weeks following the injection, donor insertion was observed with targeting efficiencies in the range of 1–3% (Li et al. 2011a). This resulted in an increase of FIX levels to 2–3% of normal that was sufficient to reduce clotting time (Li et al. 2011a). Stable genomic correction was assessed by partial hepatectomy, where mice treated with both ZFN and donor retained FIX expression when compared to animals injected with episomal vector (Li et al. 2011a). No clinical trial has yet been approved to treat children suffering from this condition. Potential of neonatal genome editing using CRISPR/Cas9 has also been evaluated in a mouse model of X-linked deficiency of the OTC enzyme (Yang et al. 2016). A liver specific two-vector approach was used to inject 2 days neonates with the RNA-guided nuclease and a donor template (Yang et al. 2016). HDR-mediated repair was observed in 8.3% of alleles, which led to a 100-fold increase in OTC-expressing hepatocytes (Yang et al. 2016). One concern about genome editing is that persistence of the nuclease could lead to off-target activity. When injected in neonates, nuclease levels were undetectable by 8 weeks, suggesting that cell proliferation in the newborn liver could be responsible for the desired decline (Yang et al. 2016). Further analyses revealed that OTC enzyme activity in the liver was 20% of normal (Yang et al. 2016). When submitted to a high-protein diet challenge, treated mice showed a decrease of 40% in circulating ammonia levels as compared to untreated controls, and showed no clinical signs of the disease (Yang et al. 2016). Interestingly, these effects were not recapitulated when adult mice were injected, suggesting that treatment of newborn could represent a significant advantage in genome editing. These reports highlight the feasibility of genome editing in the neonatal liver using AAV delivery and support the hypothesis that permanent DNA modification is a potential therapeutic avenue for the treatment of children.

Genome editing was also tested in animal models of tyrosinemia. First, nuclease and donor template were administered through intravenous hydrodynamic injection in adult mice (Yin et al. 2014). Mice were removed from NTBC 3 days following injection and weight loss was prevented. At sacrifice, liver damage was substantially less in treated mice as assessed by liver enzymes levels and histology, suggesting a rescue of the FAH phenotype (Yin et al. 2014). Indeed, genome editing generated FAH producing hepatocytes with an initial repair frequency estimated to 0.4% (Yin et al. 2014). In vivo selection via NTBC removal led to the presence of widespread patches of FAH^+ liver cells that were sufficient to prevent weight loss (Yin et al. 2014). Similarly to what was observed with gene therapy, only a fraction of hepatocytes was repaired leading to a correction of the phenotype, but without alleviating concerns about cancer. This report describes the potential for gene therapy in tyrosinemia, however hydrodynamic injection is not likely to be used in humans. Using an alternative approach, the same group delivered CRISPR nuclease mRNA via lipid nanoparticles and gRNA as well as donor template in an AAV (Yin et al. 2016). The rationale behind this strategy was to limit nuclease expression in time and to reduce AAV load, since two different AAV would be needed to package the nuclease, gRNA and repair template. Following 30 days of NTBC removal, treated mice displayed expression of FAH in up to 6% of hepatocytes, which was paralleled by weight gain and a decrease in liver enzymes (Yin et al. 2016). This report demonstrates the feasibility of combining viral and non-viral delivery for the modification of liver cells, and confirms the feasibility of hepatic genome editing.

21.7 Limits of Genome Editing

Specificity has been an ongoing concern since the development of engineered nucleases (Doyon et al. 2011; Slaymaker et al. 2016). A lot of efforts are currently directed towards the discovery and validation of more specific nucleases. Nucleases from the CRISPR system can tolerate a certain number of mismatches in targeted sequence leading to off-target activity (Slaymaker et al. 2016). It was suggested that the most commonly used enzyme variant (SpCas9) can cleave an average of 90 sites in vitro (Kim et al. 2016b), leading to concerns about oncogenesis. To circumvent this issue, more specific variants have been developed (Slaymaker et al. 2016), and other nucleases from the CRISPR system displaying increased specificity have been discovered or engineered (Zetsche et al. 2015; Kim et al. 2016a). Further in vivo studies will be needed to compare on and off-target activities of these nucleases.

Using engineered nucleases to cut out part of a gene, hence invalidating its function, yields relatively high success rates (Wang et al. 2013). This strategy is in fact already being tested in clinical trials for HIV (Wang and Cannon 2016). On the opposite, when these nucleases are used to correct a gene using a repair template, efficacy drastically decreases. This phenomenon can be explained by the fact that integration of DNA uses HDR, which is only active in dividing cells (Shrivastav et al. 2008). Considering the current inability to select for corrected cells in vivo, developments aimed at increasing HDR frequency will be essential for the success of future clinical trials.

To date, in vivo editing has largely been achieved through the use of viral vectors with defined, tissue-specific activity. Such vectors are currently limited in terms of cargo carrying capacity and tropism, restricting this mode of therapy to organs where transduction with clinically useful vectors is efficient, such as the liver, muscle and eye (Maguire et al. 2008; Grieger and Samulski 2012; Wang et al. 2012a). Nonetheless, treatment of most genetic diseases requires the targeting of a large subset of organs, and in some cases the entire body. Techniques for targeting numerous systems are not optimal and will require further development. Importantly, viral cargo capacity needs to be improved, either by developing smaller nucleases or by finding alternatives to the AAV vectors currently used. Considering that two or three different AAV are required in order to package the nuclease, gRNA and repair template and that high doses of AAV are necessary for therapeutic effects, an increase in viral cargo capacity could lead to a significant reduction of total viral vectors injected. As for gene therapy, immunogenicity remains a concern in genome editing. In addition to a potential reaction against AAV capsid, it was suggested that an immune response could be mounted towards the nuclease as well (Wang et al. 2015). Delivering nuclease mRNA or protein could limit the exposure, thus reducing potential immunogenicity.

21.8 Concluding Remark

Current advances support the idea that gene therapy and genome editing could have a positive impact on the phenotype of patients suffering from tyrosinemia. Of note, most strategies lead to weight gain and a reduction of liver enzymes, however toxic metabolites levels were not always measured. Still, correction frequency remains low and even with in vivo selection using NTBC withdrawal uncorrected cells remain, which could lead to neoplasm concerns. The feasibility and consequences of NTBC removal to stimulate liver repopulation by corrected cells in humans will need to be assessed before clinical trials are designed. At the current state of the technology, one could hypothesize that tyrosinemic patients treated by gene therapy/genome editing could eventually lessen diet restriction due to a partial rescue of FAH activity. Nonetheless, until a strategy leading to the repair of 100% hepatocytes is developed, NTBC regimen would need to be continued to lessen cancer risks. Genome editing offers a greater degree of precision in how specific DNA sequences are changed. While ethical concerns have previously been raised over genetic manipulations, the advent of engineered nucleases has prompted recent calls for a genome editing moratorium over germ-line modification. Some have also raised the possibility that genome editing of somatic cells for "enhancement" (intelligence, muscular function, height, etc) poses a specific ethical issue. Benefits of genome editing have included better targeted gene therapy in animal models of some diseases; these tools are also hoped to lead to a better understanding of the structure, function and regulation of genes. Currently modifying plants, animals, and non-inheritable cells in humans is allowed under strict controls. However, modifications that alter the human germ-line are not allowed, with the exception of the decision in the UK to allow mitochondrial replacement. Gene therapy strategies diverge in efficacy and potential adverse effects, and their use is established on defining a risk-benefit ratio. As with any other medical procedure, this balance will be individually defined depending on the type of cell being edited, the nature and extent of the editing procedure, and the therapeutic application. Given our still rudimentary knowledge of the genome and its regulation, the issue of identifying unintended/unanticipated consequences should be put forefront.

References

Bainbridge JW, Smith AJ, Barker SS, Robbie S, Henderson R, Balaggan K, Viswanathan A, Holder GE, Stockman A, Tyler N, Petersen-Jones S, Bhattacharya SS, Thrasher AJ, Fitzke FW, Carter BJ, Rubin GS, Moore AT, Ali RR (2008) Effect of gene therapy on visual function in Leber's congenital amaurosis. N Engl J Med 358(21):2231–2239

Bennett J, Wellman J, Marshall KA, McCague S, Ashtari M, DiStefano-Pappas J, Elci OU, Chung DC, Sun J, Wright JF, Cross DR, Aravand P, Cyckowski LL, Bennicelli JL, Mingozzi F, Auricchio A, Pierce EA, Ruggiero J, Leroy BP, Simonelli F, High KA, Maguire AM (2016) Safety and durability of effect of contralateral-eye administration of AAV2 gene therapy in patients with childhood-onset blindness caused by RPE65 mutations: a follow-on phase 1 trial. Lancet 388(10045):661–672

Boutin S, Monteilhet V, Veron P, Leborgne C, Benveniste O, Montus MF, Masurier C (2010) Prevalence of serum IgG and neutralizing factors against adeno-associated virus (AAV) types 1, 2, 5, 6, 8, and 9 in the healthy population: implications for gene therapy using AAV vectors. Hum Gene Ther 21(6):704–712

Brantly ML, Chulay JD, Wang L, Mueller C, Humphries M, Spencer LT, Rouhani F, Conlon TJ, Calcedo R, Betts MR, Spencer C, Byrne BJ, Wilson JM, Flotte TR (2009) Sustained transgene expression despite T lymphocyte responses in a clinical trial of rAAV1-AAT gene therapy. Proc Natl Acad Sci U S A 106(38):16363–16368

Bryant LM, Christopher DM, Giles AR, Hinderer C, Rodriguez JL, Smith JB, Traxler EA, Tycko J, Wojno AP, Wilson JM (2013) Lessons learned from the clinical development and market authorization of Glybera. Hum Gene Ther Clin Dev 24(2):55–64

Calcedo R, Morizono H, Wang L, McCarter R, He J, Jones D, Batshaw ML, Wilson JM (2011) Adeno-associated virus antibody profiles in newborns, children, and adolescents. Clin Vaccine Immunol 18(9):1586–1588

Cotugno G, Annunziata P, Tessitore A, O'Malley T, Capalbo A, Faella A, Bartolomeo R, O'Donnell P, Wang P, Russo F, Sleeper MM, Knox VW, Fernandez S, Levanduski L, Hopwood J, De Leonibus E, Haskins M, Auricchio A (2011) Long-term amelioration of feline Mucopolysaccharidosis VI after AAV-mediated liver gene transfer. Mol Ther 19(3):461–469

Cox DB, Platt RJ, Zhang F (2015) Therapeutic genome editing: prospects and challenges. Nat Med 21(2):121–131

Cunningham SC, Dane AP, Spinoulas A, Logan GJ, Alexander IE (2008) Gene delivery to the juvenile mouse liver using AAV2/8 vectors. Mol Ther 16(6):1081–1088

Cunningham SC, Spinoulas A, Carpenter KH, Wilcken B, Kuchel PW, Alexander IE (2009) AAV2/8-mediated correction of OTC deficiency is robust in adult but not neonatal Spf(ash) mice. Mol Ther 17(8):1340–1346

Dolgin E (2016) Early clinical data raise the bar for hemophilia gene therapies. Nat Biotechnol 34(10):999–1001

Doyon Y, Vo TD, Mendel MC, Greenberg SG, Wang J, Xia DF, Miller JC, Urnov FD, Gregory PD, Holmes MC (2011) Enhancing zinc-finger-nuclease activity with improved obligate heterodimeric architectures. Nat Methods 8(1):74–79

Ehrhardt A, Haase R, Schepers A, Deutsch MJ, Lipps HJ, Baiker A (2008) Episomal vectors for gene therapy. Curr Gene Ther 8(3):147–161

Espejel S, Roll GR, McLaughlin KJ, Lee AY, Zhang JY, Laird DJ, Okita K, Yamanaka S, Willenbring H (2010) Induced pluripotent stem cell-derived hepatocytes have the functional and proliferative capabilities needed for liver regeneration in mice. J Clin Invest 120(9):3120–3126

Friedmann T, Roblin R (1972) Gene therapy for human genetic disease? Science 175(4025):949–955

Gaspar HB, Parsley KL, Howe S, King D, Gilmour KC, Sinclair J, Brouns G, Schmidt M, Von Kalle C, Barington T, Jakobsen MA, Christensen HO, Al Ghonaium A, White HN, Smith JL, Levinsky RJ, Ali RR, Kinnon C, Thrasher AJ (2004) Gene therapy of X-linked severe combined immunodeficiency by use of a pseudotyped gammaretroviral vector. Lancet 364(9452):2181–2187

Grieger JC, Samulski RJ (2012) Adeno-associated virus vectorology, manufacturing, and clinical applications. Methods Enzymol 507:229–254

Grompe M (2001) The pathophysiology and treatment of hereditary tyrosinemia type 1. Semin Liver Dis 21(4):563–571

Hacein-Bey-Abina S, Le Deist F, Carlier F, Bouneaud C, Hue C, De Villartay JP, Thrasher AJ, Wulffraat N, Sorensen R, Dupuis-Girod S, Fischer A, Davies EG, Kuis W, Leiva L, Cavazzana-Calvo M (2002) Sustained correction of X-linked severe combined immunodeficiency by ex vivo gene therapy. N Engl J Med 346(16):1185–1193

Hickey RD, Mao SA, Amiot B, Suksanpaisan L, Miller A, Nace R, Glorioso J, O'Connor MK, Peng KW, Ikeda Y, Russell SJ, Nyberg SL (2015) Noninvasive 3-dimensional imaging of liver regeneration in a mouse model of hereditary tyrosinemia type 1 using the sodium iodide symporter gene. Liver Transpl 21(4):442–453

Hickey RD, Mao SA, Glorioso J, Elgilani F, Amiot B, Chen H, Rinaldo P, Marler R, Jiang H, DeGrado

TR, Suksanpaisan L, O'Connor MK, Freeman BL, Ibrahim SH, Peng KW, Harding CO, Ho CS, Grompe M, Ikeda Y, Lillegard JB, Russell SJ, Nyberg SL (2016) Curative ex vivo liver-directed gene therapy in a pig model of hereditary tyrosinemia type 1. Sci Transl Med 8(349):349ra399

Joung JK, Sander JD (2013) TALENs: a widely applicable technology for targeted genome editing. Nat Rev Mol Cell Biol 14(1):49–55

Kaeppel C, Beattie SG, Fronza R, van Logtenstein R, Salmon F, Schmidt S, Wolf S, Nowrouzi A, Glimm H, von Kalle C, Petry H, Gaudet D, Schmidt M (2013) A largely random AAV integration profile after LPLD gene therapy. Nat Med 19(7):889–891

Kaplitt MG, Feigin A, Tang C, Fitzsimons HL, Mattis P, Lawlor PA, Bland RJ, Young D, Strybing K, Eidelberg D, During MJ (2007) Safety and tolerability of gene therapy with an adeno-associated virus (AAV) borne GAD gene for Parkinson's disease: an open label, phase I trial. Lancet 369(9579):2097–2105

Kim D, Kim J, Hur JK, Been KW, Yoon SH, Kim JS (2016a) Genome-wide analysis reveals specificities of Cpf1 endonucleases in human cells. Nat Biotechnol 34:863

Kim D, Kim S, Kim S, Park J, Kim JS (2016b) Genome-wide target specificities of CRISPR-Cas9 nucleases revealed by multiplex Digenome-seq. Genome Res 26(3):406–415

Kok CY, Cunningham SC, Carpenter KH, Dane AP, Siew SM, Logan GJ, Kuchel PW, Alexander IE (2013) Adeno-associated virus-mediated rescue of neonatal lethality in argininosuccinate synthetase-deficient mice. Mol Ther 21(10):1823–1831

Leone P, Shera D, McPhee SW, Francis JS, Kolodny EH, Bilaniuk LT, Wang DJ, Assadi M, Goldfarb O, Goldman HW, Freese A, Young D, During MJ, Samulski RJ, Janson CG (2012) Long-term follow-up after gene therapy for canavan disease. Sci Transl Med 4(165):165ra163

Li H, Haurigot V, Doyon Y, Li T, Wong SY, Bhagwat AS, Malani N, Anguela XM, Sharma R, Ivanciu L, Murphy SL, Finn JD, Khazi FR, Zhou S, Paschon DE, Rebar EJ, Bushman FD, Gregory PD, Holmes MC, High KA (2011a) In vivo genome editing restores haemostasis in a mouse model of haemophilia. Nature 475(7355):217–221

Li H, Malani N, Hamilton SR, Schlachterman A, Bussadori G, Edmonson SE, Shah R, Arruda VR, Mingozzi F, Wright JF, Bushman FD, High KA (2011b) Assessing the potential for AAV vector genotoxicity in a murine model. Blood 117(12):3311–3319

Lieber MR (2010) The mechanism of double-strand DNA break repair by the nonhomologous DNA end-joining pathway. Annu Rev Biochem 79:181–211

Lisowski L, Lau A, Wang Z, Zhang Y, Zhang F, Grompe M, Kay MA (2012) Ribosomal DNA integrating rAAV-rDNA vectors allow for stable transgene expression. Mol Ther 20(10):1912–1923

Ljung RC (1998) Can haemophilic arthropathy be prevented? Br J Haematol 101(2):215–219

Lofqvist T, Nilsson IM, Berntorp E, Pettersson H (1997) Haemophilia prophylaxis in young patients – a long-term follow-up. J Intern Med 241(5):395–400

Maguire AM, Simonelli F, Pierce EA, Pugh EN Jr, Mingozzi F, Bennicelli J, Banfi S, Marshall KA, Testa F, Surace EM, Rossi S, Lyubarsky A, Arruda VR, Konkle B, Stone E, Sun J, Jacobs J, Dell'Osso L, Hertle R, Ma JX, Redmond TM, Zhu X, Hauck B, Zelenaia O, Shindler KS, Maguire MG, Wright JF, Volpe NJ, McDonnell JW, Auricchio A, High KA, Bennett J (2008) Safety and efficacy of gene transfer for Leber's congenital amaurosis. N Engl J Med 358(21):2240–2248

Maguire AM, High KA, Auricchio A, Wright JF, Pierce EA, Testa F, Mingozzi F, Bennicelli JL, Ying GS, Rossi S, Fulton A, Marshall KA, Banfi S, Chung DC, Morgan JI, Hauck B, Zelenaia O, Zhu X, Raffini L, Coppieters F, De Baere E, Shindler KS, Volpe NJ, Surace EM, Acerra C, Lyubarsky A, Redmond TM, Stone E, Sun J, McDonnell JW, Leroy BP, Simonelli F, Bennett J (2009) Age-dependent effects of RPE65 gene therapy for Leber's congenital amaurosis: a phase 1 dose-escalation trial. Lancet 374(9701):1597–1605

Manno CS, Pierce GF, Arruda VR, Glader B, Ragni M, Rasko JJ, Ozelo MC, Hoots K, Blatt P, Konkle B, Dake M, Kaye R, Razavi M, Zajko A, Zehnder J, Rustagi PK, Nakai H, Chew A, Leonard D, Wright JF, Lessard RR, Sommer JM, Tigges M, Sabatino D, Luk A, Jiang H, Mingozzi F, Couto L, Ertl HC, High KA, Kay MA (2006) Successful transduction of liver in hemophilia by AAV-Factor IX and limitations imposed by the host immune response. Nat Med 12(3):342–347

Markusic DM, Herzog RW (2012) Liver-directed adeno-associated viral gene therapy for hemophilia. J Genet Syndr Gene Ther 1:1–9

McKay TR, Rahim AA, Buckley SM, Ward NJ, Chan JK, Howe SJ, Waddington SN (2011) Perinatal gene transfer to the liver. Curr Pharm Des 17(24):2528–2541

Mingozzi F, High KA (2013) Immune responses to AAV vectors: overcoming barriers to successful gene therapy. Blood 122(1):23–36

Moynahan ME, Jasin M (2010) Mitotic homologous recombination maintains genomic stability and suppresses tumorigenesis. Nat Rev Mol Cell Biol 11(3):196–207

Nakai H, Yant SR, Storm TA, Fuess S, Meuse L, Kay MA (2001) Extrachromosomal recombinant adeno-associated virus vector genomes are primarily responsible for stable liver transduction in vivo. J Virol 75(15):6969–6976

Nakai H, Montini E, Fuess S, Storm TA, Grompe M, Kay MA (2003) AAV serotype 2 vectors preferentially integrate into active genes in mice. Nat Genet 34(3):297–302

Nathwani AC, Davidoff AM, Hanawa H, Hu Y, Hoffer FA, Nikanorov A, Slaughter C, Ng CY, Zhou J, Lozier JN, Mandrell TD, Vanin EF, Nienhuis AW (2002) Sustained high-level expression of human factor IX (hFIX) after liver-targeted delivery of recombinant

adeno-associated virus encoding the hFIX gene in rhesus macaques. Blood 100(5):1662–1669
Nathwani AC, Gray JT, Ng CY, Zhou J, Spence Y, Waddington SN, Tuddenham EG, Kemball-Cook G, McIntosh J, Boon-Spijker M, Mertens K, Davidoff AM (2006) Self-complementary adeno-associated virus vectors containing a novel liver-specific human factor IX expression cassette enable highly efficient transduction of murine and nonhuman primate liver. Blood 107(7):2653–2661
Nathwani AC, Rosales C, McIntosh J, Rastegarlari G, Nathwani D, Raj D, Nawathe S, Waddington SN, Bronson R, Jackson S, Donahue RE, High KA, Mingozzi F, Ng CY, Zhou J, Spence Y, McCarville MB, Valentine M, Allay J, Coleman J, Sleep S, Gray JT, Nienhuis AW, Davidoff AM (2011a) Long-term safety and efficacy following systemic administration of a self-complementary AAV vector encoding human FIX pseudotyped with serotype 5 and 8 capsid proteins. Mol Ther 19(5):876–885
Nathwani AC, Tuddenham EG, Rangarajan S, Rosales C, McIntosh J, Linch DC, Chowdary P, Riddell A, Pie AJ, Harrington C, O'Beirne J, Smith K, Pasi J, Glader B, Rustagi P, Ng CY, Kay MA, Zhou J, Spence Y, Morton CL, Allay J, Coleman J, Sleep S, Cunningham JM, Srivastava D, Basner-Tschakarjan E, Mingozzi F, High KA, Gray JT, Reiss UM, Nienhuis AW, Davidoff AM (2011b) Adenovirus-associated virus vector-mediated gene transfer in hemophilia B. N Engl J Med 365(25):2357–2365
Nowrouzi A, Penaud-Budloo M, Kaeppel C, Appelt U, Le Guiner C, Moullier P, von Kalle C, Snyder RO, Schmidt M (2012) Integration frequency and intermolecular recombination of rAAV vectors in non-human primate skeletal muscle and liver. Mol Ther 20(6):1177–1186
Overturf K, Al-Dhalimy M, Tanguay R, Brantly M, Ou CN, Finegold M, Grompe M (1996) Hepatocytes corrected by gene therapy are selected in vivo in a murine model of hereditary tyrosinaemia type I. Nat Genet 12(3):266–273
Overturf K, al-Dhalimy M, Ou CN, Finegold M, Tanguay R, Lieber A, Kay M, Grompe M (1997) Adenovirus-mediated gene therapy in a mouse model of hereditary tyrosinemia type I. Hum Gene Ther 8(5):513–521
Paulk NK, Wursthorn K, Wang Z, Finegold MJ, Kay MA, Grompe M (2010) Adeno-associated virus gene repair corrects a mouse model of hereditary tyrosinemia in vivo. Hepatology 51(4):1200–1208
Pennisi E (2013) The CRISPR craze. Science 341(6148):833–836
Rittelmeyer I, Rothe M, Brugman MH, Iken M, Schambach A, Manns MP, Baum C, Modlich U, Ott M (2013) Hepatic lentiviral gene transfer is associated with clonal selection, but not with tumor formation in serially transplanted rodents. Hepatology 58(1):397–408
Sharma R, Anguela XM, Doyon Y, Wechsler T, DeKelver RC, Sproul S, Paschon DE, Miller JC, Davidson RJ, Shivak D, Zhou S, Rieders J, Gregory PD, Holmes MC, Rebar EJ, High KA (2015) In vivo genome editing of the albumin locus as a platform for protein replacement therapy. Blood 126(15):1777–1784
Shrivastav M, De Haro LP, Nickoloff JA (2008) Regulation of DNA double-strand break repair pathway choice. Cell Res 18(1):134–147
Slaymaker IM, Gao L, Zetsche B, Scott DA, Yan WX, Zhang F (2016) Rationally engineered Cas9 nucleases with improved specificity. Science 351(6268):84–88
Smith BK, Collins SW, Conlon TJ, Mah CS, Lawson LA, Martin AD, Fuller DD, Cleaver BD, Clement N, Phillips D, Islam S, Dobjia N, Byrne BJ (2013) Phase I/II trial of adeno-associated virus-mediated alpha-glucosidase gene therapy to the diaphragm for chronic respiratory failure in Pompe disease: initial safety and ventilatory outcomes. Hum Gene Ther 24(6):630–640
Testa F, Maguire AM, Rossi S, Pierce EA, Melillo P, Marshall K, Banfi S, Surace EM, Sun J, Acerra C, Wright JF, Wellman J, High KA, Auricchio A, Bennett J, Simonelli F (2013) Three-year follow-up after unilateral subretinal delivery of adeno-associated virus in patients with Leber congenital Amaurosis type 2. Ophthalmology 120(6):1283–1291
Travis J (2015) Making the cut. Science 350(6267):1456–1457
Urnov FD, Rebar EJ, Holmes MC, Zhang HS, Gregory PD (2010) Genome editing with engineered zinc finger nucleases. Nat Rev Genet 11(9):636–646
Wang CX, Cannon PM (2016) The clinical applications of genome editing in HIV. Blood 127(21):2546–2552
Wang L, Bell P, Lin J, Calcedo R, Tarantal AF, Wilson JM (2011) AAV8-mediated hepatic gene transfer in infant rhesus monkeys (*Macaca mulatta*). Mol Ther 19(11):2012–2020
Wang L, Wang H, Bell P, McMenamin D, Wilson JM (2012a) Hepatic gene transfer in neonatal mice by adeno-associated virus serotype 8 vector. Hum Gene Ther 23(5):533–539
Wang Z, Lisowski L, Finegold MJ, Nakai H, Kay MA, Grompe M (2012b) AAV vectors containing rDNA homology display increased chromosomal integration and transgene persistence. Mol Ther 20(10):1902–1911
Wang H, Yang H, Shivalila CS, Dawlaty MM, Cheng AW, Zhang F, Jaenisch R (2013) One-step generation of mice carrying mutations in multiple genes by CRISPR/Cas-mediated genome engineering. Cell 153(4):910–918
Wang D, Mou H, Li S, Li Y, Hough S, Tran K, Li J, Yin H, Anderson DG, Sontheimer EJ, Weng Z, Gao G, Xue W (2015) Adenovirus-mediated somatic genome editing of Pten by CRISPR/Cas9 in mouse liver in spite of Cas9-specific immune responses. Hum Gene Ther 26(7):432–442
Weinstein DA, Correia CE, Conlon T, Specht A, Verstegen J, Onclin-Verstegen K, Campbell-Thompson M, Dhaliwal G, Mirian L, Cossette H, Falk DJ, Germain S, Clement N, Porvasnik S, Fiske L, Struck M,

Ramirez HE, Jordan J, Andrutis K, Chou JY, Byrne BJ, Mah CS (2010) Adeno-associated virus-mediated correction of a canine model of glycogen storage disease type Ia. Hum Gene Ther 21(7):903–910

Wierzbicki AS, Viljoen A (2013) Alipogene tiparvovec: gene therapy for lipoprotein lipase deficiency. Expert Opin Biol Ther 13(1):7–10

Worgall S, Sondhi D, Hackett NR, Kosofsky B, Kekatpure MV, Neyzi N, Dyke JP, Ballon D, Heier L, Greenwald BM, Christos P, Mazumdar M, Souweidane MM, Kaplitt MG, Crystal RG (2008) Treatment of late infantile neuronal ceroid lipofuscinosis by CNS administration of a serotype 2 adeno-associated virus expressing CLN2 cDNA. Hum Gene Ther 19(5):463–474

Wu G, Liu N, Rittelmeyer I, Sharma AD, Sgodda M, Zaehres H, Bleidissel M, Greber B, Gentile L, Han DW, Rudolph C, Steinemann D, Schambach A, Ott M, Scholer HR, Cantz T (2011) Generation of healthy mice from gene-corrected disease-specific induced pluripotent stem cells. PLoS Biol 9(7):e1001099

Yang Y, Wang L, Bell P, McMenamin D, He Z, White J, Yu H, Xu C, Morizono H, Musunuru K, Batshaw ML, Wilson JM (2016) A dual AAV system enables the Cas9-mediated correction of a metabolic liver disease in newborn mice. Nat Biotechnol 34(3):334–338

Yin H, Xue W, Chen S, Bogorad RL, Benedetti E, Grompe M, Koteliansky V, Sharp PA, Jacks T, Anderson DG (2014) Genome editing with Cas9 in adult mice corrects a disease mutation and phenotype. Nat Biotechnol 32(6):551–553

Yin H, Song CQ, Dorkin JR, Zhu LJ, Li Y, Wu Q, Park A, Yang J, Suresh S, Bizhanova A, Gupta A, Bolukbasi MF, Walsh S, Bogorad RL, Gao G, Weng Z, Dong Y, Koteliansky V, Wolfe SA, Langer R, Xue W, Anderson DG (2016) Therapeutic genome editing by combined viral and non-viral delivery of CRISPR system components in vivo. Nat Biotechnol 34(3):328–333

Zetsche B, Gootenberg JS, Abudayyeh OO, Slaymaker IM, Makarova KS, Essletzbichler P, Volz SE, Joung J, van der Oost J, Regev A, Koonin EV, Zhang F (2015) Cpf1 is a single RNA-guided endonuclease of a class 2 CRISPR-Cas system. Cell 163(3):759–771

Zhang L, Shao Y, Li L, Tian F, Cen J, Chen X, Hu D, Zhou Y, Xie W, Zheng Y, Ji Y, Liu M, Li D, Hui L (2016) Efficient liver repopulation of transplanted hepatocyte prevents cirrhosis in a rat model of hereditary tyrosinemia type I. Sci Rep 6:31460

Index

© Springer International Publishing AG 2017

R.M. Tanguay (ed.), *Hereditary Tyrosinemia*, Advances in Experimental Medicine and Biology 959,
DOI 10.1007/978-3-319-55780-9

MIX
Papier aus verantwortungsvollen Quellen
Paper from responsible sources
FSC® C105338

If you have any concerns about our products,
you can contact us on
ProductSafety@springernature.com

In case Publisher is established outside the EU,
the EU authorized representative is:
Springer Nature Customer Service Center GmbH
Europaplatz 3, 69115 Heidelberg, Germany

Printed by Libri Plureos GmbH
in Hamburg, Germany